T0297535

Mushroom Biotechnology

Developments and Applications

Mushroom Biotechnology

Developments and Applications

Edited by

Marian Petre

University of Pitesti, Faculty of Sciences,
1 Targul din Vale Street, Arges County, Romania

ELSEVIER

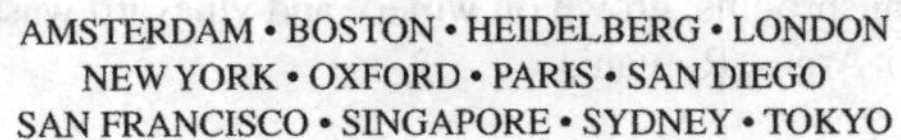

AMSTERDAM • BOSTON • HEIDELBERG • LONDON
NEW YORK • OXFORD • PARIS • SAN DIEGO
SAN FRANCISCO • SINGAPORE • SYDNEY • TOKYO

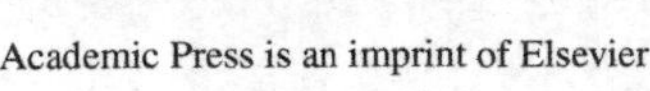

Academic Press is an imprint of Elsevier

Academic Press is an imprint of Elsevier
125, London Wall, EC2Y 5AS.
525 B Street, Suite 1800, San Diego, CA 92101-4495, USA
225 Wyman Street, Waltham, MA 02451, USA
The Boulevard, Langford Lane, Kidlington, Oxford OX5 1GB, UK

First published 2016

ISBN: 978-0-12-802794-3

British Library Cataloguing-in-Publication Data

A catalogue record for this book is available from the British Library.

Library of Congress Cataloging-in-Publication Data

A catalog record for this book is available from the Library of Congress.

For Information on all Academic Press publications
visit our website at http://store.elsevier.com/

Typeset by MPS Limited, Chennai, India
www.adi-mps.com

Printed and bound in the United States

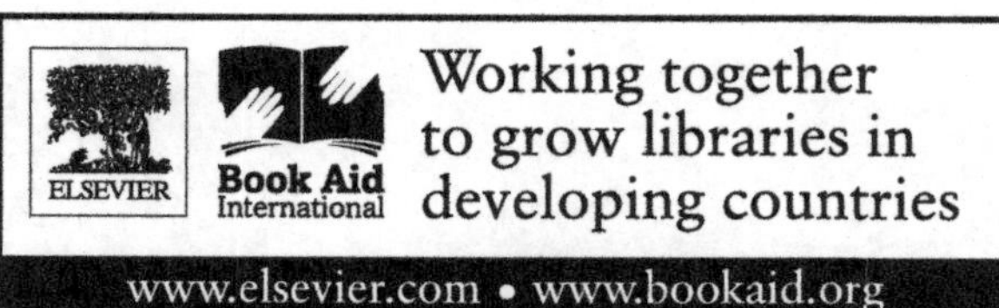

Cover image: *Pleurotus ostreatus* mushrooms, grown on winery and vineyard wastes, in the research laboratory Stefanesti-Arges, Romania

Publisher: Nikki Levy
Acquisitions Editor: Patricia Osborn
Editorial Project Manager: Jaclyn A. Truesdell
Production Project Manager: Lisa Jones
Designer: Matthew Limbert

Dedication

To my whole family, who understood my passion for mushrooms and supported me all the time!

Contents

Editor Biography

Marian Petre, Ph.D. Habil. in Biological Sciences, is Professor of Biotechnology for Environmental Protection, Microbial Biotechnology, Bioremediation, Microbial Ecology and Bioengineering in the Faculty of Sciences at University of Pitesti. Since he graduated the Faculty of Biology from University of Bucharest, in 1981, he has published over 150 scientific articles, 73 of them in international journals and proceeding volumes. In the last decade, he has written and edited 25 books on applied biotechnology, environmental biotechnology, microbiology, bioremediation, as well as microbial ecology. As first author, he has also registered 10 Romanian patents in the field of mushroom biotechnology, being awarded for them with gold and silver medals at international exhibitions for inventions, research, and new technologies in Brussels, Geneva, SuZhou (China), and Bucharest. So far, he has been designated as chairman of five international congresses and symposia on mushroom biotechnology and he has managed 14 research projects financially supported by the Romanian Ministry of Education and Research, being invited as an active expert to bring his contribution to the scientific evaluation of research project proposals registered in European academic contests.

List of Contributors

Noorlidah Abdullah
Mushroom Research Centre, Institute of Biological Sciences, Faculty of Science, University of Malaya, Kuala Lumpur, Malaysia

Georgia Cezara Avram
Faculty for Engineering and Management of Technological Systems, Politehnica University of Bucharest, Bucharest, Romania

Dapeng Bao
Institute of Edible Fungi, Shanghai Academy of Agricultural Sciences, Shanghai City, People's Republic of China

Avner Barazani
Institute of Evolution, Haifa University, Mt. Carmel Haifa, Israel

Gayane S. Barseghyan
Institute of Evolution, Haifa University, Mt. Carmel Haifa, Israel

Marin Berovic
Faculty of Chemistry and Chemical Technology, University of Ljubljana, Ljubljana, Slovenia

Reyna L. Camacho-Morales
El Colegio de la Frontera Sur, Tapachula, Chiapas, México

José M. Domínguez
Department of Chemical Engineering, Faculty of Sciences, University of Vigo (Campus Ourense), Ourense, Spain; Laboratory of Agro-food Biotechnology, CITI (University of Vigo)-Tecnópole, Technological Park of Galicia, San Cibrao das Viñas, Ourense, Spain

Michèle Largeteau
INRA, UR1264 MycSA, Villenave d'Ornon, France

Beng Fye Lau
Mushroom Research Centre, Institute of Biological Sciences, Faculty of Science, University of Malaya, Kuala Lumpur, Malaysia

Dan Andrei Marinescu
EDAG Engineering GmbH, Wolfsburg–Westhagen, Germany

Gerardo Mata
Instituto de Ecología, A.C., Red de Manejo Biotecnólogico de Recursos, Xalapa, Veracruz, Mexico

Florin Adrian Nicolescu
Faculty for Engineering and Management of Technological Systems, Politehnica University of Bucharest, Bucharest, Romania

Florin Pătrulescu
Faculty of Sciences, University of Pitesti, Pitesti, Romania

Noelia Pérez-Rodríguez
Department of Chemical Engineering, Faculty of Sciences, University of Vigo (Campus Ourense), Ourense, Spain; Laboratory of Agro-food Biotechnology, CITI (University of Vigo)-Tecnópole, Technological Park of Galicia, San Cibrao das Viñas, Ourense, Spain

Marian Petre
Faculty of Sciences, University of Pitesti, Pitesti, Romania

Violeta Petre
Department of Biology, Sfântul Sava College, Bucharest, Romania

Bojana Boh Podgornik
Faculty of Natural Sciences and Engineering, University of Ljubljana, Ljubljana, Slovenia

Ionela Rusea
Faculty of Sciences, University of Pitesti, Pitesti, Romania

José E. Sánchez
El Colegio de la Frontera Sur, Tapachula, Chiapas, México

Jean-Michel Savoie
INRA, UR1264 MycSA, Villenave d'Ornon, France

Florin Stănică
Faculty of Horticulture, University of Agronomic Sciences and Veterinary Medicine, Bucharest, Romania

Răzvan Ionuţ Teodorescu
Faculty of Land Reclamation and Environmental Engineering, University of Agronomic Sciences and Veterinary Medicine, Bucharest, Romania

Ana Torrado-Agrasar
Bromatology Group, Department of Analytical and Food Chemistry, Faculty of Sciences, University of Vigo (Campus Ourense), As Lagoas, Ourense, Spain

Hong Wang
Institute of Edible Fungi, Shanghai Academy of Agricultural Sciences, Shanghai City, People's Republic of China

Solomon P. Wasser
Institute of Evolution, Haifa University, Mt. Carmel Haifa, Israel

Foreword

Mushroom Biotechnology – Developments and Applications focuses our attention on the highly diversified attributes of a fascinating group of fungi whose contributions to economic and technological development have often been greatly under-estimated. The volume is edited by Professor Marian Petre, organizer in 2012 of the International School of Advanced Studies on Mushroom Biotechnology and Bioengineering at the University of Pitesti. Contributing authors are distinguished mushroom biologists who are all actively engaged in research at prestigious universities and research institutes worldwide. Compiling a publication of this kind is a demanding and complicated exercise, and all concerned are to be congratulated on a highly successful outcome.

Mushrooms impact on human welfare in many ways. For centuries, edible varieties have been treasured for their high nutritive value and desirable organoleptic qualities. Over 60 species are now cultivated on a commercial scale, and this figure is increasing every year as more species are domesticated. Also, the health-promoting properties of mushrooms have long been recognized in some cultures, especially in China, although this perception has, until recently, largely depended on empirical observations. However, latter-day application of modern analytical techniques has identified various mushroom-derived compounds, polysaccharides and triterpenoids for example, which exhibit a wide range of medicinal properties including immuno-enhancing, anti-tumor, anti-viral and hypocholesterolemic activities. There is growing experimentally-based evidence to suggest that dietary supplements based on bioactive compounds extracted from mushrooms (mushroom nutriceuticals) increase resistance to disease and, in some cases, cause regression of a diseased state. Mushroom cultivation also impacts positively on the environment since the lignocellulosic waste materials generally used as growth substrates are often disposed of using less environmentally-friendly methods. Moreover, the metabolic diversity of mushrooms is integral to bioremediation and biocontrol functions.

One important facet of mushroom biotechnology is focused on mushroom products obtained by fermentation or extraction from fruiting bodies, fungal mycelium, or spent culture liquor. It is this sector of the mushroom industry, currently estimated to be worth in excess of 20 billion US dollars annually, that is expanding most rapidly. Therefore, it is appropriate that two contributions to this book (Chapters 1 and 9) focus on fungal biomass production using fermenter-based systems. Depending upon the mushroom species, traditional mushroom cultivation periods can extend to several months, during which time microbial contamination and/or insect infestation may occur and adversely effect quality and yield. Production of bioactive mushroom metabolites through the controlled cultivation of fungal biomass in bioreactors, using both submerged and solid–state processes, allows system parameters to be easily and accurately manipulated to maximize product yields in the shortest time.

The two chapters on the use of winery and fruit tree wastes will appeal to the more entrepreneurial wine producer/fruit grower. Although some basic principles apply, mushroom cultivation does not necessarily require highly automated growth facilities and the heavy capital investment associated with *Agaricus bisporus* (white button mushroom) production in Europe and North America. Low-technology cultivation systems also have the potential to increase profit margins by generating an added-value cash crop from plentiful supplies of locally-available agricultural waste materials that would otherwise require cost-incurring disposal.

Major mushroom growing enterprises, especially those producing *A. bisporus*, are already using highly automated, computer-controlled systems. It will be interesting to see how long it takes for the virtual robotic prototype for safe and efficient cultivation of mushrooms, introduced in Chapter 4, to become reality.

Since it is generally accepted that *A. bisporus* cultivation was first undertaken in France, it is perhaps fitting that French researchers are associated with the two chapters in the book that are focused solely on "Le champignon." The first emphasizes the role of the mushroom in sustainable agricultural development and the importance of conserving and improving germplasm resources. The second pertains to the major pathogens affecting the mushroom and the need to adopt environmentally-friendly solutions.

The chapter on sclerotium-forming mushrooms is a helpful addition to the relatively sparse literature describing this interesting group of fungi. Sclerotium-forming *Pleurotus tuber-regium* is economically important in Africa, both as a food and a medicine, while *Inonotus obliquus* (chaga mushroom) has a long history of use as a tonic and for the treatment of various ailments.

Mushrooms are non-photosynthetic and instead produce a battery of extracellular enzymes (e.g., cellulases, hemicellulases, and ligninases) in order to convert the lignocellulosic residues that normally serve as the growth substrate into products that can be assimilated for fungal nutrition. Although *Aspergillus niger*, the fungal subject of Chapter 10, is not a mushroom, it is feasible to extrapolate the methodology used to produce xylanases by this fungus to high xylanase-producing mushroom species.

The penultimate chapter describes the identification of mating type loci and genes in the straw mushroom, *Volvariella volvacea*, and the various techniques adopted for molecular marker–assisted breeding. The methodologies described are again generally applicable but this chapter will be of special interest to breeders and growers located in tropical and subtropical regions where the straw mushroom is widely cultivated.

Two other contributions are both concerned with "mycorestoration" – the use of fungi to restore degraded environments. Mushrooms as a source of various biocontrol agents are the subject of the former, while the latter describes the significance of fungal ligninolytic enzymes in the degradation of recalcitrant agro-pesticides (mycoremediation).

Mushroom Biotechnology – Developments and Applications covers a wide range of topics which highlight the versatility of mushrooms and their fundamental importance to the welfare of humankind. It will appeal to both specialists and non-specialists alike, and I am confident it will enjoy a wide readership and provide a stimulus for future research.

John Buswell

Visiting Professor, Institute of Edible Fungi,

Shanghai Academy of Agricutural Sciences

September, 2015

Preface

Mushrooms are considered one of the most diversified groups of biological species adapted for living in extreme environmental conditions all over the Earth. For centuries, many mushroom species have been used as outstanding sources of food and medicine, but in the recent past humans have discovered some of their powerful features to clean the environment through the bioconversion of organic residues from the habitats where they live and continuous recycling of chemical elements.

Nevertheless, for humankind, there is as an urgent need to sustain the efforts to change the current status of serious crises in food, human health, and environmental pollution through the beneficial applications of mushroom biotechnology!

In this respect, a better understanding of the main interactions between biological, biophysical, and biochemical phenomena and processes involved in biotechnological applications of using mushrooms as one of the most important biologic tools for maintaining environmental health will be a key solution for the future progress of humanity.

Mushroom biotechnology is defined as a component discipline of mushroom biology applications including mushroom cultivation, mushrooms for biocontrol of phytopathogens, and mushrooms as bioremediation agents. In this respect, a new field of using mushrooms in cleansing organic and inorganic wastes from the environment has been developed as mycoremediation.

The book *Mushroom Biotechnology—Developments and Applications* has been conceived as a synthetic mirror of recent scientific achievements in the fields of controlled cultivation of culinary and medicinal mushrooms as organic sources of food and medicines, automatic cultivation and processing of mushrooms, biocontrol of pathogens and pests, improvement of mushroom breeding by genetic methods, as well as biodegradation of recalcitrant contaminants through the application of advanced mycological biotechnology.

The content is divided into 12 chapters, each of which provides detailed information regarding scientific experiments carried out in various countries of the world to test novel applications designed to shed light on the beneficial effects of mushroom biotechnology.

The first three chapters are focused on biotechnology for conversion of organic agricultural wastes, both through submerged and solid-state cultivation of culinary and medicinal mushroom species. Chapter 4 has as its main subject the automatic cultivation and processing of mushrooms through a modular robotic prototype designed to produce both fruit bodies and sterilized and inoculated bags filled with mycelium of culinary and medicinal mushrooms. The next two chapters describe the biotechnology of *Agaricus bisporus* cultivation as well as specific methods for pathogen control in this button mushroom species. The seventh chapter presents current perspectives on sclerotia-forming mushrooms as an emerging source of medicines. Then, the next two chapters characterize medicinal mushrooms regarding their specific antiphytopathogenic and insecticidal properties as well as their cultivation in different types of bioreactors.

Chapter 10 relates to the use of *Aspergillus niger* extracts obtained by solid-state fermentation for enzyme production, and the next chapter highlights the identification and application of *Volvariella volvacea* mating type genes in mushroom breeding. The last chapter focuses on the biotechnological use of fungi for degradation of recalcitrant agro-pesticides.

This book is especially addressed to researchers, students, and specialists in mushroom biotechnology, mycological research, food biotechnology, environmental biotechnology, bioengineering, and bioremediation, but also all readers who want to improve their knowledge of biotechnological applications of mushrooms for the well-being of human society.

In conclusion, after a whole year of tremendous editorial activity, I would like to thank each of the contributors for their considerable efforts to present the most valuable achievements in their fields, and I really hope that readers will be interested in the scientific content of these chapters.

In addition, I take real pleasure in expressing my sincere gratitude toward Patricia Osborn, the Senior Acquisitions Editor of Elsevier Books Division, for her remarkable professionalism and kindness in support of this book project from the beginning of our cooperation in order to achieve such outstanding work!

Last but not least, my warm and sincere thanks are forwarded to Editorial Project Managers Jaclyn Truesdell, Lisa Jones and Carrie Bolger for their careful assistance and great patience during our joint work, as well as to whole staff of Elsevier Inc. for their professional involvement in publishing this book!

Marian Petre
Editor
University of Pitesti, Pitesti, Romania
May, 2015

CHAPTER 1

BIOTECHNOLOGY OF MUSHROOM GROWTH THROUGH SUBMERGED CULTIVATION

Marian Petre[1] **and Violeta Petre**[2]

[1]*Faculty of Sciences, University of Pitesti, Pitesti, Romania*

[2]*Department of Biology, Sfântul Sava College, Bucharest, Romania*

1.1 INTRODUCTION

From the beginning of this century, the submerged cultivation of culinary and medicinal mushrooms has received a great deal of attention as a promising and reproducible alternative for the efficient production of mycelia biomass and fungal metabolites. Due to economic reasons, the submerged cultivation of mushrooms (SCM) has gained an ascending attention due to its significant potential for industrial applications, but its prospective success on a commercial scale depends on increasing product yields and development of novel production systems that address the problems associated with this biotechnology of mushroom cultivation.

In the recent literature, there are described several methods of growing strains of Basidiomycetes in submerged cultures, which provide an opportunity to get a huge production of biomass containing high concentrations of bioactive compounds with healthful effects on humans, such as proteins, essential amino acids, vitamins, and polysaccharides (Verstraete and Top, 1992; Smith, 1998; Stamets, 2000; Sanchez, 2004; Wasser, 2010).

Any technology for bioprocessing raw materials or their constituents into bioproducts requires the following three steps: process design, system optimization, and model development. To achieve all these steps, a biotechnological proceeding involves the use of biocatalysts, as whole microorganisms or their enzymes, to synthesize or bioconvert raw materials into new and useful products. At the same time, optimization of any submerged cultivation bioprocess is essential for biotechnology development in an industrial-scale application. In this respect, it should be taken into consideration that physical and chemical factors interact and affect the efficacy of the bioprocess regarding mycelia growth within the liquid medium. However, for the time being, in spite of research into optimizing the production of bioactive metabolites by synthesis by mushrooms, the physiological and engineering aspects of all submerged cultures are still far from being thoroughly studied (Wood, 1992; Wedde et al., 1999; Elisashvili, 2012).

Mushroom Biotechnology. DOI: http://dx.doi.org/10.1016/B978-0-12-802794-3.00001-1

1.2 THE CONCEPT OF SCM

First of all, it is necessary to point out that the SCM has an exclusive and specific character concerning fungal cell growth and development in totally different conditions compared with the natural environment where all native mushrooms exist.

This means that the concept of SCM refers to a biotechnological process of mushroom growth inside an artificial environment represented by the volume of a liquid medium in which all physical and chemical factors needed for optimal development of mycelium are provided without any risk of chemical or biological contamination.

The specific status of all mushroom species as native or indigenous fungi is to grow and develop in natural habitats in terrestrial ecosystems; in other words, they are species adapted to colonize only solid substrates containing a certain amount of water and involving living organisms or organic structures accumulated outside or inside the soil (Vournakis and Runstadler, 1989; Wedde et al., 1999; Uphoff, 2002).

More precisely, no known mushroom species has any capability of growing and developing in natural aquatic habitats; more than that, none of them is adapted to form fruiting bodies inside a liquid medium. This is a restrictive living condition for all native mushroom species of planet Earth, by which they are compelled to live only inside terrestrial ecosystems from the natural environment due to their strictly specific adaptation to aerobic respiration.

The cellular metabolic processes of any mushroom species require permanent oxygen intake in appropriate concentrations, supplied from the outer environment of the mycelia, and this cannot be achieved inside a liquid volume of any natural environment where there does not exist the proper concentration of dissolved oxygen (DO) in order to maintain the mushroom's life!

Mushroom species have great potential for adapting to any habitat which provides a solid support and containing only a small amount of water to sustain their natural life cycle. If this support is entirely formed by water, there is no chance for a mushroom strain to survive due to the lack of DO intake to the membrane surface of fungal cells. In such circumstances, the only way to artificially grow mushroom species inside a liquid medium is to keep the dissolved oxygen concentrations (DOC) at required levels to maintain the mushroom's metabolic activities by using special devices to force oxygen penetration inside the liquid volume.

Thus, despite both the shear forces and turbulence generated by oxygen intake inside the liquid cultivation medium from the culture vessel of a bioreactor, the mycelium is forced to move circularly according to the specific rheology of such a medium. During the cultivation process, the fungal cells are able to grow in submerged conditions and, due to centrifugal force, these cells metabolize the nutritive particles from the cultivation medium and develop as a biomass containing many mycelium pellets of different sizes and almost rounded shapes.

1.3 METHODS AND TECHNIQUES USED FOR SCM

As a general matter, SCM requires full control of the cultivation bioprocess regarding the automatic tracking of all chemical and physical parameters and keeping them at optimal values.

This biotechnological method permits fully standardized production of the fungal biomass with high nutritional value or the biosynthesis of mushroom metabolites with a predictable composition. At the same time, the downstream processing after submerged cultivation is very feasible and easier to carry out as compared with the classical procedure of solid-state cultivation. Inside the cultivation vessel of a bioreactor, it is possible to control the culture conditions, such as temperature, agitation, DOC, temperature, substrate and metabolite concentration, as well as the pH inside the liquid culture substrate (Kim et al., 2007; Elisashvili, 2012; Turlo, 2014; Homolka, 2014).

It is well known that the morphology of mycelia in submerged cultures has a significant influence on the rheology of the culture broth. At the same time, the initial viscosity of the liquid medium, as well as the stirring speed and air intake pressure, have important effects on fungal pellet formation during the cultivation cycle of mushroom spawn. Thus, the agitation rate and dispersion effect induced by shear forces upon the fragile structure of the mycelium, especially in the first period of time during a cultivation process, have determinant influence upon the fragile structure of the mycelium which is to develop inside the liquid culture medium as fungal pellets with different shapes and sizes. After many experiments to study the effects of stirring rate and share forces, it was noticed that an inverse relationship exists between agitation speed and pellet features. In fact, increased agitation determines the formation of small and very compact pellets; on the other hand, a vigorous agitation seems to prevent pellet formation (Park et al., 2001; Papagianni, 2004; Turlo, 2014).

Along the evolution of submerged cultures, the mushroom mycelia generate globular shaped aggregates called pellets. The morphological forms of pellets are characteristic of each mushroom species. In any submerged culture, the pellet size determines the oxygen and nutrient transport into its center. In the core region of a large pellet, the fungal cells stop their growth because of low DOC and nutrients, and for this reason the smaller pellet diameter could be advantageous in terms of increased mycelia biomass (Lee et al., 2004; Kim et al., 2007; Elisashvili et al., 2009; Xu et al., 2011; Turlo, 2014).

However, pellet size is influenced by various variables, such as agitation regime, density of the inoculums, and sugar concentration in the culture medium (Petre et al., 2010).

During the cultivation process, the culture viscosity increases significantly, and sometimes, mushroom mycelia start to wrap around impellers, spreading into the sampling devices and feed tubing with nutrients, causing functional blockages. All these problems limit the operation time of bioreactors, and they must be avoided by constant control and correction of the culture density (Shih et al., 2008; Elisashvili, 2012; Turlo, 2014).

While the SCM mycelia induce relatively high energy costs required for agitation, oxygen supply, and constant control of the temperature of the liquid medium during the whole cultivation process, this biotechnological method has significant industrial potential due to the possibility of process upscaling and operation of large-scale bioreactors.

There are many biotechnological methods for cultivating the mycelia of edible mushrooms in liquid media by applying various strategies. In this respect, batch culture is one of the most frequently used biotechnological methods for the SCM. In this cultivation method, no fresh nutritive elements are added to the culture composition and no end products of fungal metabolism are discharged during the process.

The simplest technique used for this kind of cultivation is based on shake flask cultures in order to get relatively small quantities of mycelia that can be used as inocula for the larger production of mycelia biomass by growing in the culture vessels of laboratory-scale bioreactors designed for batch cultures (Porras-Arboleda et al., 2009; Lin, 2010; Xu et al., 2011; Elisashvili, 2012; Petre and Petre, 2013; Homolka, 2014).

1.4 BIOTECHNOLOGY FOR SUBMERGED CULTIVATION OF *PLEUROTUS OSTREATUS* AND *LENTINULA EDODES*

The main problem that needs to be solved for the intensive biotechnological process of submerged cultivation of edible and medicinal mushrooms on nutrient substrates made of agricultural wastes resulting from cereal grain processing is to convert these natural waste products of organic agriculture into nutritive biomass to be used as food supplements that are made only through biological means (Petre et al., 2014a; Petre and Petre, 2013).

In our recent studies on the application of laboratory-scale biotechnology for submerged cultivation of culinary mushrooms, we tested two Basidiomycetes species, described in the following lines.

Lentinula edodes (Berkeley) Pegler is a heterothallic mushroom species belonging to Basidiomycetes group. The optimum temperature for spore germination is 22–25°C, but for mycelial growth temperature can range from 5°C to 35°C. The species of the genus *Lentinula* can grow on various culture media, both natural and synthetic, depending on the cultivation procedure, and they have certain morphological and physiological characteristics that distinguish them from other types of mushrooms (Carlile and Watkinson, 1994; Hawksworth et al., 1995; Jones, 1995; Hobbs, 1995).

Pleurotus ostreatus (Jacquin ex Fries) Kummer, also known by its popular name as the oyster mushroom, is a Basidiomycetes species belonging to the family Pleurotaceae (Agaricales, Agaricomycetes). The species have carpophores with eccentric pileus and decurente blades showing white or hyaline enhanced with cylindrical or oval forms (Chahal and Hachey, 1990; Carlile and Watkinson, 1994; Hawksworth et al., 1995).

The pure cultures of these mushroom species, which were tested in our experiments, are represented by two strains, *L. edodes* LE 07 and *P. ostreatus* PO 14, belonging to the mushroom collection of the University of Pitesti.

Before starting the application of submerged mushroom cultivation, the pure mycelial cultures were inoculated into 250-mL flasks containing 100 mL of MEYE (malt extract 20%, yeast extract 2%) medium, and then they were placed in a rotary shaker incubator set to keep the temperature level at 23°C with a stirring speed of 110 rpm (rotations per minute) for 5–7 days. Then the fungal cultures were placed by aseptic inoculation inside the bioreactor vessel for submerged cultivation.

The main stages of biotechnology to get high nutritive mycelial biomass by controlled submerged fermentation are as follows: (i) preparation of culture substrates, (ii) steam sterilization of the bioreactor culture vessel, (iii) aseptic inoculation of sterilized culture media with the pure cultures of selected mushroom strains, (iv) running the submerged cultivation cycles under controlled conditions, and (v) collecting, washing, and filtering the fungal pellets that were obtained.

Such cellulosic wastes as apple marc and winery wastes were chosen as the main components of mushroom cultivation substrates; these were mixed with cereal bran, such as wheat, barley, and oats, which were weighed before mixing with limestone powder and tap water in different ratios, as shown in Table 1.1.

The first stage of biotechnology for submerged cultivation of *P. ostreatus* and *L. edodes* was achieved by preparation of culture substrates represented by agricultural wastes resulting from industrial processing of cereal grains (wheat bran, barley bran, oat bran) and apple and winery wastes (grape marc and apple marc), with pure water (in the composition shown in Table 1.1), which were then poured into the cultivation vessel of the bioreactor.

Table 1.1 Variants of Substrates Used for Submerged Cultivation of Mushrooms (SCM)

Variants of Substrates	Substrate Composition (% d.m.)
I	Apple marc 40%, wheat bran 5%, limestone powder 5%
II	Apple marc 40%, barley bran 5%, limestone powder 5%
III	Apple marc 40%, oat bran 5%, limestone powder 5%
IV	Grape marc 30%, wheat bran 15%, limestone powder 10%
V	Grape marc 30%, barley bran 15%, limestone powder 10%
VI	Grape marc 30%, oat bran 15%, limestone powder 10%
Control	Pure cellulose (Merck)

In the next stage, steam sterilization of the bioreactor vessel containing the culture substrates was performed at 121°C and 1.1 atm. for 30 min. The third stage was aseptic inoculation of the sterilized culture media with pure cultures of selected mushroom strains inside the bioreactor vessel, using a sterile air hood with laminar flow.

In the next stage, the submerged cultivation cycles were carried out under controlled conditions: temperature 23 ± 2°C, stirring speed 70 rpm, and dissolved oxygen tension of 25–35%. Finally, the last step was accomplished by collecting, washing, and filtering the fungal pellets obtained by the submerged fermentation of substrates made of by-products resulting from cereal grains processing (Petre and Petre, 2008; Petre and Petre, 2012).

The biotechnological experiments inside the bioreactor vessel were conducted under the following conditions: temperature, 25°C; stirring speed, 120–180 rpm; dissolved oxygen tension (DOT) of 35%, and an initial pH of 4.5–5.5. After 10–12 days of cultivation, the mycelial pellets were collected from the bioreactor vessel where they were formed, having various shape and size characteristics. All experiments were carried out in three replicates. Then we proceeded to filtration and concentration of the fungal pellets dispersed inside the cultivation vessel to get a pure and fresh mycelial biomass, as can be seen in Figure 1.1A and B.

At the end of the fermentation process in submerged conditions, we obtained a mushroom biomass with a brownish color, composed of mycelial formations consisting of compact and spherical anastomosed hyphae that were concentrically distributed in their internal structure. Finally, the fungal pellets that were collected from the cultivation vessels were chemically analyzed for the percentage of dry matter, sugar content, Kjeldahl nitrogen, and total protein, as shown in Table 1.2.

The experiments were carried out three times for each mushroom species and substrate variant. The reported results are the means of the three repeated experiments. Samples for analysis were collected at the end of the submerged cultivation process, when mycelia pellets took on specific shapes and characteristic sizes. The mycelia biomass was washed repeatedly with double distilled water in a sieve with a 2 mm diameter eye to remove the remained bran in each culture medium. Biochemical analyses of mycelia biomass samples obtained through SCM were carried out for the solid fractions after their separation from the remaining fluid by pressing and filtering. In each experimental variant, the amount of fresh biomass mycelia was determined. Percentage of dry biomass was determined by dehydration obtained at a temperature of 70°C until constant weight was obtained.

The total protein content was determined by a biuret method, whose principle is similar to the Lowry method, this method being recommended for protein content ranging from 0.5 to 20 mg/100 mg

FIGURE 1.1

Mycelial pellets of *Lentinula edodes* (A) and *Pleurotus ostreatus* (B), dispersed inside the culture vessels of laboratory-scale bioreactor.

Table 1.2 Chemical Composition of Fungal Pellets Belonging to *Lentinula edodes* and *Pleurotus ostreatus*

Variants of Substrates	Dry Biomass (%)		Sugar Content (mg/g)		Kjeldahl Nitrogen (%)		Total Protein (g% d.m.)	
	L. e 07	P. o 14	L. e 07	P. o 14	L. e 07	P. o 14	L. e 07	P. o 14
I	7.97	10.70	4.95	5.15	5.95	6.50	3.53	3.67
II	7.05	9.35	5.65	4.93	6.75	5.35	3.55	3.35
III	6.75	9.95	5.55	5.55	6.53	6.70	2.95	2.95
IV	7.45	9.15	4.70	4.35	5.05	5.75	3.25	3.30
V	6.50	10.05	5.30	5.50	6.10	6.90	2.75	3.55
VI	6.95	9.95	4.90	5.30	5.50	6.10	2.80	3.10
Control	0.50	0.70	0.45	0.55	0.35	0.30	0.20	0.30

sample (Raaska, 1990; Park et al., 2001; Sanchez, 2010; Shih et al., 2008). This method requires only one sample incubation period for 20 min. In this way, interference from various chemical agents (e.g., ammonium salts) is eliminated. The principal method is based on the reaction that takes place between copper salts and compounds with two or more peptides, which results in a red-purple complex whose absorbance is read in the visible domain ($\lambda = 550$ nm) of a spectrophotometer. Using the Dubois method (1956), the sugar content of dried mycelia pellets collected after the biotechnological experiments was determined (Petre and Petre, 2008; Papaspyridi et al., 2012).

Of the two mushroom species which were tested during the biotechnological experiments, *L. edodes* cultivated on substrate variant II showed the best value of sugar content; the next best, in order, was *L. edodes* in culture variant III, and *P. ostreatus* in culture variant III could be mentioned. Regarding Kjeldahl nitrogen, most favorable amounts were registered in the case of *P. ostreatus* cultivated on substrate variant V, followed by *L. edodes* cultivated on substrate variant II and also by *P. ostreatus* on substrate I. Total protein content was the greatest in the case of *P. ostreatus* cultivated on substrate variant I, followed by the same species cultivated on substrate variant II and *L. edodes* cultivated on substrate variant V. However, the registered results concerning sugar and total nitrogen contents showed higher values than those obtained by other researchers (Beguin and Aubert, 1994; Carlile and Watkinson, 1994; Moo-Young, 1993; Confortin et al., 2008; Lin and Yang, 2006; Kim et al., 2004).

As a matter of fact, the nitrogen and protein contents of mycelia biomass represent key factors for assessing its nutritive potential, but the assessment of different protein nitrogen compounds requires additional investigation. Comparing all registered data, it can be noticed that the correlation between the dry weight of mycelia pellets and their protein, sugar, and nitrogen content is kept at a balanced ratio, as in the case of the fungal pellets of each tested mushroom species, as mentioned in a few scientific works (Park et al., 2001; Papagianni, 2004; Shih et al., 2008; Papaspyridi et al., 2012).

The optimization of any submerged cultivation bioprocess is essential for biotechnology development in an industrial-scale application. In this respect, it should be taken into consideration that the physical and chemical factors interact and affect the efficacy of the bioprocess regarding mycelia growth inside the liquid medium. One of the best methods to optimize the culture medium composition and physical culture parameters involves the changing of one independent parameter (physical or chemical) while keeping the other factors constant.

Such a method allows one to determine the optimal parameter (e.g., carbon or nitrogen sources) but it does not provide information on interactions and correlations between parameters. The most appropriate way is to use statistical techniques that permit the simultaneous optimization of multiple factors, thereby obtaining considerable quantitative information by only a few experimental trials (Subramaniyam and Vimala, 2012; Turlo, 2014).

1.5 PHYSICAL AND CHEMICAL FACTORS THAT INFLUENCE THE SCM

The physicochemical characteristics of the submerged cultivation process may influence cell growth and, consequently, the number of viable cells per unit volume of cultivation media. These factors include temperature, pH index, the availability of oxygen (aeration), stirring, and inoculum quantity. The optimal timing for adding fresh nutritive medium in the case of batch cultivation is determined by the enzyme amount divided by the number of viable cells existing in the culture medium (Chahal, 1994; Bae et al., 2000).

Usually, the ideal moment to add fresh culture medium in the cell culture is in the course of the exponential growth phase, especially toward the middle or end of it, ensuring in this way the exposure of the cells growing on the contact surface of recently added substrate (Glazebrook et al., 1992; Zarnea, 1984).

1.5.1 CHEMICAL FACTORS

In the establishment of any biotechnological application, the most important key factors that must be taken into consideration are the chemical factors, because these could influence the running of bioprocesses to the highest degree.

1.5.1.1 Carbon sources

The carbon sources, particularly important due to the inducible characteristics of cellulases, must be accessible and at competitive manufacturing costs; this will result in a major production of enzymes. To select the optimal carbon source in order to increase the cellulose production of *P. ostreatus*, the strain PO 14 was grown on nutrient media containing soluble sugars (maltose, glucose, and xylose) in concentration ranging from 3% to 15%. The amount of soluble sugars determined in the fluid culture did not inhibit the total cellulose production of the *P. ostreatus* strain (van den Twell et al., 1994; Zhang and Cheung, 2011).

In order to achieve certain experiments regarding the biodegradation of cellulose constituents by cultivating *P. ostreatus* in a batch system, we tested different carbon sources in the form of suspensions in a concentration of 5–10 g% dry matter (d.m.), and a pure cellulose solution (Schuhardt) was used as the control. Monitoring the enzyme biosynthesis intensity, there were results that showed much higher values of enzymatic activity in the case of using natural carbon sources, represented by wastes resulting from industrial processing of cereal grains rather than the ones belonging to the control. In this respect, depending on the concentration of natural agricultural wastes, the cultivation substrates can be classified into the following categories: solid substrates (80–90 g% d.m.), semisolid substrates with high viscosity (30–40 g% d.m.), and colloidal suspension (5–10 g% d.m.).

There are many mushroom species that develop on cultivation substrates made of carbohydrates. It was noticed that fungal growth in media containing polysaccharides, such as starch, is much slower, since the depolymerization reaction of the carbohydrate compound was a limiting factor in the dynamic of cell growth process. As a result, the mushroom grew much faster on the equivalent substrate consisting of a single monosaccharide (Beguin, 1990; Petre et al., 2010). The results showed a direct correlation between particle size of cultivation substrate and the final amount of fungal biomass, which is explained by the influence of the ratio of volume/surface conversion during the dynamic biochemical processes of submerged cultivation.

1.5.1.2 Nitrogen sources

When the nitrogen sources were tested in experimental work, some of them (e.g., yeast extracts, malt extracts, peptones, triptones) did not provide nitrogen molecules which were specifically accessible to the cells, but they had provided vitamins and essential minerals for the proper development of the mushroom species used in certain cycles of cultivation.

For many mushroom species, these organic compounds may be used as combined sources of carbon and nitrogen at the same time. The nitrogen source capable of inducing the synthesis of cellulases at the optimum level during the cultivation of *P. ostreatus* strain PO 14 was determined by assaying organic (urea, yeast extract, peptone) or inorganic (ammonium nitrate, nitrite potassium) compounds, and the total amount of nitrogen registered in the culture fluid was 0.5 g/L. This demonstrated a need for the presence of organic compounds such as urea, peptone, yeast, or meat extracts and inorganic salts such as ammonium nitrate, as preferred sources of nitrogen to produce cellulolytic enzymes (Tanaka and Matsuno, 1985; Songulashvili et al., 2005).

Regarding the influence of some nitrogen sources upon the level of enzyme activity of cellulolytic fungi, a series of experimental tests were carried out using inorganic sources, such as ammonium nitrate in concentrations of 0.01%, 0.05%, 0.10%, 0.5%, and 1%. The stimulating effects on the cellulolytic enzyme biosynthesis, such as endoglucanase in batch cultures of *L. edodes* and *P. ostreatus* mushrooms, were registered at ammonium nitrate concentrations of 0.5% and 1%, respectively. The highest enzyme activities were recorded in terms of using a mixture of beef extract, peptone, and inorganic salts such as ammonium nitrate or potassium nitrate. In each case, the enzyme activity, dosed in the culture supernatants, reached the maximum values (Zarnea, 1994; Cohen et al., 2002).

1.5.1.3 pH index

The influence of pH values upon the selected substrates for edible mushroom cultivation has been studied in a relatively broad range of variation, between 3.0 and 7.5 pH. For the mushroom species of *L. edodes* and *P. ostreatus*, the maximum cellulase activity was registered at an initial pH of 5.5–6.0. The highest values of enzymatic activities during the cultivation cycle of the mentioned mushrooms were registered between 2.80 and 3.70 U/mL.

Adjusting the pH index to 5.5 during the whole cycle of submerged cultivation of the PO 14 strain, an increase in enzyme activity of this species of approximately 20–23% was registered compared with the uncorrected pH variant of blank. Thus, the variation of pH from 4 to 7 units resulted in a decrease of endoglucanase activity of the L.E.07 strain of 30–40% compared to the blank, which was maintained under standard conditions at a pH value of 7.0. Other results registered during cultivation cycles ranged from 2.80 to 3.70 U/mL, such variation being dependent on the time consumed from the beginning of mushroom growth. The results confirm the data from the literature, according to which the production of cellulase by the fungal strains is carried out in a weakly acidic pH range (Finkelstein and Ball, 1992; Gregg and Saddler, 1996).

1.5.1.4 Oxygen intake

Fungi in general are strictly aerobic microorganisms, and therefore, aeration is an extremely important factor in the development of metabolic processes. Experiments conducted under certain conditions, with different average values of surface/volume ratio (S/V), had provided relevant data on the sensitivity of oyster mushroom strains depending on the air flow, as well as about how proper is the actual implementation of its uniform dispersion throughout the mass of nutrient medium (Mikiashvili et al., 2006).

Concerning the mycelial growth of *P. ostreatus* in a batch cultivation system, a significant increase was determined in the amounts of dried protein biomass at levels that ranged from 15.9 to 23.8 g% d.m., registering a difference of 8.9 to 15.5 g% d.m., compared to the blank represented by the same species grown in a cultivation system without aeration, as compared to the initial protein content, which registered between 2.1 and 4.0 g% d.m.

It is worth pointing out that at flow rates of the air volume at 5 and 10 m^3/h, the differences are negligible, which shows that an additional air intake (over 5 m^3/h) has no significant influence on the metabolic activity of cultivated mushrooms. The registered data showed that the maximum levels of mycelial biomass belonging to PO 14 (15.35 g/L) and LE 07 (14.20 g/L) were noticed at aeration rates of 0.5 and 1.5 vvm, respectively. Also, the biometric analysis of mycelial morphology showed that mycelia formed fungal pellets even in the early stages of culture. Subsequently, during submerged cultivation the fungal pellets increased in their sizes.

In addition to these results, when the same species were grown in a chemostat-based system, an increase in the yield protein biomass conversion process was reflected in the final composition of mycelia biomass, which registered between 14.5 and 15.9 g% d.m., versus the initial content of 2.0–4.0 g% d.m. at the beginning of the cultivation cycle. The optimal specific growth of mycelia biomass was significantly enhanced, from 3.50 to 7.30 g/L, when the aeration rate increased from 0.5 to 1.5 vvm, but it dropped to 5.10 g/L at an aeration rate of 1.7 vvm. Finally, it was established that the optimal aeration rate is between 0.7 and 1.2 vvm.

1.5.2 PHYSICAL FACTORS THAT INFLUENCE THE SCM

1.5.2.1 Temperature

In a biosynthetic process, the temperature factor is of exceptional importance, as this parameter increases in its optimum value affecting the specific growth rate, according to the Arrhenius relationship. Experiments relating to the monitoring of temperature, to see how they influence enzymatic activity when the value of this biochemical parameter is different from that required for adequate cell growth, have been performed mainly by cultivation of *L. edodes*. The intensity of *L. edodes* enzyme activities was determined for temperatures in the range of 10–30°C, under conditions of stationary cultivation on Mandels and Sternberg medium with cellulose 1%; a variation was noted in the range of 0.9–2.50 U/mL, the maximum activity being recorded at 23°C, while under a stirring regime it ranged between 1.50 and 4.50 U/mL, with a maximum at a temperature of 28°C. Experiments around the effect of temperature on the stationary cultivation of *L. edodes* have revealed a decrease in enzymatic activity of about 30–40% compared with those results obtained by cultivation under a stirring regime. This highlighted the positive impact that was exercised by the stirring process, in close correlation with the carbon source, as well as its state of dispersion inside the culture medium (Jiang, 2010; Petre et al., 2014a).

1.5.2.2 Fragmentation degree of cultivation substrates

Increasing the degree of fragmentation of cellulosic materials used for submerged cultivation of several species of mushrooms is directly proportional to the extension of contact surfaces of cellulolytic enzymes. Mechanical fragmentation at dimensions of microns ensured an enhanced enzymatic activity by increasing the coefficient of adhesion of the particles to the substrate on the external surface of the cell walls of hyphae from the mycelium structure (Leahy and Colwell, 1990; Frankland, 1992; Howard et al., 2003).

1.5.2.3 Stirring rate

The method of achieving a uniform dispersion of nutrient compounds contained in a cultivation liquid substrate demonstrated the important role of using a stirring regime, properly adapted to the morpho-physiological requirements of fungal cultures (Glazebrook et al., 1992; Davitashvili et al., 2008). Without a controlled stirring regime, significant changes in the intensity of cellular metabolic activity were noticed, particularly as a result of the emergence of some areas deficient in nutrients, while there were others that kept their concentration within appropriate limits. These disturbances with entropic effects on the fungal microhabitat were caused by an improper stirring regime, which may occur when certain shear forces are generated due to the swirling effects that occur in the area adjacent to the impeller blades. In certain experiments, the intensity of such effects increased exponentially, as the optimal rheology of the cultivation medium decreased during the development of fungal biomass (Carlile and Watkinson, 1994; Cocker, 1980).

Data recorded during the experiments regarding simulation of growing cycles under various stirring speeds demonstrated that each of the species tested in cultures needed a certain stirring regime, through a well-defined number of rotations per minute, allowing optimal deployment of morphogenesis processes, and eliminating the risk of cell damage (Petre et al., 2010).

1.6 THE BIOLOGICAL FACTORS THAT INFLUENCE THE SCM

Many species of Basidiomycetes mushrooms produce extracellular enzymes that give them the ability to break down polysaccharides such as pulps and convert these organic compounds into polymeric carbohydrates and other low molecular weight substances. The metabolic characteristics of cellulolytic fungi require the use of a cell culture with high enzymatic potential (Beguin and Aubert, 1994; Boddy, 1992, Hawksworth, 1992; Carlile and Watkinson, 1994).

Determination of the optimal influence of spore inoculum upon the submerged cultures was performed by inoculation of the spore suspensions with the following titers: 3%, 5%, and 7%. The experimental results showed that in small culture volumes, the amount of inoculum had no significant influence on enzyme biosynthesis (Kirk and Eriksson, 1990; Nevalainen and Pentilla, 1995; Baker et al., 1995). However, submerged cultivations in chemostat and batch systems have demonstrated the need for certain types of inoculum with morphological and physiological characteristics corresponding to the specificity of the cultivation process for which they are used as biocatalysts.

The optimal age of fungal inoculums to be used for submerged cultures was determined by testing the spore suspensions of *L. edodes* LE 07 through cultivation in petri dishes on MEYEA (malt extract-yeast extract-agar) medium; the fungal cultures were maintained for 5 days at 28°C and a pH value of 4.5. When the PO 14 strain was tested, the variant of the mycelial inoculum having an age of 72h showed a stimulation of enzyme activity compared with another variant of 120h. The influence of inoculum amount was obviously much more important when the cultivation processes were carried out in large volumes of culture medium (Wainwright, 1992; Trinci, 1992; Tsivileva et al., 2005). The use of an appropriate volume of inoculum has the advantage of inducing mycelia growth in the shortest period of time, starting at an optimal development level that ensures the reproducibility of the cultivation process (Baker et al., 1995; Beveridge et al., 1997).

1.7 NEW BIOTECHNOLOGY FOR SUBMERGED CO-CULTIVATION OF MUSHROOM SPECIES

Preparation of substrates for the cultivation of edible mushrooms to convert the apple marc by submerged fermentation in order to produce protein biomass to be used as feed products was carried out in two different compositions after prior grinding and hydration of apple marc.

In the first variant of the cultivation, as substrate 1, the apple wastes were mixed with the following organic and inorganic ingredients as dried matter: apple marc (30%), wheat bran (5%), $CaCO_3$ (1%), NH_4NO_3 (0.5%), and $MgSO_4{\cdot}5H_2O$ (0.5%). The second alternative substrate for the cultivation of edible mushrooms (substrate 2) was prepared with the same sort of apple wastes, supplemented with organic and inorganic following components: apple marc (30%), barley bran (5%), NH_4NO_3 (0.5%), and $MgSO_4{\cdot}5H_2O$ (0.3%).

Both of these substrates were used in experiments of fungal fermentation of winery wastes through mono- and co-cultures of *L. edodes* and *P. ostreatus* mushroom species. Optimal temperatures for the growth and development of mycelium in both monocultures and co-cultures of these mushroom species were recorded in the range of 23–25°C, at an initial pH index of 5.5–6.5, and a stirring speed which ranged between 60 and 90 rpm.

The composition of the cultivation substrate, the index levels of pH, incubation temperature, agitation speed, inoculum age, and volume of samples are all the same physical, chemical, and biological factors that influence the evolution of the submerged fermentation process in its different stages up to the end of substrate conversion into useful biological products (Petre et al., 2014b; Ropars et al., 1992).

Experiments in the cultivation of both mushroom species, *L. edodes* and *P. ostreatus*, in monocultures and co-cultures as submerged fermentation of winery wastes, were performed using a laboratory-scale bioreactor equipped with main control system operational parameters, incorporating a device for maintaining a constant temperature, a device for the supply of sterile air, a mechanical stirrer, an inoculum tank, a pH index correction device, and an automation system for driving the biotechnological process.

To establish the efficiency of submerged fermentation processes for the winery wastes for conversion into fungal biomass to be used as feed products, the mushroom species have been used in pairs, as co-cultures, and separated as monocultures, for periods of time between 20 and 30 days, using and inoculum age of 3 days and having a volume size between 3% and 9% (v/v).

The reducing sugar content was determined during the biotechnological experiments by using the method laid down by Kubicek et al. (1993), and the total amount of nitrogen that was accumulated in the fungal biomass obtained by culturing the two species of mushrooms on two different substrates was analyzed by the Kjeldahl method (Table 1.3).

The results recorded during experiments showed an increase in the amount of reducing sugars in conjunction with a corresponding increase in the total nitrogen content when using co-cultures as compared to monocultures belonging to the same species of edible mushrooms.

The optimum temperatures for the growth and development of the mycelium in both monocultures and co-cultures of both species of fungi have been recorded in the range from 23°C to 25°C, at an initial level of pH index between 5.5 and 6.5 and a stirring speed which ranged between 60 and 90 rpm. The analysis of total nitrogen content of fungal biomass (g% d.m.) in the conversion process, for substrate S1 and substrate S2, strictly depending on the type of culture and time period, highlights the constant accumulation of nitrogen in the total fungal biomass in the three culture types (Table 1.4).

The results of the determination of reducing sugars (Kubicek et al., 1993) were correlated with those on lowering the amount of dry matter contained in cultivation substrates, during the course of conversion processes fungal, and then compared between the monocultures and co-cultures of *L. edodes* and *P. ostreatus*, as shown in Table 1.5. The progress of weight loss in the total amount of dry matter (%) of the composition of the two types of the cultivation substrates (S1 and S2) is in direct correlation to the rate of their decomposition.

Table 1.3 Total Reducing Sugar Concentration (mg/g) of Fungal Biomass during the Bioconversion of Substrates Depending on the Culture Type and Time of Cultivating

	Total Reducing Sugar Concentration (mg/g)					
	L. edodes (Monoculture)		*P. ostreatus* (Monoculture)		*L. edodes-P. ostreatus* (Co-Culture)	
Time (h)	Substrate 1	Substrate 2	Substrate 1	Substrate 2	Substrate 1	Substrate 2
72	2.10	2.80	4.50	6.90	9.30	12.80
144	4.10	4.90	5.80	8.10	11.10	15.50
216	5.70	6.80	7.90	10.40	14.90	18.30
288	7.80	8.10	10.70	12.80	18.30	21.80
360	9.50	10.90	14.10	15.50	21.90	25.30
432	10.70	12.50	16.30	18.20	24.50	27.50
504	11.45	15.30	19.70	21.50	26.30	30.10
576	12.50	17.70	21.80	23.30	28.80	32.50
648	14.80	19.30	23.50	25.80	30.10	33.90
720	15.10	20.50	25.10	28.30	30.50	35.10

Table 1.4 Total Nitrogen Content of Fungal Biomass (g% d.m.) during the Bioconversion of Substrates, Depending on the Culture Type and the Time for Cultivating

	Total Nitrogen Content of Fungal Biomass (g% d.m.)					
	L. edodes (Monoculture)		*P. ostreatus* (Monoculture)		*L. edodes-P. ostreatus* (Co-culture)	
Time (h)	Substrate 1	Substrate 2	Substrate 1	Substrate 2	Substrate 1	Substrate 2
72	3.50	3.90	4.50	5.10	7.90	9.50
144	4.10	4.75	5.80	6.40	9.30	12.10
216	5.70	6.55	7.70	8.50	14.10	15.80
288	7.80	7.90	9.80	10.10	15.80	18.10
360	9.50	9.80	12.10	12.50	18.30	21.90
432	10.70	11.10	14.00	14.40	21.50	23.30
504	11.45	12.70	16.70	17.30	23.60	25.70
576	12.10	13.50	18.50	20.10	25.90	27.10
648	12.80	14.30	20.80	21.80	27.20	28.90
720	12.90	14.50	21.30	23.20	28.10	30.30

The bioprocess to obtain the nutritional biomass of *P. ostreatus* was carried out using the cultivation medium composed of all natural ingredients, which provide the development of submerged fermentation induced by fungal enzymatic activity, and was much faster and showed far greater economic efficiency compared to currently used methods.

Table 1.5 The Weight Loss of Dried Matter Amount (%) from the Substrates, Depending on the Culture Type and Time for Cultivating

	The Weight Loss of Dry Matter Amount from the Substrate (g%)					
	L. edodes (Monoculture)		*P. ostreatus* (Monoculture)		*L. edodes-P. ostreatus* (Co-Culture)	
Time (h)	Substrate 1	Substrate 2	Substrate 1	Substrate 2	Substrate 1	Substrate 2
72	2.50	4.30	5.10	6.40	7.30	10.40
144	3.10	5.50	5.90	7.00	8.50	12.80
216	3.90	6.70	6.70	8.50	9.70	14.10
288	4.80	7.90	7.90	9.40	10.80	15.30
360	5.50	8.80	8.80	10.50	12.50	16.50
432	6.40	9.50	10.90	12.80	14.80	18.30
504	7.30	10.90	12.70	14.30	16.30	20.10
576	8.50	12.10	13.50	15.90	17.70	21.80
648	9.30	13.70	14.90	17.40	18.50	23.70
720	10.10	14.30	15.80	18.50	20.80	25.50

The biotechnological process of controlled SCM showed the following advantages:

1. uses nutrient media consisting of fully natural ingredients for culturing the strain *P. ostreatus*, in order to obtain a food supplement with high nutritional value;
2. removes the technological processes and does not require expensive cultivation substrates and auxiliary materials, which could increase production costs;
3. provides short production time with increased amounts of fungal biomass, which contains biologically active substances with nutritional properties significantly higher than other farming methods employed to date.

1.8 CONCLUDING REMARKS

The optimization of any submerged cultivation bioprocess is essential for biotechnological development in an industrial-scale application. In this respect, it should be taken into consideration that the physical and chemical factors interact and affect the efficacy of the bioprocess regarding mycelia growth inside the liquid medium. Comparing all registered data, it may be noticed that the correlation between the dry weight of mycelia pellets and their protein, sugar, and nitrogen contents is kept at a balanced ratio, as in case of the fungal pellets of each tested mushroom species.

The processing of the recorded data on variations in the concentration of total reducing sugars (mg/g) in the fungal biomass grown on substrates S1 and S2, depending on the culture type and for different periods of time, showed significant differences between the amounts of total reducing sugars in the biomass accumulated obtained during the three types of fungal cultures. The co-culture of *L. edodes* and *P. ostreatus* showed the highest rate of accumulation of reducing sugars, with values of 30–32 mg/g, significantly higher than specific monocultures of *L. edodes* and *P. ostreatus*, which recorded maximum amounts of 26–28.5 mg/g and 19–21 mg/g, respectively.

The data relating to the variation of total nitrogen content in the fungal biomass synthesized during the bioconversion of substrate S1, depending on the time and type of culture, shows a highly significant percentage difference between the type of co-culture and the monocultures, in particular with that belonging to *L. edodes*. Statistically analyzing the change in the total nitrogen content of the fungal biomass (g% dry matter) during the bioconversion of substrate S2, depending on the culture type and the specific period of time, showed significant differences among all mushroom cultures, which reflect the lower efficiency of nitric biosynthesis for monocultures compared to the type of co-culture.

The results concerning the reducing sugars were correlated with those concerning the lowering of dry matter contained in cultivation substrates during the bioconversion processes, and then they were comparatively analyzed in case of the monocultures and co-culture of *L. edodes* and *P. ostreatus*. The weight loss of the total amount of dry matter (%) from the composition of the cultivation substrates (S1 and S2) is in direct correlation to the rate of their decomposition. The analysis of total nitrogen content of fungal biomass (g% d.m.) in the conversion process, substrate S1 and substrate S2, strictly depending on the type of culture and time period, highlights the constant accumulation of nitrogen in the total fungal biomass in the three culture types.

By comparing the registered results concerning the use of both substrates for SCM, there are significant differences between them whatever the type of fungal cultures and time periods used for each phase of the bioconversion, which highlights the important influence of substrate composition in the efficient cultivation of mushroom for getting a mycelial biomass with high nutritional value and economic benefit.

REFERENCES

Bae, J.T., Sinha, J., Park, J.P., Song, C.H., Yun, J.W., 2000. Optimization of submerged culture conditions for exo-biopolymer production by *Paecilomyces japonica*. J. Microbiol. Biotechnol. 10, 482–487.

Baker, J.O., Adney, W.S., Thomas, S.R., Nives, R.A., 1995. Synergism between purified bacterial and fungal cellulases. ACS Symp. Ser. 618, 114–141.

Beguin, P., 1990. Molecular biology of cellulose degradation. Ann. Rev. Microbiol. 44, 219–248.

Beguin, P., Aubert, J.P., 1994. The biological degradation of cellulose. FEMS Microbiol. Rev. 13, 25–58.

Beveridge, T.J., Makin, S.A., Kaduregamuwa, J.L., Li, Z., 1997. Interactions between biofilms and the environment. FEMS Microbiol. Rev. 20, 291–303.

Boddy, L., 1992. Fungal communities in wood decomposition. In: Carroll, G.C., Wicklow, D.T. (Eds.), The Fungal Community: Its Organization and Role in the Ecosystem, second ed. Marcel Dekker, New York, NY, pp. 749–782.

Carlile, M.J., Watkinson, S.C., 1994. The Fungi. Academic Press, London, UK.

Chahal, D.S., 1994. Biological disposal of lignocellulosic wastes and alleviation of their toxic effluents. In: Chaudry, G.R. (Ed.), Biological Degradation and Bioremediation of Toxic Chemicals. Chapman & Hall, London, pp. 364–385.

Chahal, D.S., Hachey, J.M., 1990. Use of hemicellulose and cellulose system and degradation of lignin by *Pleurotus sajor-caju* grown on corn stalks. Am. Chem. Soc. Symp. 433, 304–310.

Cocker, R., 1980. Interactions between fermenter and microorganism: tower fermenter. In: Smith, J.E., Berry, D.R., Kristiansen, B. (Eds.), Fungal Biotechnol. Academic Press, London, pp. 112–127.

Cohen, R., Persky, L., Hadar, Y., 2002. Biotechnological applications and potential of wood-degrading mushrooms of the genus *Pleurotus*. Appl. Microbiol. Biotechnol. 58, 582–594.

Confortin, F.G., Marchetto, R., Bettin, F., Camassola, M., Salvador, M., Dillon, A.J., 2008. Production of *Pleurotus sajor-caju* strain PS-2001 biomass in submerged culture. J. Ind. Microbiol. Biotechnol. 35 (10), 1149–1155.

Davitashvili, E., Kapanadze, E., Kachlishvili, E., Khardziani, T., Elisashvili, V., 2008. Evaluation of higher Basidiomycetes mushroom lectin activity in submerged and solid-state fermentation of agro-industrial residues. Int. J. Med. Mushrooms 10, 173–178.

Dubois, M., Gilles, K.A., Hamilton, J.K., Rebers, P.A., Smith, F., 1956. Colorimetric method for determination of sugars and related substances. Anal. Chem. 28, 350–356.

Elisashvili, V., 2012. Submerged cultivation of medicinal mushrooms: bioprocesses and products (review). Int. J. Med. Mushrooms 14 (3), 211–239.

Elisashvili, V., Kachlishvili, E., Wasser, S., 2009. Carbon and nitrogen source effects on *Basidiomycetes* exopolysaccharide production. Appl. Biochem. Microbiol. 45, 531–535.

Finkelstein, D.B., Ball, C., 1992. Biotechnology of Filamentous Fungi: Technology and Products. Butterworth-Heinemann, Boston, MA, pp. 15–56.

Frankland, J.C., 1992. Mechanisms in fungal succesions. In: Wicklow, D.T., Carroll, G.C. (Eds.), The Fungal Community: Its Organisation and Role in the Ecosystem Marcel Dekker, New York, NY, pp. 383–410.

Glazebrook, M.A., Vining, L.C., White, R.L., 1992. Growth morphology of *Streptomyces akiyoshiensis* in submerged culture: influence of pH, inoculum, and nutrients. Can. J. Microbiol. 38, 98–103.

Gregg, D.J., Saddler, J.N., 1996. Factors affecting cellulose hydrolysis and the potential of enzyme recycle to enhance the efficiency of an integrated wood to ethanol production. Biotechnol. Bioeng. 51 (4), 375–381.

Hawksworth, D.L., 1992. Biodiversity in microorganisms and its role in ecosystem function. In: Solbrig, O.T., van Emden, H.M., van Oordt, P.G.W.J. (Eds.), Biodiversity and Global Change. IUBB, Paris, pp. 83–93.

Hawksworth, D.L., Kirk, P.M., Sutton, Pegler, D.N., 1995. Ainsworth & Bisby's Dictionary of the Fungi, eighth ed. Wallingford, pp. 56–59, 211–214, 575–579.

Hobbs, C., 1995. Medicinal Mushrooms. An Exploration of Tradition, Healing & Culture. Interweave Press, Inc., Loveland, CO, USA, pp. 14–56.

Homolka, L., 2014. Preservation of live cultures of *Basidiomycetes* – recent methods. Fungal Biology 118, 107–125.

Howard, R.L., Abotsi, E., Jansen van Rensburg, E.L., Howard, S., 2003. Lignocellulose biotechnology: issues of bioconversion and enzyme production. Afr. J. Biotechnol. 2 (12), 602–619.

Jiang, L.F., 2010. Optimization of fermentation conditions for pullulan production by *Aureobasidium pullulans* using response surface methodology. Carbohydr. Polym. 79, 414–417.

Jones, K., 1995. Shiitake – The Healing Mushroom. Healing Arts Press, Rochester, VT, pp. 3–15.

Kim, H.O., Lim, J.M., Hwang, H.J., Choi, J.W., Yun, J.W., 2007. Optimization of submerged culture condition for the production of mycelial biomass and exopolysaccharides by *Agrocybe cylindracea*. Bioresour. Technol. 96 (10), 1175–1182.

Kim, S.W., Hwang, H.J., Park, J.P., Cho, Y.J., Song, C.H., Yun, J.W., 2004. Mycelial growth and exo-biopolymer production by submerged culture of various edible mushrooms under different media. Lett. Appl. Microbiol. 34, 56–61.

Kirk, T.K., Eriksson, K.E., 1990. Roles of biotechnology in manufacture. In: Robert, F. (Ed.), World Pulp & Paper Technology The Sterling Publishing Group, London, pp. 23–28.

Kubicek, C.P., Messner, R., Guber, F., Mach, R.L., 1993. The *Trichoderma* cellulase regulatory puzzle: from the interior life of a secretory fungus. Enzyme Microb. Technol. 15, 90–98.

Leahy, J.G., Colwell, R.R., 1990. Microbial degradation of hydrocarbons in the environment. Microbiol. Rev. 54, 305–315.

Lee, B.C., Bae, J.T., Pyo, H.B., Choe, T.B., Kim, S.W., Hwang, H.J., et al., 2004. Submerged culture conditions for the production of mycelial biomass and exopolysaccharides by the edible Basidiomycete *Grifola frondosa*. Enzyme Microb. Technol. 35, 369–376.

Lin, E.S., 2010. Submerged culture medium composition for the antioxidant activity by *Grifola frondosa* TFRI1073. Food Sci. Biotechnol. 19, 917–922.

Lin, J.H., Yang, S.S., 2006. Mycelium and polysaccharide production of *Agaricus blazei* Murrill by submerged fermentation. J. Microbiol. Immunol. Infect. 39 (2), 98–108.

Mikiashvili, N.A., Elisashvili, V., Wasser, S.P., Nevo, E., 2006. Comparative study of lectin activity of higher Basidiomycetes. Int. J. Med. Mushrooms 8, 31–38.

Moo-Young, M., 1993. Fermentation of cellulose materials to mycoprotein foods. Biotechnol. Adv. 11 (3), 469–482.

Nevalainen, H., Pentilla, M., 1995. Molecular biology of cellulolytic fungi. In: Kuck, H. (Ed.), The Mycota. Genetics and Biotechnology, vol. 2. Springer-Verlag, Berlin-Heidelberg, pp. 303–319.

Papagianni, M., 2004. Fungal morphology and metabolite production in submerged mycelial processes. Biotechnol. Adv. 22 (3), 189–259.

Papaspyridi, L.M., Aligiannis, N., Topakas, E., Christakopoulos, P., Skaltsounis, A.L., Fokialakis, N., 2012. Submerged fermentation of the edible mushroom *Pleurotus ostreatus* in a batch stirred tank bioreactor as a promising alternative for the effective production of bioactive metabolites. Molecules 17 (3), 2714–2724.

Park, J.P., Kim, S.W., Hwang, H.J., Yun, J.W., 2001. Optimization of submerged culture conditions for the mycelial growth and exo-biopolymer production by *Cordyceps militaris*. Lett. Appl. Microbiol. 33 (1), 76–81.

Petre, M., Petre, V., 2008. Environmental biotechnology to produce edible mushrooms by recycling the winery and vineyard wastes. J. Environ. Protect. Ecol. 9 (1), 87–97.

Petre, M., Petre, V., 2012. The semi-solid state cultivation of edible mushrooms on agricultural organic wastes. Scientific Bull. Ser. F. Biotechnol. vol. XVI, 36–40.

Petre, M., Petre, V., 2013. Environmental biotechnology for bioconversion of agricultural and forestry wastes into nutritive biomass. In: Petre, M. (Ed.), Environmental Biotechnology – New Approaches and Prospective Applications, InTech, Rijeka, Croatia, pp. 3–23.

Petre, M., Petre, V., Duţă, M., 2014a. Mushroom biotechnology for bioconversion of fruit tree wastes into nutritive biomass. Rom. Biotech. Lett. 19 (6), 9952–9958.

Petre, M., Petre, V., Rusea, I., 2014b. Microbial composting of fruit tree wastes through controlled submerged fermentation. Italian J. Agron. 9 (4), 152–156.

Petre, M., Teodorescu, A., Tuluca, E., Andronescu, A., 2010. Biotechnology of mushroom pellets producing by controlled submerged fermentation. Rom. Biotech. Lett. 15 (2), 50–56.

Porras-Arboleda, S.M., Valdez-Cruz, N.A., Rojano, B., Aguilar, C., Rocha-Zavaleta, L., Trujillo-Roldán, M.A., 2009. Mycelial submerged culture of new medicinal mushroom, *Hum-phreya coffeata* (Berk.) Stey. (Aphyllophoromycetideae) for the production of valuable bioactive metabolites with cytotoxicity, genotoxicity, and antioxidant activity. Int. J. Med. Mushrooms 11, 335–350.

Raaska, L., 1990. Production of *Lentinus edodes* mycelia in liquid media: improvement of mycelial growth by medium modification. Mushroom J. Tropics 8, 93–98.

Ropars, M., Marchal, R., Pourquie, J., Vandercasteele, J.P., 1992. Large scale enzymatic hydrolysis of agricultural lignocellulosic biomass. Bioresour. Technol. 42, 197–203.

Sanchez, C., 2004. Modern aspects of mushroom culture technology. Appl. Microbiol. Biotechnol. 64 (6), 756–762.

Sanchez, C., 2010. Cultivation of *Pleurotus ostreatus* and other edible mushrooms. Appl. Microbiol. Biotechnol. 85 (5), 1321–1337.

Shih, I.L., Chou, B.W., Chen, C.C., Wu, J.Y., Hsieh, C., 2008. Study of mycelial growth and bioactive polysaccharide production in batch and fed-batch culture of *Grifola frondosa*. Bioresour. Technol. 99 (4), 785–793.

Smith, J.E., 1998. Biotechnology, third ed. Cambridge University Press, UK, pp. 56–70.

Songulashvili, G., Elisashvili, V., Penninckx, M., Metreveli, E., Hadar, Y., Aladashvili, N., et al., 2005. Bioconversion of plant raw materials in value-added products by *Lentinus edodes* (Berk.) Singer and *Pleurotus* spp. Int. J. Med. Mushrooms 7 (3), 467–468.

Stamets, 2000. Growing Gourmet and Medicinal Mushrooms, third ed. Ten Speed Press, Berkeley, CA, pp. 123–127.

Subramaniyam, R., Vimala, R., 2012. Solid state and submerged fermentation for the production of bioactive substances: a comparative study. Int. J. Sci. Nat. 3 (3), 480–486.

Tanaka, M., Matsuno, R., 1985. Conversion of lignocellulosic materials to single-cell protein (SCP): review developments and problems. Enzyme Microbial. Technol. 7, 197–207.

Trinci, A.P.J., 1992. Myco-protein: a twenty-year overnight success story. Mycol. Res. 96, 1–13.

Tsivileva, O.M., Nikitina, V.E., Garibova, L.V., 2005. Effect of culture medium composition on the activity of extracellular lectins of *Lentinus edodes*. Appl. Biochem. Microbiol. 41, 174–176.

Turlo, J., 2014. The biotechnology of higher fungi—current state and perspectives. Folia Biol. Oecol. 10, 49–65.

Uphoff, N., 2002. Agroecological Innovations: Increasing Food Production with Participatory Development. Earthscan, London, pp. 153–160.

van den Twell, W.J.J., Leak, D., Bielicki, S., Petersen, S., 1994. Biocatalysts production. In: Cabral, J.M.S., Boros, D.B.L., Tramper, J. (Eds.), Applied Biocatalysis Harwood Acad. Publ. GmbH, Chur., Switzerland, pp. 157–235.

Verstraete, W., Top, E., 1992. Holistic Environmental Biotechnology. Cambridge University Press, UK, pp. 1–18.

Vournakis, J.N., Runstadler, P.W., 1989. Microenvironment: the key to improve cell culture products. J. Biotechnol. 7, 143–145.

Wainwright, M., 1992. An Introduction to Fungal Biotechnology. Wiley, Chichester, West Sussex, pp. 56–73.

Wasser, S.P., 2010. Medicinal mushroom science: history, current status, future trends, and unsolved problems. Int. J. Med. Mushrooms 12 (1), 1–16.

Wedde, M., Iacobs, M., Stahl, U., 1999. Fungi: important organisms in history and today. In: Oliver, R.P., Schweizer, M. (Eds.), Molecular Fungal Biology. Cambridge University Press, UK, pp. 21–35.

Wood, T.M., 1992. Fungal cellulases. Biochem. Soc. Trans. 20, 46–52.

Xu, X., Yan, H., Chen, J., Zhang, X., 2011. Bioactive proteins from mushrooms. Biotechnol. Adv. 29, 667–674.

Zarnea, G., 1984. The physiology of microorganisms Treaty of Microbiology, vol. 2. Romanian Academy Publishing House, Bucharest, pp. 28–65.

Zarnea, G., 1994. Theoretical bases of microbial ecology Treaty of Microbiology, vol. 5. Romanian Academy Publishing House, Bucharest, pp. 154–163.

Zhang, B.B., Cheung, P.C.K., 2011. A mechanistic study of the enhancing effect of Tween 80 on the mycelial growth and exopolysaccharide production by *Pleurotus tuber-regium*. Bioresour. Technol. 102, 8323–8326.

CHAPTER 2

BIOTECHNOLOGICAL RECYCLING OF FRUIT TREE WASTES BY SOLID-STATE CULTIVATION OF MUSHROOMS

Violeta Petre[1], Marian Petre[2], Ionela Rusea[2] and Florin Stănică[3]

[1]Department of Biology, Sfântul Sava College, Bucharest, Romania [2]Faculty of Sciences, University of Pitesti, Pitesti, Romania [3]Faculty of Horticulture, University of Agronomic Sciences and Veterinary Medicine, Bucharest, Romania

2.1 INTRODUCTION

The woody wastes which are produced every year during fruit tree pruning in all Romanian orchards represent in total a huge amount of redundant materials that need to be recycled through their use as main substrates for solid-state mushroom cultivation. These organic materials coming from the fruit trees are composed of dried trunks, branches, leaves, and even fruit seeds. Statistical data showed that in 2012 between 1.2 and 1.5 tons/ha of dried trunks and branches of fruit tree wastes are produced on around 75.000 ha in Romania, resulting a total amount of 90,000 tons up to 112,500 tons of such lignocellulosic materials (www.eubia.org; www.insse.ro).

Taking into consideration that almost 90% of all these redundant organic wastes are used as the cheapest fuels for heating of fruit tree farm owners and only about 10% are used as raw materials for furniture manufacturing, it is obviously a great challenge for fruit tree farmers to apply the biotechnology of recycling the fruit tree wastes as natural substrates for mushroom cultivation.

The biotechnology of lignocellulosic material conversion into high-value products normally requires multistep processes, which include pretreatment (mechanical, chemical, or biological), polymer hydrolysis to produce readily metabolizable molecules (e.g., hexose or pentose sugars), the use of such molecules to support microbial growth or to produce biochemical compounds, and the separation and purification of final products (Chang and Miles, 2004; Chahal, 1994; Breene, 1990).

The laboratory experiments which are presented in this chapter were carried out on testing and optimization of fruit tree waste recycling through controlled cultivation of edible and medicinal mushroom species *Ganoderma lucidum* and *Pleurotus ostreatus* in order to retrieve their carpophores to be used as food and nutraceuticals. To achieve these goals, a new and innovative environmental biotechnology was applied for full recovery and valorization of all fruit tree wastes (leaves, branches, dried trunks), usable as raw materials of hitherto untapped economic value, to prepare nutritive substrates for mushroom growth. In this way, the lignocellulosic wastes of fruit trees may be integrated extremely quickly into the main cycles of organic matter in nature, as new links in the natural food chain made by the

Mushroom Biotechnology. DOI: http://dx.doi.org/10.1016/B978-0-12-802794-3.00002-3

cultivation of edible and medicinal mushrooms belonging to the mentioned species (Beguin, 1990; Verstraete and Top, 1992; Smith, 1998; Wedde et al., 1999).

Among the Basidiomycetes group, the genus *Pleurotus* includes a group of ligninolytic mushrooms with medicinal properties and important biotechnological and environmental applications. Thus, the solid-state cultivation of *Pleurotus* species is an economically important food industry worldwide, which has expanded in the past few years (Arjona et al., 2009; Das and Mukherjee, 2007; Kurt and Buyukalaca, 2010). *Pleurotus ostreatus* is the second most cultivated edible mushroom worldwide after *Agaricus bisporus*, according to Sanchez (2010). Normally, *P. ostreatus*, known as oyster mushroom, requires a shorter growth time in comparison to other edible mushrooms (Ropars et al., 1992; Salmones et al., 2005; Kachlishvili et al., 2006).

The substrates used for the cultivation of this mushroom species do not require complex ingredients to be added, and hence are less expensive (Leahy and Colwell, 1990; Moser, 1994; Rani et al., 2008). By growing *P. ostreatus*, the oyster mushrooms, the lignocellulosic materials are converted in a high percentage into fruiting bodies, increasing profitability (McIntyre, 1987; Obodai et al., 2003; Sanchez, 2004; Songulashvili et al., 2005).

Pleurotus ostreatus demands few environmental controls, their fruiting bodies are not often attacked by pathogens or pests, and they can be cultivated in a simple and cheap way. All this makes *P. ostreatus* cultivation an excellent alternative for production of mushrooms as compared to other mushroom species (Chahal and Hachey, 1990; Eichlerova et al., 2000; Uphoff, 2002; Holker et al., 2004).

Cultivation of *P. ostreatus* strains for commercial purposes could be performed in plastic bags, and the tree leaves appear to be an excellent growth substrate for conversion into fruiting bodies, with a biological efficiency of 108–118% (Stamets, 2000; Robinson et al., 2001; Elisashvili et al., 2008).

2.2 THE SOLID-STATE CULTIVATION OF MUSHROOMS (SSCM) ON LIGNOCELLULOSIC WASTES OF FRUIT TREES

For the development and application of biotechnology for conversion of fruit tree wastes, experiments were set up for the cultivation of the mushroom species previously mentioned on substrates made of wastes from performing operations involving the cutting of dry or unproductive branches of tree species of apple, plum, and cherry, and the dried leaves after their collection. The application of such biotechnology allows a consistent reduction of pollutant effects induced by constant accumulation of huge quantities of lignocellulosic wastes, which are produced annually on every fruit tree farm in Romania.

In our experiments, two mushroom species, namely *G. lucidum* (Wm. Curtis: Fries) Karsten and *P. ostreatus* (Jacquin ex Fries) Kummer, belonging to the mushroom collection of the University of Pitesti, were used for testing as biodegrading agents of lignocellulosic wastes of fruit trees. The stock cultures were maintained on Difco malt-extract agar (MEA) slants (Difco), incubated at 25°C for 5–7 days, and then stored at 4°C. To prepare the inoculum for solid-state cultivation on fruit tree wastes, the seed cultures were grown in 250-mL Erlenmayer flasks containing 100 mL of Difco malt-extract broth (MEB) at 23°C on rotary shaker incubators at 150 rpm (rotations per minute) for 7 days (Jiang, 2010; Park et al., 2001; Wagner et al., 2004; Elisashvili, 2012).

The mushroom cultures were prepared by inoculating 100 mL of MEB culture medium using 3–5% (v/v) of the seed culture and then were grown at 23–25°C inside the Erlenmayer flasks of 250 mL, mounted on a rotary shaking incubator. The experiments were conducted under the following conditions: temperature,

Table 2.1 The Composition of Substrate Variants Used for Controlled Cultivation of Mushroom Species *Ganoderma lucidum* and *Pleurotus ostreatus*

Variants of Substrates	Composition of Cultivation Substrates
S1	Dried apple branches 60%, dried apple leaves 15%, barley bran 5%, wheat bran 5%
S2	Dried plum branches 60%, dried plum leaves 15%, barley bran 5%, wheat bran 5%
S3	Dried cherry branches 60%, dried cherry leaves 15%, barley bran 5%, wheat bran 5%
S4	Dried apple branches 20%, dried plum branches 20%, dried cherry branches 20%, barley bran 5%, wheat bran 5%
S5	Dried apple leaves 25%, dried plum leaves 25%, dried cherry leaves 25%, barley bran 5%, wheat bran 5%
Control	Pure cellulose

25°C; agitation speed, 120–180 rev/min; initial pH, 4.5–5.5. After 10–12 days of incubation, all the mushroom cultures were ready to be inoculated with mycelium in aseptic conditions.

2.2.1 PREPARATION OF SUBSTRATES FOR SSCM

Before starting the proper bioprocess of mushroom cultivation, all lignocellulosic wastes, mainly composed of trunks, branches, and leaves belonging to apple, plum, and cherry trees, were dried at least 6 months at 25–30°C after they were collected from the fruit tree farms. Then all woody dried materials were chopped and split into relatively equal-sized fragments of 3–5 cm.

Five variants of mushroom cultivation substrates were set up, made of lignocellulosic wastes belonging to apple, plum, and cherry trees, mixed with cereal grain wastes from the milling industry, such as wheat bran (5% w/w) and barley bran (5% w/w). These natural ingredients were added with the role of enhancing the processes of growth and development of the mycelium of the fungi used, in order to stimulate enzymatic activity of the fungal species, as well as the processes of growth and development of mycelia biomass, according to Table 2.1.

2.2.2 MAIN STAGES OF SSCM

In the first phase, all variants of substrates for mushroom cultivation were soaked in a tap water–based synthetic medium containing 5 g/L yeast extract for 20 h at room temperature. After leaching, 1 kg of the substrate was placed in polypropylene gas-permeable bags for sterilization by autoclaving at 121°C for 1 h (Saddler et al., 1993; Wainwright, 1992). After cooling at room temperature (23–25°C), in the second phase of the experiments, the bags containing lignocellulosic wastes of apple, plum, and cherry trees were inoculated with 10% (wet weight) of mycelium from pure cultures of the species *G. lucidum* and *P. ostreatus*, using the hood with sterile air laminar flow for aseptic handling of biological materials during the inoculation. The inoculated bags were incubated in the dark at 23°C. Immediately after inoculation of substrates with the pure mushroom cultures of mentioned species, the inoculated plastic bags (three replicates for each strain/substrate) were placed in growth chambers to be kept at a constant temperature of 21 or 23°C, during an incubation period lasting between 20 and 30 days, depending on the cultivated mushroom species (van den Twell et al., 1994; Stamets, 2000).

After incubation, the bags were exposed over 3 days at 4°C to experience a cold shock necessary for the stimulation of fruit body formation. Then the blocks were kept in the fruiting room at 15–18°C, having a relative humidity around 90%, an aeration volume of 3 shifts/h, and under illumination of about 1500 lux (Arjona et al., 2009; Cohen et al., 2002; Stamets, 2000). When the first primordia appeared, the bags were removed from the blocks formed by the mycelium colonization on the whole surface and inside the substrate volume. Thus, the carpophores emerged outside the plastic bags and they were collected during three consecutive flushes (Rosado et al., 2002; Songulashvili et al., 2005; Sainos et al., 2006).

Regarding the amounts of fresh fungal biomass produced by cultivation of *G. lucidum* and *P. ostreatus* mycelia, the experimental results, which are presented in Figures 2.1 and 2.2, reflect the variation of fungal biomass growth on the substrates consisting of fruit tree wastes, supplemented with natural ingredients, according to Table 2.1, depending on both substrate type and fungal species.

Taking into consideration the statistical processing of the recorded data after performing chemical analysis, there was a variation in the amount of fresh fungal biomass produced by *G. lucidum* mycelia, depending on the substrates made of apple, plum, and cherry tree wastes, which reached an average value of 21% w/w; in the case of *P. ostreatus*, the average quantity of biomass was significantly greater, registering 28.75% w/w, as shown in Figure 2.1.

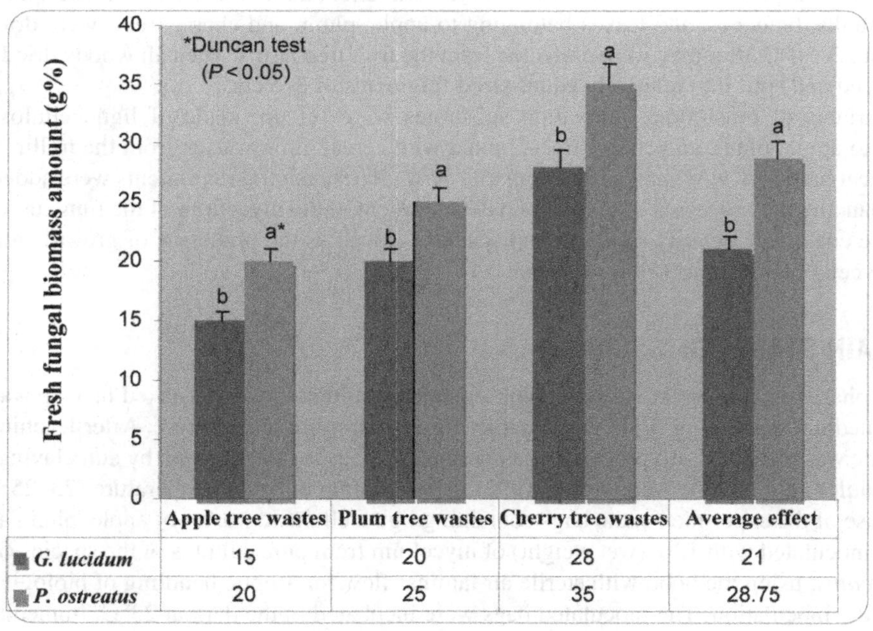

	Apple tree wastes	Plum tree wastes	Cherry tree wastes	Average effect
G. lucidum	15	20	28	21
P. ostreatus	20	25	35	28.75

FIGURE 2.1

Variation of fungal biomass amount produced by cultivation of *Ganoderma lucidum* and *Pleurotus ostreatus* on substrates made of fruit tree wastes depending on the substrate type.

On the other hand, the ratio of the average quantity of mycelial biomass obtained by the solid-state cultivation of mushroom (SSCM) species *G. lucidum* and *P. ostreatus* registered only a difference of 5 g% in the substrate variants made of apple and plum tree wastes, while in the case of cherry tree wastes, the difference was of 9 g% w/w in comparison with plum tree wastes and 14% w/w compared to apple tree wastes (Figure 2.2).

2.2.3 CHEMICAL ANALYSIS OF THE COLLECTED MUSHROOMS

To prove the effectiveness of the biotechnological procedure which was applied to test the biochemical ability of mushroom species to convert the fruit tree wastes into useful products, chemical analyses were carried out at the end of the mushroom cultures.

The fungal biomass produced after the cultivation cycles was investigated to determine the dry matter, carbohydrates, total nitrogen, and protein, both for secondary mycelium and for carpophores belonging to the cultivated fungal species. Initially, the contents of dry matter of mycelia and carpophores were analyzed for both mushroom species *G. lucidum* and *P. ostreatus*, using a drying oven at a temperature of 105°C and weighing the dried sample directly. The results showed a higher dried matter content for *G. lucidum* species compared with *P. ostreatus*. The differences between dry matter contents of the biomass of fungi tested in experiments demonstrated significant differences of 15–16 g%

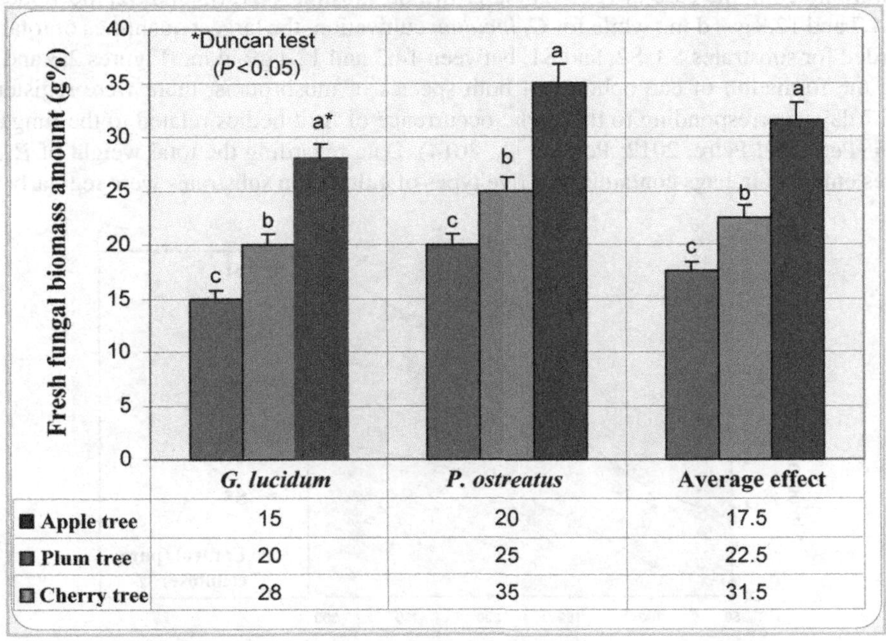

	G. lucidum	P. ostreatus	Average effect
Apple tree	15	20	17.5
Plum tree	20	25	22.5
Cherry tree	28	35	31.5

FIGURE 2.2

Variation of fungal biomass amount produced by cultivation of *Ganoderma lucidum* and *Pleurotus ostreatus* on substrates made of fruit tree wastes, depending on the fungal species.

between secondary mycelium and the carpophores (fruit bodies) of each of the mentioned species (Petre and Petre, 2013).

To determine the soluble carbohydrates in carpophores as well as the secondary mycelium of *G. lucidum* and *P. ostreatus*, the colorimetric method with anthrone reagent was used (Dubois et al., 1956), modified by the phenol-sulfuric acid method in microplate format Masuko et al., 2005). Comparing the results, there are no significant differences between the two cultivated species. Regarding the variation of soluble carbohydrate content as determined by the analysis of mushroom biomass belonging to *G. lucidum* and *P. ostreatus*, depending on the stage of mycelium development, there is a significant difference of 4.1 g% d.m. between the secondary and tertiary stage (corresponding to carpophore formation) of *G. lucidum* but only 3.35 g% d.m. in the case of *P. ostreatus*. At the end of the cultivation cycle, the total protein content was determined by the Lowry method (1951), for both the secondary mycelium and carpophores of mushrooms species *G. lucidum* and *P. ostreatus*. There were significant differences between the total protein contents of secondary mycelia compared to those of carpophores for both mushroom species. The total protein content of the carpophores belonging to *G. lucidum* was registered as 17.1 g/kg d.m., which means a significant improvement, while for *P. ostreatus* carpophores, it was only 15.4 g/kg d.m.

The quantitative analyses of total nitrogen in the fungal biomass during the controlled cultivation cycles of *G. lucidum* and *P. ostreatus* on each of cultivation substrates were carried out by the Kjeldahl method. The variations in total nitrogen content, calculated in relation to dry matter of the obtained biomass, were noted in the case of *P. ostreatus* grown on substrates S1, S3 and S2, the values ranging between 14.7 and 12.3 g% d.m., while for *G. lucidum* cultivation, the largest quantities of total nitrogen were recorded for substrates S3, S2, and S1, between 14.3 and 12.1 g% d.m. (Figures 2.3 and 2.4).

During the formation of carpophores of both species of mushrooms, there were registered three periods of 7 days, corresponding to the cyclic occurrence of fruit bodies related to the fungal species in question (Petre and Petre, 2012; Petre et al., 2014). Data regarding the total weight of *P. ostreatus* carpophores emerged in bags containing all five types of cultivation substrates were regularly recorded

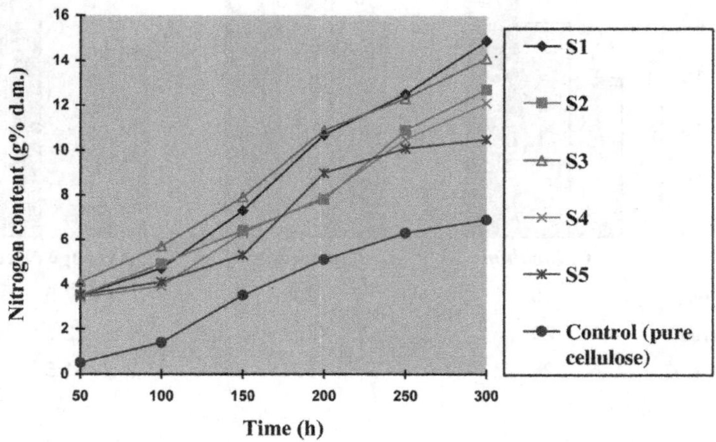

FIGURE 2.3

Variation of total nitrogen amount of *Pleurotus ostreatus* biomass.

and are presented in Table 2.2. The entire fruiting period was conducted over 30 days under constant temperature (15–18°C), relative humidity (80–85%), and air intake (3 shifts/h).

The cultivation of both mushroom species on the same cultivation substrates showed that the mushroom yield was strain dependent (Table 2.2). Biological efficiency (BE) was estimated as the ratio of the weight of fresh fruiting bodies and weight of dry substrate, multiplied by 100. *P. ostreatus* cultivated on substrate S1 and S3 as well as *G. lucidum* grown on S1 appeared to be the most productive mushroom/substrate variant ratio compared with the rest of the cultivation variants, registering a BE between 95% and 90.5%. All

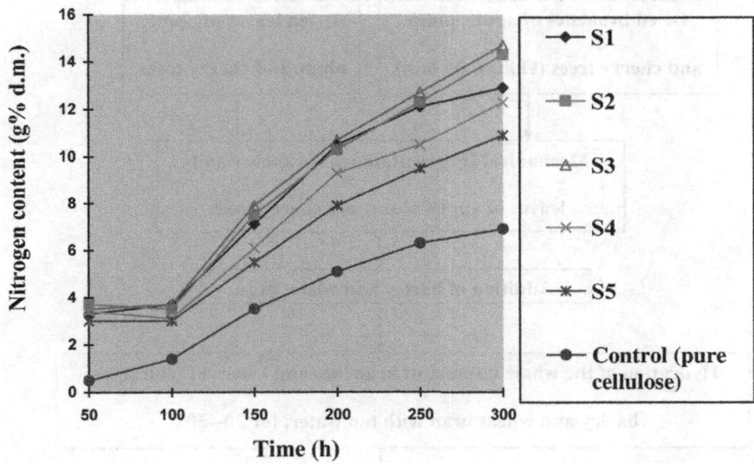

FIGURE 2.4

Variation of total nitrogen amount of *Ganoderma lucidum* biomass.

Table 2.2 The Variation of Carpophores Production during the Cultivation of Mushroom Species *Ganoderma lucidum* and *Pleurotus ostreatus*, on Five Different Cultivation Substrates

Mushroom/Substrate Variant	Fruiting Body Yield (g/kg Substrate)				
	Flush I	Flush II	Flush III	Total	BE (%)
G. lucidum/S1	495	270	140	905	90.5
G. lucidum/S2	437	215	175	827	82.7
G. lucidum/S3	435	205	155	795	79.5
G. lucidum/S4	370	195	105	670	67.0
G. lucidum/S5	350	175	85	610	61.0
P. ostreatus/S1	560	250	140	950	95.0
P. ostreatus/S2	495	230	120	850	84.5
P. ostreatus/S3	530	270	125	925	92.5
P. ostreatus/S4	470	215	125	810	81.0
P. ostreatus/S5	450	190	115	755	75.5

the cultivated mushroom species were more productive, especially during the first flush. Statistical analysis of the results regarding the weight of carpophores, revealed the highest values in the first three stages of cultivation (Petre et al., 2014). At the same time, the most productive substrates were found to be S1 and S2, followed, finally, by substrate S3. Based on the experiments carried out, the results of the laboratory-scale biotechnology for recycling of fruit tree wastes by controlled cultivation of mushroom species *G. lucidum* and *P. ostreatus* are shown in Figure 2.5.

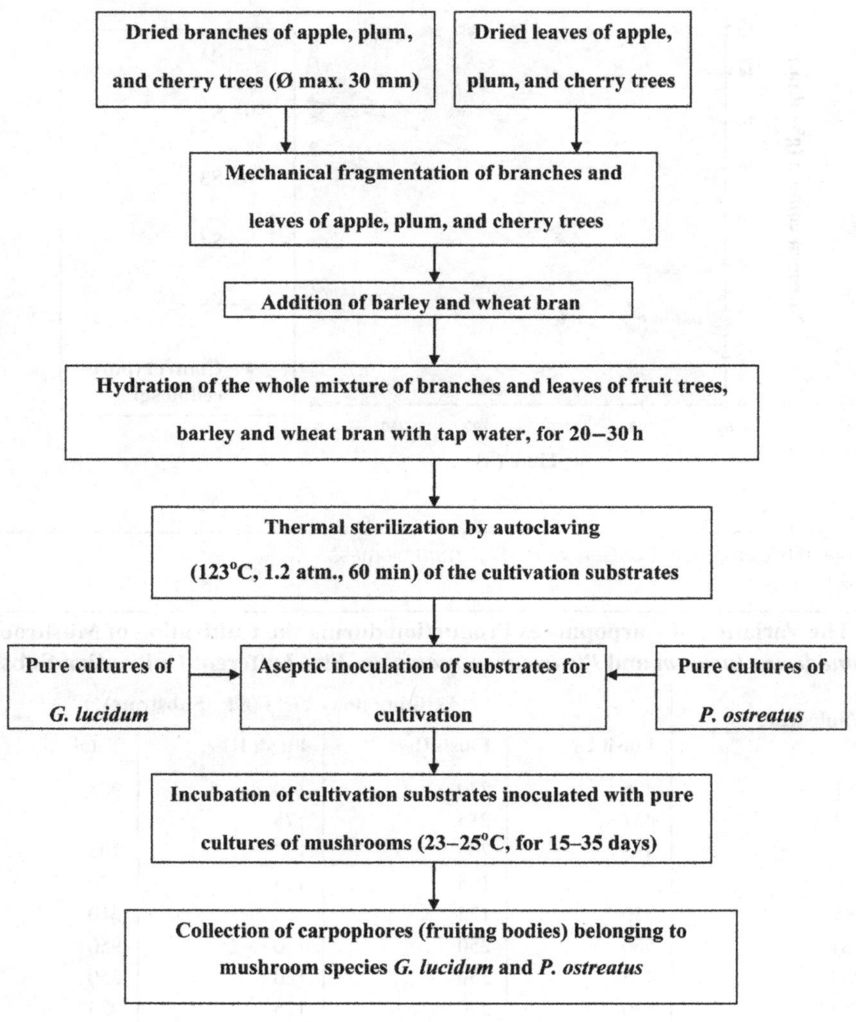

FIGURE 2.5

The scheme of biotechnology for ecological recycling of lignocellulosic fruit tree wastes through controlled cultivation of mushroom species *Ganoderma lucidum* and *Pleurotus ostreatus*.

2.3 CONCLUSIONS

The results showed a higher dried matter content for *G. lucidum* species compared with *P. ostreatus*. The differences between dry matter contents in the biomass of fungi tested in experiments demonstrated significant differences of 15–16 g% between secondary mycelium and the carpophores (fruiting bodies) of each of the mentioned species.

Regarding the variation of soluble carbohydrate content determined by the analysis of mushroom biomass belonging to *G. lucidum* and *P. ostreatus*, depending on the stage of mycelium development, there is a significant difference of 4.1 g% d.m. between the secondary and tertiary stages (corresponding to carpophore formation) of *G. lucidum* and only 3.35 g% d.m. in case of *P. ostreatus*. It was noticed that there are significant differences between total protein contents of secondary mycelia compared to those of carpophores for the two mushroom species. Thus, the total protein content of the carpophores belonging to *G. lucidum* was registered as 17.1 g/kg d.m., which is a significant improvement, while for *P. ostreatus* carpophores, it was only 15.4 g/kg d.m.

The variations in total nitrogen content, calculated in relation to the dry matter of the obtained biomass, were noticed in the case of *P. ostreatus* species grown on the substrates S1, S3, and S2, the values ranging between 14.7 and 12.3 g% d.m., while for *G. lucidum* cultivation, the largest quantities of total nitrogen were recorded for substrates S3, S2, and S1, between 14.3 and 12.1 g% d.m.

Statistical analysis of the results regarding the weight of carpophores harvested at the end of the cultivation cycle on fruit tree wastes used as substrates revealed that the highest values were registered in the first three stages of cultivation, which were recorded as the most significant results, and the most productive substrates were found to be S1 and S2, followed, finally, by the substrate S3.

ACKNOWLEDGMENTS

This research was carried out in the framework of Project No. 201/28.10.2013 from the Research Program "Innovation"—Subprogram "Checks of Innovation," funded by Romanian Ministry of Education and Research.

REFERENCES

Arjona, D., Aragon, C., Aguilera, J.A., Ramirez, L., Pisabarro, A.G., 2009. Reproducible and controllable light induction of *in vitro* fruiting of the white-rot Basidiomycete *Pleurotus ostreatus*. Mycol. Res. 113 (5), 552–558.

Beguin, P., 1990. Molecular biology of cellulose degradation. Ann. Rev. Microbiol. 44, 219–248.

Breene, W.M., 1990. Nutritional and medicinal value of specialty mushrooms. J. Food Protect. 53, 833–894.

Chahal, D.S., 1994. Biological disposal of lignocellulosic wastes and alleviation of their toxic effluents. In: Chaudry, G.R. (Ed.), Biological Degradation and Bioremediation of Toxic Chemicals. Chapman & Hall, London, pp. 156–173.

Chahal, D.S., Hachey, J.M., 1990. Use of hemicellulose and cellulose system and degradation of lignin by *Pleurotus sajor-caju* grown on corn stalks. Am. Chem. Soc. Symp. 433, 304–310.

Chang, S.T., Miles, P.G., 2004. Mushrooms: Cultivation, Nutritional Value, Medicinal Effect, and Environmental Impact, second ed. CRC Press, Boca Raton, FL, pp. 470–481.

Cohen, R., Persky, L., Hadar, Y., 2002. Biotechnological applications and potential of wood-degrading mushrooms of the genus *Pleurotus*. Appl. Microbiol. Biotechnol. 58, 582–594.

Das, N., Mukherjee, M., 2007. Cultivation of *Pleurotus ostreatus* on weed plants. Bioresour. Technol. 98 (14), 2723–2726.

Dubois, M., Gilles, K.A., Hamilton, J.K., Rebers, P.A., Smith, F., 1956. Colorimetric method for determination of sugars and related substances. Anal. Chem. 28, 350–356.

Eichlerova, I., Homolka, L., Nerud, F., Zadrazil, F., Baldrian, P., Gabriel, J., 2000. Screening of *Pleurotus ostreatus* isolates for their ligninolytic properties during cultivation on natural substrates. Biodegradation 11 (5), 279–287.

Elisashvili, V., 2012. Submerged cultivation of medicinal mushrooms: bioprocesses and products (Review). Int. J. Med. Mushrooms 14 (3), 211–239.

Elisashvili, V., Penninckx, M., Kachlishvili, E., Tsiklauri, N., Metreveli, E., Kharziani, T., et al., 2008. *Lentinus edodes* and *Pleurotus* species lignocellulolytic enzymes activity in submerged and solid-state fermentation of lignocellulosic wastes of different composition. Bioresour. Technol. 99, 457–462.

Holker, U., Hofer, M., Lenz, J., 2004. Biotechnological advantages of laboratory-scale solid state fermentation with fungi. Appl. Microbiol. Biotechnol. 64, 175–186.

Jiang, L.F., 2010. Optimization of fermentation conditions for pullulan production by *Aureobasidium pullulans* using response surface methodology. Carbohydr. Polym. 79, 414–417.

Kachlishvili, E., Penninckx, M.J., Tsiklauri, N., Elisashvili, V., 2006. Effect of nitrogen source on lignocellulolytic enzyme production by white-rot Basidiomycetes under solid-state cultivation. World J. Microbiol. Biotechnol. 22 (4), 391–397.

Kurt, S., Buyukalaca, S., 2010. Yield performances and changes in enzyme activities of *Pleurotus* spp. (*P. ostreatus* and *P. sajor-caju*) cultivated on different agricultural wastes. Bioresour. Technol. 101 (9), 3164–3169.

Leahy, J.G., Colwell, R.R., 1990. Microbial degradation of hydrocarbons in the environment. Microbiol. Rev. 54, 305–315.

Masuko, T., Minami, A., Iwasaki, N., Majima, T., Nishimura, S.I., Lee, Y.C., 2005. Carbohydrate analysis by a phenol-sulfuric acid method in microplate format. Anal. Biochem. 339, 69–72.

McIntyre, T.C., 1987. An overview of the environmental impacts anticipated from large scale biomass/energy systems. In: Moo-Young, M. (Ed.), Biomass Conversion Technology: Principles and Practice. Pergamon Press, Toronto, ON, pp. 45–52.

Moser, A., 1994. Sustainable biotechnology development: from high-tech to eco-tech. Acta Biotechnol. 12, 2–6.

Obodai, M., Cleland-Okine, J., Vowotor, K.A., 2003. Comparative study on the growth and yield of *Pleurotus ostreatus* mushroom on different lignocellulosic by-products. J. Ind. Microbiol. Biotechnol. 30 (3), 146–149.

Park, J.P., Kim, S.W., Hwang, H.J., Yun, J.W., 2001. Optimization of submerged culture conditions for the mycelial growth and exo-biopolymer production by *Cordyceps militaris*. Lett. Appl. Microbiol. 33 (1), 76–81.

Petre, M., Petre, V., 2012. The semi-solid state cultivation of edible mushrooms on agricultural organic wastes. Scientific Bull. Ser. F. Biotechnol. XVI, 36–40.

Petre, M., Petre, V., 2013. Environmental biotechnology for bioconversion of agricultural and forestry wastes into nutritive biomass. In: Petre, M. (Ed.), Environmental Biotechnology—New Approaches and Prospective Applications. InTech, Rijeka, Croatia, pp. 3–23.

Petre, M., Petre, V., Rusea, I., 2014. Ecotechnology for fully recovery of fruit tree wastes through controlled cultivation of eatable mushrooms. Scientific Bull. Ser. F. Biotechnol. XVIII, 48–54.

Rani, P., Kalyani, N., Prathiba, K., 2008. Evaluation of lignocellulosic wastes for production of edible mushrooms. Appl. Biochem. Biotechnol. 2–3, 151–159.

Robinson, T., Singh, D., Nigam, P., 2001. Solid-state fermentation: a promising microbial technology for secondary metabolite production. Appl. Microbiol. Biotechnol. 55, 284–289.

Ropars, M., Marchal, R., Pourquie, J., Vandercasteele, J.P., 1992. Large scale enzymatic hydrolysis of agricultural lignocellulosic biomass. Bioresour. Technol. 42, 197–203.

Rosado, F.R., Kemmelmeier, C., Da Costa, S.M., 2002. Alternative method of inoculum and spawn production for the cultivation of the edible Brazilian mushroom *Pleurotus ostreatoroseus* SING. J. Basic Microbiol. 42 (1), 37–44.

Saddler, J.N., Khan, A.W., Martin, S.M., 1993. Steam pretreatment of lignocellulosic residues. In: Saddler, J.N. (Ed.), Bioconversion of Forest and Agricultural Plant Residues, pp. 73–92.

Sainos, E., Díaz-Godínez, G., Loera, O., Montiel-González, A.M., Sánchez, C., 2006. Growth of *Pleurotus ostreatus* on wheat straw and wheat-grain-based media: biochemical aspects and preparation of mushroom inoculum. Appl. Microbiol. Biotechnol. 72 (4), 812–815.

Salmones, D., Mata, G., Waliszewski, K.N., 2005. Comparative culturing of *Pleurotus* spp. on coffee pulp and wheat straw: biomass production and substrate biodegradation. Bioresour. Technol. 96 (5), 537–544.

Sanchez, C., 2004. Modern aspects of mushroom culture technology. Appl. Microbiol. Biotechnol. 64 (6), 756–762.

Sanchez, C., 2010. Cultivation of *Pleurotus ostreatus* and other edible mushrooms. Appl. Microbiol. Biotechnol. 85 (5), 1321–1337.

Smith, J.E., 1998. Biotechnology, third ed. Cambridge University Press, UK, pp. 23–30.

Songulashvili, G., Elisashvili, V., Penninckx, M., Metreveli, E., Hadar, Y., Aladashvili, N., et al., 2005. Bioconversion of plant raw materials in value-added products by *Lentinus edodes* (Berk.) singer and *Pleurotus* spp. Int. J. Med. Mushrooms 7 (3), 467–468.

Stamets, P., 2000. Growing Gourmet and Medicinal Mushrooms, third ed. Ten Speed Press, Berkeley, CA, pp. 56–73.

Uphoff, N., 2002. Agroecological Innovations: Increasing Food Production with Participatory Development. Earthscan, London, UK, pp. 153–160.

van den Twell, W.J.J., Leak, D., Bielicki, S., Petersen, S., 1994. Biocatalysts production. In: Cabral, J.M.S., Boros, D.B.L., Tramper, J. (Eds.), Applied biocatalysis. Harwood Acad. Publ. GmbH, Chur., Switzerland, pp. 157–235.

Verstraete, W., Top, E., 1992. Holistic Environmental Biotechnology. Cambridge University Press, UK, pp. 1–18.

Wagner, R., Mitchell, D.A., Sassaki, G.L., De Almeida Amazonas, M.A., 2004. Links between morphology and physiology of *Ganoderma lucidum* in submerged culture for the production of exopolysaccharide. J. Biotechnol. 114 (1–2), 153–164.

Wainwright, M., 1992. An Introduction to Fungal Biotechnology. Wiley, Chichester, West Sussex, pp. 123–156.

Wedde, M., Iacobs, M., Stahl, U., 1999. Fungi: important organisms in history and today. In: Oliver, R.P., Schweizer, M. (Eds.), Molecular Fungal Biology. Cambridge University Press, UK, pp. 27–56.

www.eubia.org

www.insse.ro

CHAPTER

CONTROLLED CULTIVATION OF MUSHROOMS ON WINERY AND VINEYARD WASTES

3

Marian Petre[1], Florin Pătrulescu[1] and Răzvan Ionuţ Teodorescu[2]

[1]Faculty of Sciences, University of Pitesti, Pitesti, Romania [2]Faculty of Land Reclamation and Environmental Engineering, University of Agronomic Sciences and Veterinary Medicine, Bucharest, Romania

3.1 INTRODUCTION

Many of the agricultural lignocellulosic wastes which are produced every day in the world cause serious environmental pollution effects if they are allowed to accumulate in the agro-ecosystems or, much worse, burned for uncontrolled domestic purposes. Every year, large amounts of lignocellulosic wastes are generated through forestry and agricultural practices, in timber industries and many agroindustries, generating environmental pollution problems by their burning on the soil surface or their incorporation into the soil matrix. The optimal and efficient way to solve these problems is to recycle these lignocellulosic wastes as main ingredients in nutritive compost preparations that could be used for edible mushrooms cultivation (Petre and Petre, 2012).

So far, the basis of most studies on lignocellulose-degrading fungi has been economic rather than ecological, with emphasis on the applied aspects of lignin and cellulose decomposition, including biodegradation and bioconversion (Boddy, 1992; Beguin and Aubert, 1994; Chahal, 1994; Carlile and Watkinson, 1994; Uphoff, 2002).

A wide range of biomass resources are produced all over the world, including whole plants, plant parts (e.g., seeds, roots, stems) or plant constituents (e.g., starch, lipids, protein, and fiber), processing by-products (distiller's grains, corn solubles), animal by-products, and municipal and industrial wastes (Smith et al., 1987). All these resources can be used to create new biomaterials, and for this purpose they have to be converted into bioproducts. This process of bioconversion requires a deep knowledge of raw material composition, so that the desired functional elements can be obtained for bioproduct production (Kleman-Leyer et al., 1992; Nevalainen and Pentilla, 1995; Lestan and Lamar, 1996).

Bioconversion of lignocellulosic materials to useful, higher-value products normally requires multi-step processes, which include: (i) pretreatment (mechanical, chemical, or biological) of raw materials, (ii) hydrolysis of the polymers to produce readily metabolizable molecules (e.g., hexose or pentose sugars), (iii) use of these compounds to support microbial growth or to produce chemical products, and (iv) separation and purification (Tanaka and Matsuno, 1985; Smith et al., 1987; Smith, 1998; Verstraete and Top, 1992; Zarnea, 1994).

Mushroom Biotechnology. DOI: http://dx.doi.org/10.1016/B978-0-12-802794-3.00003-5

3.2 SOLID-STATE CULTIVATION OF MUSHROOMS (SSCM) ON WINERY AND VINEYARD WASTES

The biotechnological processes for the total recovery of lignocellulosic wastes resulting from industrial processing of grapes, as well as the maintenance of vineyards, through solid-state cultivation of mushrooms were tested using two Basidiomycetes species, namely, *Ganoderma lucidum* (Curt.: Fr.) P. Karst and *Pleurotus ostreatus* (Jacquin ex Fries) Kummer (Hawksworth et al., 1995).

The experiments were achieved by *in vitro* growth of these fungal species in special rooms, where the main culture parameters were kept at optimal levels in order to get the greatest production of mushroom fruit bodies. The effects of culture compost composition (carbon, nitrogen, and mineral sources) as well as other physical and chemical factors (such as temperature, inoculum amount, pH level, incubation time) on mycelial net formation, and especially on fruit body induction, were investigated.

The stock cultures were maintained on malt-extract agar (MEA) slants. Slants were incubated at 25°C for 5–7 days and then stored at 4°C. The fungal cultures were grown in 250-mL flasks containing 100mL of MEA medium (20% malt extract, 2% yeast extract, 20% agar-agar) at 23°C on rotary shaker incubators at 110 rotations per minute (rpm) for 5–7 days. The fungal cultures designed to be used in experiments were prepared by inoculating 100mL of cultivation medium with 3–5% (v/v) of the seed culture and then cultivated at 23–25°C in rotary shake flasks of 250mL. The experiments were conducted under the following conditions: temperature, 25°C; agitation speed, 90–120rpm; initial pH, 4.5–5.5. The seed culture was then transferred to the fungal culture medium and cultivated for 7–12 days (Petre et al., 2007; Glazebrook et al., 1992; Ropars et al., 1992).

In the next step of the experiments, the incubation of prepared fungal cultures was performed by keeping them in growth rooms where all the culture parameters were maintained at optimal levels in order to get the greatest production of fruit bodies. During the experiments, the effects of culture compost composition (carbon, nitrogen, and mineral sources) as well as other physical and chemical factors (such as temperature, inoculum size and volume, and incubation time) on mycelial net formation, and especially on fruit body induction, were investigated (Petre and Petre, 2008).

During and after the run cycles conducted for proper cultivation of edible mushroom species, certain biochemical analyses were performed in order to determine the composition of plant constituents which were used as substrates for cultivation, determining initial dry matter content by dehydrating the plant material at 105°C to constant weight. Then we proceeded to determine photometrically the content of soluble sugars by the method proposed by Panczel and Eifert (1960), with anthrone reagent, after their extraction from plant tissue with 80% ethanol; Kerepesi et al.'s (1996) method was also referred to. The amount of crude fiber (residue consisting of soluble or insoluble parts of cell membranes in the plant tissue, obtained by treating the residue under specified conditions with acids and bases of certain concentrations) was determined by the method of Henneberg and Stohmann (1864), supplemented by that of Greenfield and Southgate (1992).

The total nitrogen content determined for the two types of organic wastes was determined by the Kjeldahl method. In addition, measurements were taken to establish the content of phenolic compounds, through dosage of UV (λ-275nm) using tannins extracted in ethyl alcohol 80% (Hagerman and Butler, 1978).

The lipids were dosed according to the protocol proposed by Zollner and Kirsch (Gunstone et al., 1994), involving extraction in methanol:chloroform= 1:1 (v/v). Mineral components were quantified by atomic absorption spectroscopy, after solubilization of ash with 5N HCl (Ihnat, 1982). Also, measurements were taken of the pH of the aqueous solution of grape marc, where the ratio of grape marc extract to water was 1:10; the potentiometric pH values were recorded indicating an acid reaction of grape marc.

Due to the particularities of structure and chemical composition of these vegetable wastes, the pretreatment methods proved to be a particularly important application before their being used as substrates to grow mushrooms (McIntyre, 1987; Leahy and Colwell, 1990; Gregg and Saddler, 1996). In addition, the physical and chemical properties of grape marc differ according to the proportion of stalks, seeds, and skins of the grapes, and they vary as well depending on the variety (McCarty, 1988; Chahal and Hachey, 1990; Lamar et al., 1992).

Chemical analysis of grape marc showed that total nitrogen was present in a relatively high proportion (13.7%), which allows its use as a source of nitrogen for macromycetes which are grown on substrates made of such winery wastes. Also, the cellulose and soluble carbohydrate contents in grape marc have registered appropriate levels for use as carbon sources during the conversion process of substrates for mushroom cultivation.

The relatively high content of phenolic compounds in the composition of grape marc was due to primarily to the anthocyanin pigments that exist in red grapes. The amount of ash resulting from the incineration of plant waste at 550°C was almost four times higher for grape marc compared to the sample resulting from the calcination of vineyard wastes. The results were recorded per 100 g dry matter, as shown in Figures 3.1 and 3.2.

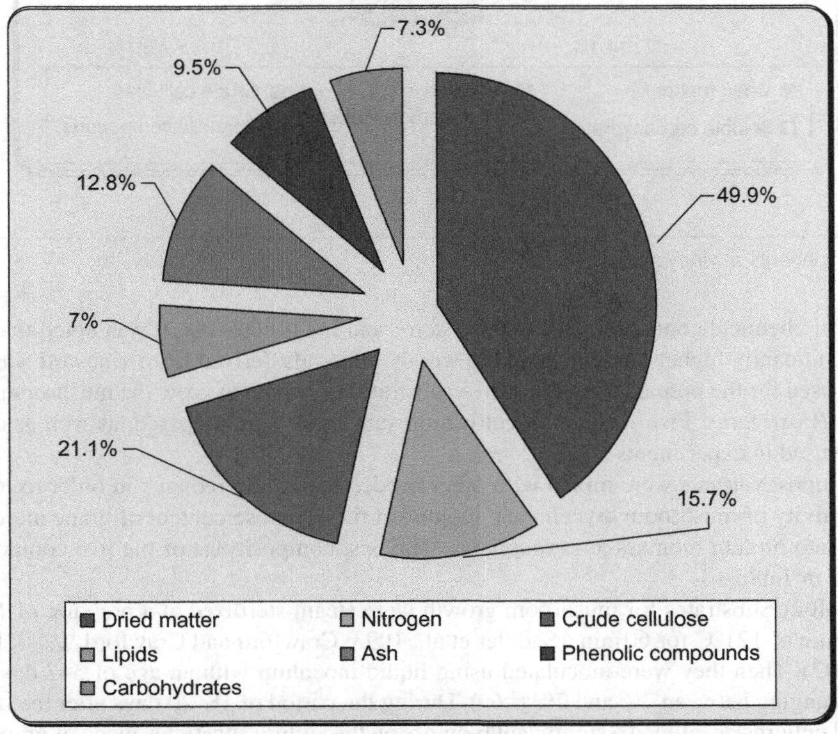

FIGURE 3.1

Chemical composition of grape marc.

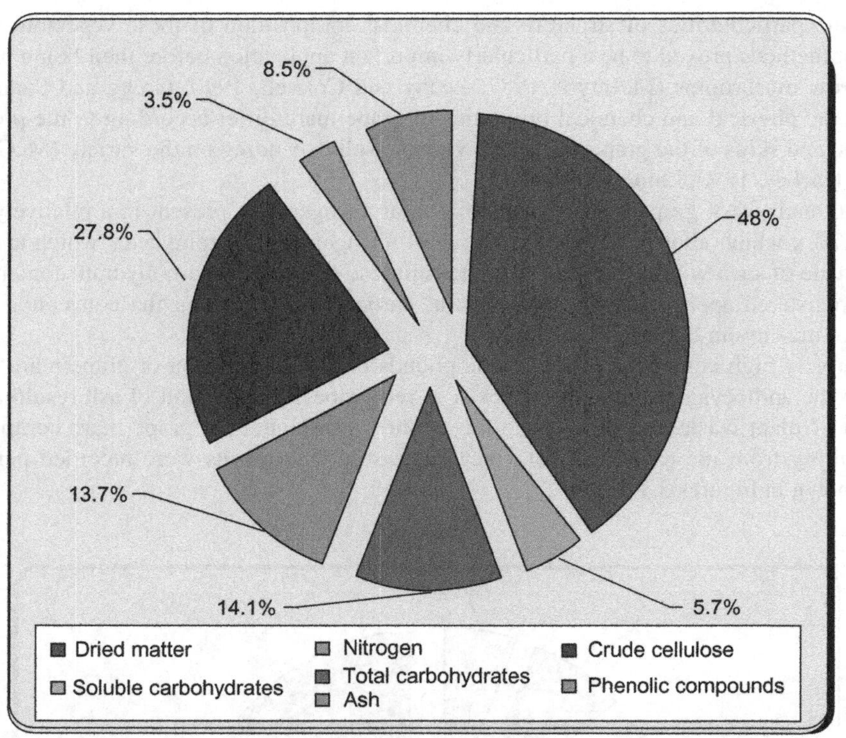

FIGURE 3.2

Chemical components of vineyard wastes.

In terms of chemical content regarding the macro- and micronutrients, it was noted that the grape marc had significantly higher amounts than the woody materials derived from vineyard wastes. These wastes were used for the preparation of nutritive substrates on which to grow the mushroom species *G. lucidum* and *P. ostreatus*. Five variants of cultivation substrates were prepared, as well as one control sample to be used in experiments.

These compost variants were mixed with other needed natural ingredients in order to improve the enzymatic activity of mushroom mycelia and to convert the cellulose content of grape marc and vineyard wastes into protein biomass as carpophores. The best compositions of the five compost variants are presented in Table 3.1.

All the culture substrates for mushroom growth were steam sterilized at a pressure of 1.1 atm and at a temperature of 121°C for 60 min. (Saddler et al., 1993; Crawford and Crawford, 1980; Finkelstein and Ball, 1992). Then they were inoculated using liquid inoculum with an age of 5–7 days, with the volume size ranging between 3% and 7% (v/w). During the period of 18–20 days after the inoculation, all the fungal cultures developed a significant biomass on the culture substrates made of grape marc and vineyard wastes. These wastes were used as the main ingredients to prepare natural substrates for mushroom growth. The optimal temperatures for incubation and mycelia growth were maintained between

Table 3.1 Compost Variants Used for Solid-state Cultivation of Mushrooms on Grape Marc and Vineyard Wastes

Compost Variants	Compost Composition
S1	Vineyard wastes + wheat bran (9:1)
S2	Vineyard wastes + rye bran (9:1)
S3	Grape marc + wheat bran (9:1)
S4	Grape marc + rye bran (9:1)
S5	Grape marc + vineyard wastes (1:1)
Control	Pure cellulose (Schuhardt)

23°C and 25°C. The whole period of mushroom growth from inoculation to fruit body formation lasted between 30 and 70 days, depending on the fungal species used in experiments (Petre et al., 2007).

The experiments were carried out inside growth rooms, where the main culture parameters (temperature, humidity, aeration) were kept at optimal levels to get the greatest production of mushroom fruit bodies (Moser, 1994; Stamets, 2000; Sanchez, 2010).

In comparison with organic nitrogen sources, the inorganic ones gave rise to relatively lower growth of mycelium and fungal biomass production (Bae et al., 2000). The influence of various mineral sources on fungal biomass production was examined at a standard concentration level of 5 mg. In comparison of KH_2PO_4 and $MgSO_4{\cdot}5H_2O$ as mineral sources tested in experiments, K_2HPO_4 yielded good mycelia growth as well as fungal biomass production, and for this reason it was recognized as a favorable mineral source. Also, K_2HPO_4 could improve productivity through its buffering action, being favorable for mycelia growth (Kirk and Eriksson, 1990; Wainwright, 1992; Moo-Young, 1993; Wedde et al., 1999; Papaspyridi et al., 2012).

In order to study the effects of the initial medium pH in correlation with the incubation temperature upon fruit body formation, experiments using *G. lucidum* and *P. ostreatus* mushrooms were carried out for 6 days at 25°C with an initial pH of 5.5. Similar observations were made by Stamets (2000), because the optimization of substrate composition and cultivation conditions are essential for enhancing efficiency in a submerged culture.

To find the optimal incubation temperature for mycelia growth, these fungal species were cultivated at different temperatures, ranging from 20°C to 25°C; in the end, the optimum temperature was found to be 23°C, and the appropriate pH level was 5.5. The best levels of pH and temperature for fruit body production were found between 5.0°C and 5.5°C and 21°C and 23°C. Among several fungal physiological properties, the age and volume of mycelia inoculum may play an important role in fungal hyphae development, as well as in fruit body formation (Petre and Petre, 2013). To examine the effect of inoculum age and volume, mushroom species *G. lucidum* and *P. ostreatus* were grown on substrates made of vineyard wastes over different time periods between 30 and 90 days, varying the inoculum volume (5–7 v/w). All the experiments were carried out at 25°C and an initial pH of 5.5. An inoculum age of 120 h as well as an inoculum volume of 6.0 (v/w) had beneficial effects on the fungal biomass production.

During the mushroom growing cycles, the specific rates of cellulose biodegradation were determined using the direct method of biomass weighing, the results being expressed as percentage of dry matter (d.m.) before and after cultivation (Songulashvili et al., 2005; Sanchez, 2010; Turlo, 2014). The data revealed that by applying this biotechnology, the grape marc and vineyard wastes could be recycled as useful raw materials for mushroom compost preparation in order to get significant mushroom production.

In order to determine the evolution of the total nitrogen content in the fungal biomass, samples were collected at precise time intervals of 50 h; these were analyzed by using the Kjeldahl method. The results concerning the evolution of total nitrogen content in *G. lucidum* biomass are presented in Figure 3.3 and the data regarding *P. ostreatus* biomass can be seen in Figure 3.4.

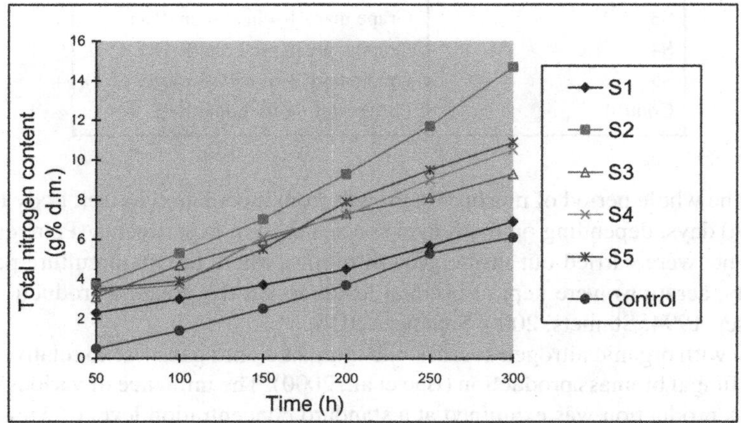

FIGURE 3.3

The evolution of total nitrogen content in biomass of *Ganoderma lucidum*.

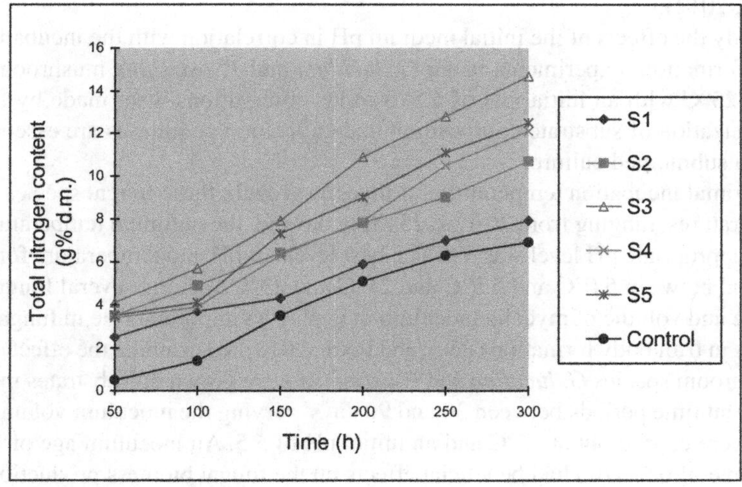

FIGURE 3.4

The evolution of total nitrogen content in biomass of *Pleurotus ostreatus*.

Based on the experimental results, the laboratory biotechnology for cultivation of edible mushrooms *G. lucidum* and *P. ostreatus* on grape marc and vineyard wastes was developed to produce nutritive biomass as carpophores (fruiting bodies).

From the point of view of total duration of biological cycle to produce carpophores, *P. ostreatus* was found to be the fastest mushroom culture (30–35 days), followed by *G. lucidum* (50–60 days). It is important to note that the experiments demonstrated the feasibility of biotechnological processes applied to edible mushroom cultivation on substrates consisting of wastes resulting from viticulture and the winemaking industry, as well as the efficiency of their use for the production of nutritive biomass as fruit bodies of mushroom species, and finally as fodder, obtained by dehydration of exhausted substrates used for mushroom cultivation.

3.3 SUBMERGED CULTIVATION OF MUSHROOMS (SCM) IN LIQUID MEDIA CONTAINING WINERY WASTES

Submerged cultivation refers to any controlled biochemical process regarding the enzymatic conversion of substrates distributed in a liquid nutrient phase, consisting mainly of carbohydrates, which promote the growth and development of microbial species (bacterial or fungal) under constant monitoring of the physical and chemical parameters (Atlas, 1984; Levinson et al., 1994; Bae et al., 2000; Park et al., 2001; Subramaniyam and Vimala, 2012).

For these experiments, three species of Basidiomycetes group were used, namely, *G. lucidum* (W. Curt.: Fr) Lloyd (popular name: reishi), *Lentinula edodes* (Berkeley) Pegler (popular name: shiitake), and *P. ostreatus* (Jaquin ex Fries) Kummer (popular name: oyster) were used (Hawksworth et al., 1995). These mushroom species were preserved on a special culture medium for the maintenance of pure cultures, specifically an average type malt-extract-agar (MEA) (20% malt extract, 2% yeast extract, 20% agar).

The nutrients absolutely necessary for the growth and development of mushroom mycelia were based on malt extract, yeast extract, peptone, $CaCO_3$, and wheat bran, to which were added certain amounts of grape marc resulting from vinification processes, as specified in Table 3.2. Each of these nutrients can affect the growth of the fungal species used in the experiments both according to the specific amount of each and by the proportionality between them.

Table 3.2 Variants of Nutritive Media for Submerged Cultivation of Mushroom Species

Nutritional Component of Cultivation Substrates	Substrate Variants		
	$M_{(Control)}$ (g/L)	V_1 (g/L)	V_2 (g/L)
Wheat bran	3	3	3
Malt extract	1	1	1
Yeast extract	0.6	0.6	0.6
Peptone	0.3	0.3	0.3
$CaCO_3$	–	0.5	0.5
Grape marc, collected after vinification	–	1	3

The pH index of the cultivation media which were prepared for these three species of mushrooms had a relatively broad range of variation over the 72h of each cultivation cycle. Analyzing the results obtained during the cultivation cycle of *P. ostreatus* species, it appeared that the version V1 presented performance results relatively similar to that of the control, namely a decrease from baseline of 6pH units to 5 units, and then a slight improvement to 5.5, while the V2 variant showed a steady downward trend in the amount from pH 5.9 to a very low value of 4.7.

In case of *L. edodes*, the pH index variation for the variants V1 and V2 of cultivation substrates was recorded as having almost identical progressions, from 7 to 7.2 to 5.2pH, compared with M version, which highlighted a constant curve downward from 5.2pH, finally reaching a value of 3.5pH.

In the case of *G. lucidum*, the pH index was recorded as a sinusoidal progress over the course of 72 hours of cultivation, based on the relative similar values for V1 and V2 variants of 7.3 to 7.4pH, reaching 5.5–5.7pH units, passing through the extreme minimum of 3.5–3.3pH. The M variant, used as the control for this species, showed a relatively balanced curve compared with the other two variants, the initial value being identical to the final one.

The Applikon bioreactor was used to carry out the biotechnological experiments in a batch cultivation system, being characterized by automatic control of bioprocesses using edible and medicinal mushrooms and permanent monitoring of all process parameters. This is an automatized bioreactor consisting of an autoclavable cultivation vessel equipped with transducers for temperature, pH, and dissolved oxygen (DO) level, and a set of systems for agitation and aeration control (Figure 3.5).

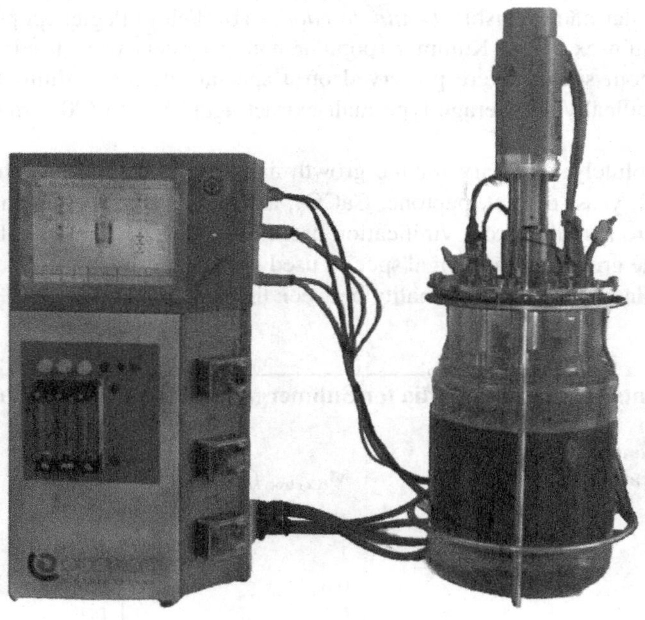

FIGURE 3.5

General overview of laboratory-scale bioreactor for submerged batch cultures.

During the course of the experiments, a system was established and developed by which the submerged cultivation developed in the bioreactor batch type can be monitored, controlled, and operated using the remote control software and hardware assemblies. This communication system allowed the video monitoring of the bioreactor vessel using a webcam and data transmission on a process controller (Packer and Thomas, 1990). A remote control system was conducted for wireless online monitoring and total control of bioprocess parameters inside the cultivation medium (Petre and Petre, 2013).

The experiments mounted for studying the fermentation processes were conducted under the following conditions:

- temperature 25°C;
- stirring speed, 120 rotations per minute (rpm);
- initial pH 4.5–5.5.

All the experiments were carried out over at least three repetitions. The analysis samples were collected at different time points during the fermentation process, centrifuged at a speed of 12,000 *g* for 15 min, and the supernatant was filtered through a Millipore membrane filter (0.45 µm). The amount of the mycelium dry matter was determined after repeated washes of the fungal pellets with distilled water and drying at 70°C overnight in a stream of air at a constant rate.

The fungal pellets belonging to *G. lucidum*, *L. edodes*, and *P. ostreatus* mushroom species which formed during the controlled submerged cultivation were analyzed using an Olympus stereomicroscope, and images taken are shown in Figures 3.6–3.8.

The fungal biomass samples were initially filtered, and then the solid filtrates were dried in an oven at 35°C for 48 h. The amount of biomass thus obtained was weighed.

In order to determine the chemical composition of fungal biomass samples collected after carrying out the submerged cultivation cycle of the mushroom species, chemical measurements were carried out

FIGURE 3.6

Fungal pellets collected after submerged cultivation of *Ganoderma lucidum* on winery wastes.

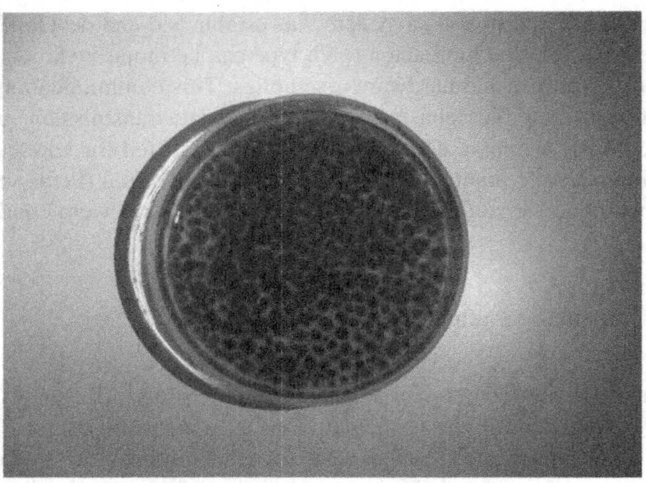

FIGURE 3.7

Fungal pellets collected after submerged cultivation of *Lentinula edodes* on winery wastes.

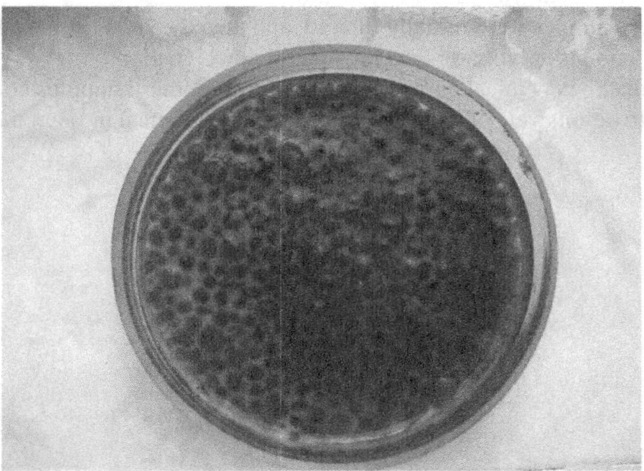

FIGURE 3.8

Fungal pellets collected after submerged cultivation of *Pleurotus ostreatus* on winery wastes.

on the solid content of dried matter, crude cellulose, ash, fat, carbohydrates, and soluble phenolic compounds (Masuko et al., 2005). The amount of total nitrogen was determined by the Kjeldahl method, and to analyze the crude fiber we used the acid hydrolysis method with an acid solution (acetic, trichloroacetic, and nitric acids) that solubilizes all the available nutrients, with the exception of the raw pulp (cellulose, hemicellulose, inlaid substances) and mineral compounds. The lipids were analyzed by the Soxhlet method, and soluble sugars were determined by the Lowry method (Lowry et al., 1951).

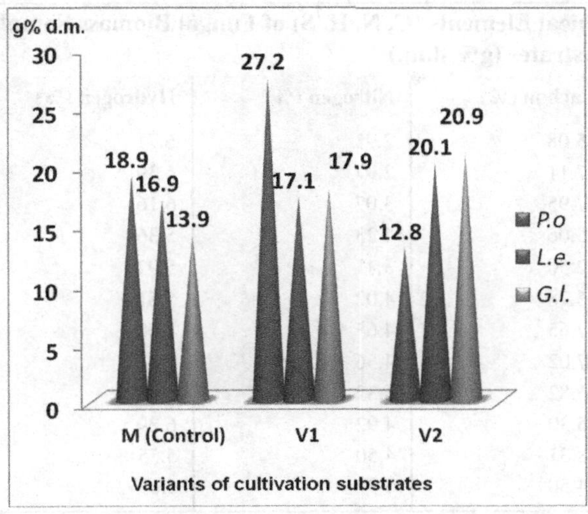

FIGURE 3.9

Final amounts of fungal biomass obtained by submerged cultivation of *Ganoderma lucidum*, *Lentinula edodes*, and *Pleurotus ostreatus*.

The chemical analyses showed that the total nitrogen in the composition of grape marc recorded a relatively high proportion (14.7%), which allows its use as a nitrogen source by the macromycetes grown on this type of waste. Both the crude fiber content (21.6%) and soluble carbohydrates (5.1%) present in grape marc contained sufficient amounts for use as a carbon source in the process of fungal conversion.

At the same time, the results showed a significant increase of biomass, with increasing amounts of grape marc in the cultivation media, with *P. ostreatus* variant V1 recording the highest value of 27.2 g% d.m. (dry matter), followed by *G. lucidum* variant V2 with 20.9 g% d.m., and then *L. edodes* variant V2 by 20 g% d.m.

In contrast to the variants that were recorded as the highest values in respect of the final amount of fungal biomass resulting after performance of controlled submerged culture processes, other variants were recorded in the range of 13.9 to 18.9 g% d.m. (Figure 3.9).

Determination of the chemical elements carbon (C%), nitrogen (N%), sulfur (S%), and hydrogen (H%) in the fungal biomass obtained by the submerged cultivation of mushrooms in the experiments was performed by the Dumas method (known as dry combustion) using the analyzer Vario Macro analyzer. This involved dehydration at 105°C, grinding, and mixing the final powder in order to obtain homogeneous samples. The combustion was done at 1150°C and, in the next step, reduction took place in the second furnace at 850°C in a helium flow, allowing the reduction of nitrogen oxides to molecular nitrogen (N_2).

The chemical elements C, N, H, and S were determined through thermoelectric cells, and the analyses were performed after previously calibrating with substances of which the composition of these elements was already known. Thus, sulfanilamide was used for calibration of the following composition: N: 16.25%, C: 41.81%, S: 18.62%, and H: 4.65%. The amount of crude protein was deducted from the amount of total nitrogen according to the relationship: Crude Protein (%) = Total Nitrogen × 4.38. The

Table 3.3 Chemical Elements (C, N, H, S) of Fungal Biomass Samples and Cultivation Substrates (g% d.m.)

Sample	Carbon (%)	Nitrogen (%)	Hydrogen (%)	Sulfur (%)
M	45.08	2.91	6.71	0.262
V_1	47.11	2.97	6.39	0.259
V_2	47.95	3.07	6.16	0.265
P.o.–M	35.06	4.23	5.36	0.360
P.o.–V_1	42.50	3.41	5.92	0.306
P.o.–V_2	45.78	4.02	5.81	0.344
L.e.–M	37.65	4.65	5.00	0.379
L.e.–V_1	37.02	4.56	5.47	0.393
L.e.–V_2	45.82	4.32	5.85	0.353
G.l.–M	45.30	4.92	6.89	0.346
G.l.–V_1	38.31	4.50	5.55	0.314
G.l.–V_2	44.50	3.99	5.63	0.370

conversion factor from total nitrogen in protein is 4.38, as described in the literature (Braaksma and Schaap, 1996).

According to the results obtained after carrying out these biochemical investigations (Table 4.6), it can be concluded that the concentration levels of the four elements analyzed were more diminished in the case of variants M, V1, and V2 compared with the corresponding cultivation substrates. This is explained by the fact that the biotechnological processes of submerged cultivation caused significant biochemical transformations of initial chemical composition of the cultivation substrates due to enzyme activities induced by the cultivated mushrooms. It may be noted that the quantitative ratio of those four chemical elements was kept relatively constant, which showed that metabolic processes induced by the enzymatic action of fungi qualitatively transformed the organic compounds from cultivation substrates in substances with a modified biochemical structure, under conditions in which the stoichiometric ratio of the elements was kept relatively constant (Table 3.3).

Regarding the protein content of fungal biomass samples belonging to *G. lucidum*, *L. edodes*, and *P. ostreatus* species, the investigation highlighted a relative decrease in the quantities of protein in variants V1 and V2 compared with the M variant as the control for the three species of cultivated mushrooms (Figure 3.10).

In the case of the V1 variant belonging to mushroom species *L. edodes* and *G. lucidum*, the protein concentration levels were fairly similar, 19.96 and 19.71 g% d.m., respectively, while that of *P. ostreatus* was only 14.94 g% d.m. Also, the V2 variant of *L. edodes* had the highest protein concentration, at 18.92 g% d.m., followed by the same variants of *P. ostreatus* with 17.62 g% d.m. and, finally, *G. lucidum* with 17.49 g% d.m. The M variants (as controls for which the substrate composition did not contain grape marc) were recorded as having the highest values of protein concentration, which proved that grape marc had an inhibitory effect on fungal enzyme activity during the bioconversion of cultivation substrates.

Regarding the nitrogen content of fungal biomass samples belonging to *G. lucidum*, *L. edodes*, and *P. ostreatus*, the results demonstrated a significant proportionality of these quantities with those of each of the proteins (Figure 3.11).

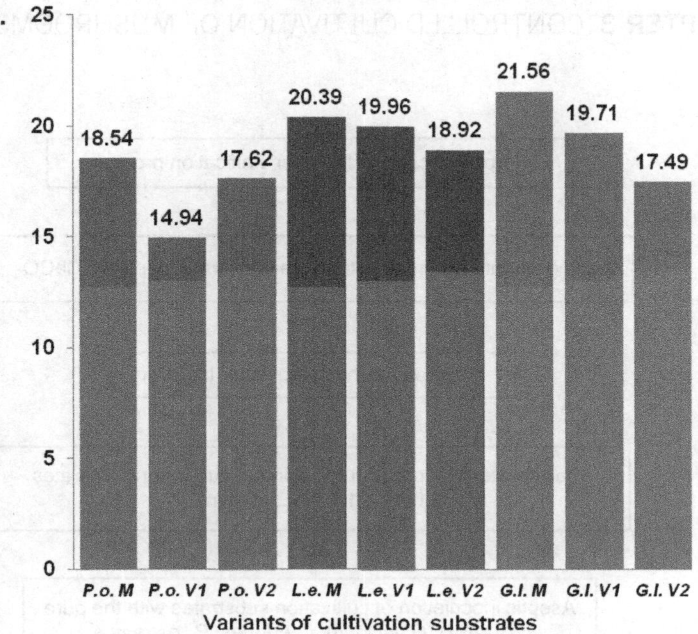

FIGURE 3.10

Protein content of fungal biomass samples belonging to *Ganoderma lucidum*, *Lentinula edodes*, and *Pleurotus ostreatus*.

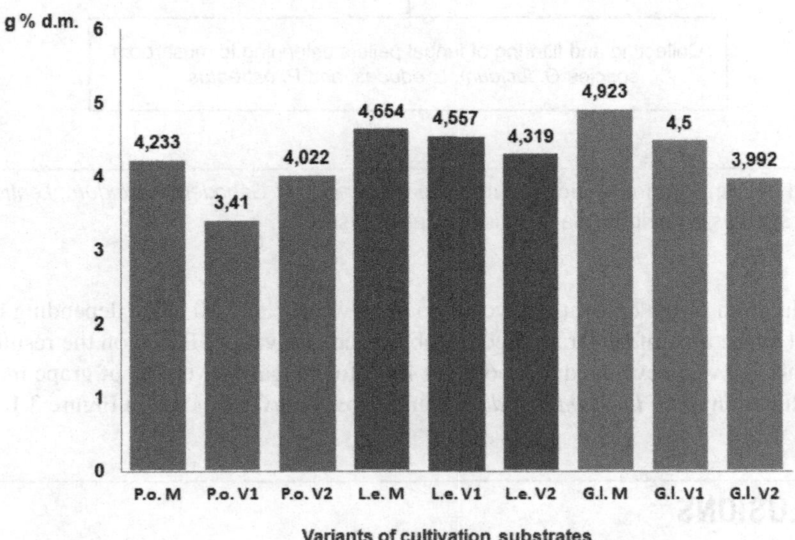

FIGURE 3.11

Nitrogen content of fungal biomass belonging to *Ganoderma lucidum*, *Lentinula edodes*, and *Pleurotus ostreatus* species.

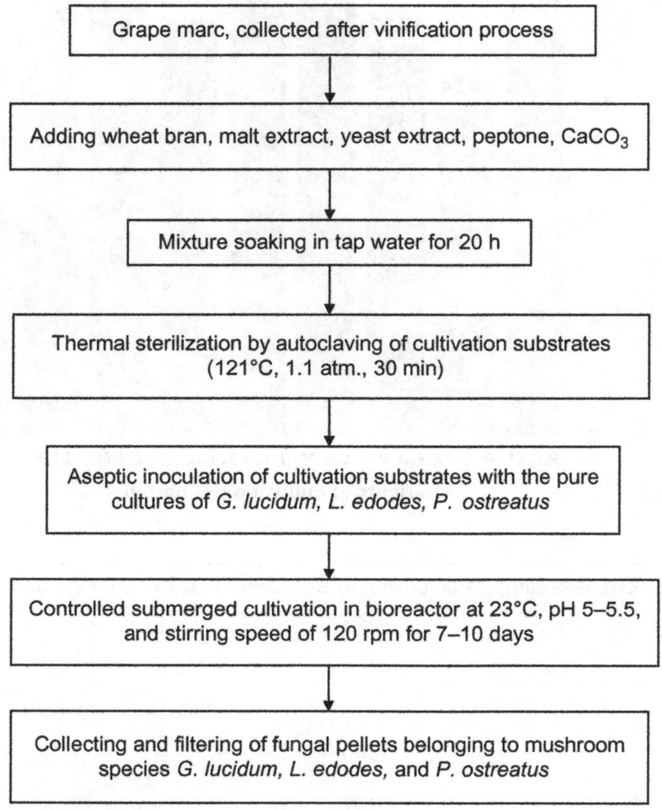

FIGURE 3.12

The scheme of biotechnology for controlled submerged cultivation of *Ganoderma lucidum, Lentinula edodes,* and *Pleurotus ostreatus* in liquid media containing winery wastes.

The total duration of the cultivation cycle varied between 7 and 10 days, depending on the fungal species as well as the type of substrate used in submerged cultivation. Based on the results, the appropriate biotechnology was developed at laboratory level for fungal conversion of grape marc wastes by submerged cultivation of *G. lucidum*, *L. edodes*, and *P. ostreatus*, as shown in Figure 3.12.

3.4 CONCLUSIONS

The results showed a significant increase of biomass with increasing amounts of grape marc in cultivation media; in this respect, *P. ostreatus* variant V1 recorded the highest value, 27.2 g% d.m. (dry matter), followed by *G. lucidum* variant V2 with 20.9 g% d.m. and *L. edodes* variant V2 with 20 g% d.m. In contrast to the variants that recorded the highest values in respect of the final amount of fungal biomass resulting after performance of controlled submerged culture processes, other variants were in the range of 13.9 to 18.9 g% d.m.

Regarding the protein content of fungal biomass samples belonging to *G. lucidum*, *L. edodes*, and *P. ostreatus*, the investigation highlights a relative decrease in the quantities of protein for variants V1 and V2 compared with the M (control) variant for all three species of cultivated mushrooms. In the case of the V1 variant belonging to *L. edodes* and *G. lucidum*, the protein concentration levels were fairly similar, at 19.96 and 19.71 g% d.m., respectively, while that for *P. ostreatus* was only 14.94 g% d.m. Also, the V2 variant of the species *L. edodes* had the highest protein concentration, 18.92 g% d.m., followed by the same variants of *P. ostreatus* with 17.62 g% d.m. and, finally, *G. lucidum* with 17.49 g% d.m.

The M variants (as controls, for which the substrate did not contain grape marc) were recorded as having the highest values of protein concentration, which proved that grape marc had an inhibitory effect on fungal enzyme activities during the bioconversion of cultivation substrates.

The nitrogen contents of fungal biomass samples belonging to *G. lucidum*, *L. edodes*, and *P. ostreatus* showed a significant proportionality of these quantities with those of each of the proteins. The contents of the chemical elements C, N, H, and S were determined through thermoelectric cells, and the analyses were performed after previous calibration with substances whose composition of these elements was known. According to the results obtained after carrying out these biochemical investigations, it can be concluded that the concentration levels of the four elements analyzed are more diminished in the case of variants M, V1, and V2 compared with the corresponding cultivation substrates.

REFERENCES

Atlas, R.M., 1984. Diversity of microbial communities. In: Marshall, K.C. (Ed.), Advances in Microbial Ecology, vol. 7. Plenum Press, New York, NY, pp. 10–47.

Bae, J.T., Sinha, J., Park, J.P., Song, C.H., Yun, J.W., 2000. Optimization of submerged culture conditions for exo-biopolymer production by *Paecilomyces japonica*. J. Microbiol. Biotechnol. 10, 482–487.

Beguin, P., Aubert, J.P., 1994. The biological degradation of cellulose. FEMS Microbiol. Rev. 13, 25–58.

Boddy, L., 1992. Fungal communities in wood decomposition. In: Carroll, G.C., Wicklow, D.T. (Eds.), The Fungal Community: Its Organization and Role in the Ecosystem, second edn. Marcel Dekker, New York, NY, pp. 749–782.

Braaksma, A., Schaap, D.J., 1996. Protein analysis of the common mushroom *Agaricus bisporus*. Postharvest Biol. Technol. 7, 119–127.

Carlile, M.J., Watkinson, S.C., 1994. The Fungi. Academic Press, London, UK.

Chahal, D.S., 1994. Biological disposal of lignocellulosic wastes and alleviation of their toxic effluents. In: Chaudry, G.R. (Ed.), Biological Degradation and Bioremediation of Toxic Chemicals. Chapman & Hall, London, UK, pp. 364–385.

Chahal, D.S., Hachey, J.M., 1990. Use of hemicellulose and cellulose system and degradation of lignin by *Pleurotus sajor-caju* grown on corn stalks. Am. Chem. Soc. Symp. 433, 304–310.

Crawford, D.L., Crawford, R.L., 1980. Microbial degradation of lignin. Enzyme Microb. Technol. 2, 11–20.

Finkelstein, D.B., Ball, C., 1992. Biotechnology of Filamentous Fungi: Technology and Products. Butterworth-Heinemann, Boston, MA, pp. 56–73.

Glazebrook, M.A., Vining, L.C., White, R.L., 1992. Growth morphology of *Streptomyces akiyoshiensis* in submerged culture: influence of pH, inoculum, and nutrients. Can. J. Microbiol. 38, 98–103.

Greenfield, H., Southgate, D.A.T., 1992. Food Composition Data: Production, Management and Use. Elsevier Applied Sciences, London, England, pp. 108.

Gregg, D.J., Saddler, J.N., 1996. Factors affecting cellulose hydrolysis and the potential of enzyme recycle to enhance the efficiency of an integrated wood to ethanol production. Biotechnol. Bioeng. 51 (4), 375–381.

Gunstone, F.D., John, L.H., Fred, B.P., 1994. The Lipid Handbook. Chapman & Hall Chemical Database, London, UK, pp. 156–164.

Hagerman, A.E., Butler, L.G., 1978. Protein precipitation method for the quantitative determination of tannins. J. Agric. Food Chem. 26, 809–812.

Hawksworth, D.L., Kirk, P.M., Sutton, Pegler, D.N., 1995. Ainsworth & Bisby's Dictionary of the Fungi, eighth ed., Wallingford, pp. 56–59, 211–214, 575–579.

Henneberg, W., Stohmann, F., 1864. Beitrage Zur Begrundung Einer Rationellen Futterung Der Wiederkauer, Book 1 (German Edition).

Ihnat, M., 1982. Application of atomic absorption spectrometry to the analysis of foodstuffs. In: Cantle, J.E. (Ed.), Atomic Absorption Spectrometry. Elsevier, Amsterdam, pp. 139–210.

Kerepesi, I., Toth, M., Boross, l, 1996. Water-soluble carbohydrates in dried plant. J. Agric. Food Chem. 44 (10), 3235–3239.

Kirk, T.K., Eriksson, K.E., 1990. Roles of biotechnology in manufacture. In: Robert, F. (Ed.), World Pulp & Paper Technology. The Sterling Publishing Group, London, UK, pp. 23–28.

Kleman-Leyer, K., Agostin, E., Conner, A.N., 1992. Changes in molecular size distribution of cellulose during attack by white-rot and brown-rot fungi. Appl. Environ. Microbiol. 58, 1267–1270.

Lamar, R.T., Glaser, J.A., Kirk, T.K., 1992. White rot fungi in the treatment of hazardous chemicals and wastes. In: Leatham, G.F. (Ed.), Frontiers in Industrial Mycology. Chapman & Hall, New York, NY, pp. 127–143.

Leahy, J.G., Colwell, R.R., 1990. Microbial degradation of hydrocarbons in the environment. Microbiol. Rev. 54, 305–315.

Lestan, D., Lamar, R.T., 1996. Development of fungal inocula for bioaugmentation of contaminated soils. Appl. Environ. Microbiol. 62 (6), 2045–2052.

Levinson, W.E., Stormo, K.E., Tao, H.L., Crawford, R.L., 1994. Hazardous waste cleanup and treatment with encapsuled or entrapped microorganisms. In: Chaudhry, G.R. (Ed.), Biological Degradation and Bioremediation of Toxic Chemicals. Chapman & Hall, London, UK, pp. 455–469.

Lowry, O.H., Rosebrough, N.J., Fan, A.L., Randall, R.J., 1951. Protein measurement with the Folin phenol reagent. J. Biol. Chem. 193, 265–273.

Masuko, T., Minami, A., Iwasaki, N., Majima, T., Nishimura, S.I., Lee, Y.C., 2005. Carbohydrate analysis by a phenol-sulfuric acid method in microplate format. Anal. Biochem. 339, 69–72.

McCarty, P.L., 1988. Bioengineering issues related to *in situ* remediation of contaminated soils and groundwater. In: Omenn, G.S. (Ed.), Environmental Biotechnology: Reducing Risks from Environmental Chemicals Through Biotechnology. Plenum Press, New York, NY, pp. 143–162.

McIntyre, T.C., 1987. An overview of the environmental impacts anticipated from large scale biomass/energy systems. In: Moo-Young, M. (Ed.), Biomass Conversion Technology: Principles and Practice. Pergamon Press, Toronto, ON, pp. 45–52.

Moo-Young, M., 1993. Fermentation of cellulose materials to mycoprotein foods. Biotech. Adv. 11 (3), 469–482.

Moser, A., 1994. Sustainable biotechnology development: from high-tech to eco-tech. Acta Biotechnol. 12, 2–6.

Nevalainen, H., Pentilla, M., 1995. Molecular biology of cellulolytic fungi. In: Kuck, H. (Ed.), The Mycota. Genetics and Biotechnology, vol. 2. Springer-Verlag, Berlin-Heidelberg, New York, pp. 303–319.

Packer, H.L., Thomas, C.R., 1990. Morphological measurements on filamentous microorganisms by fully automatic image analysis. Biotechnol. Bioeng. 35, 870–881.

Panczel, M., Eifert, J., 1960. Die Bestimung des Zuckerund Stärkegehaltes der Weinrebe mittels Anthronreagens. Mitt. Klosterneuburg 10, 102–110.

Papaspyridi, L.M., Aligiannis, N., Topakas, E., Christakopoulos, P., Skaltsounis, A.L., Fokialakis, N., 2012. Submerged fermentation of the edible mushroom *Pleurotus ostreatus* in a batch stirred tank bioreactor as a promising alternative for the effective production of bioactive metabolites. Molecules 17 (3), 2714–2724.

Park, J.P., Kim, S.W., Hwang, H.J., Yun, J.W., 2001. Optimization of submerged culture conditions for the mycelial growth and exo-biopolymer production by *Cordyceps militaris*. Lett. Appl. Microbiol. 33 (1), 76–81.

Petre, M., Petre, V., 2008. Environmental biotechnology to produce edible mushrooms by recycling the winery and vineyard wastes. J. Environ. Protect. Ecol. 9 (1), 87–97.

Petre, M., Petre, V., 2012. The semi-solid state cultivation of edible mushrooms on agricultural organic wastes. Scientific Bull. Ser. F. Biotechnol. XVI, 36–40.

Petre, M., Petre, V., 2013. Environmental biotechnology for bioconversion of agricultural and forestry wastes into nutritive biomass. In: Petre, M. (Ed.), Environmental Biotechnology—New Approaches and Prospective Applications. InTech, Rijeka, Croatia, pp. 3–23.

Petre, M., Bejan, C., Visoiu, E., Tita, I., Olteanu, A., 2007. Mycotechnology for optimal recycling of winery and vine wastes. Int. J. Med. Mushrooms 9 (3), 241–243.

Ropars, M., Marchal, R., Pourquie, J., Vandercasteele, J.P., 1992. Large scale enzymatic hydrolysis of agricultural lignocellulosic biomass. Bioresour. Technol. 42, 197–203.

Saddler, J.N., Khan, A.W., Martin, S.M., 1993. Steam pretreatment of lignocellulosic residues. In: Saddler, J.N. (Ed.), Bioconversion of Forest and Agricultural Plant Residues, pp. 73–92.

Sanchez, C., 2010. Cultivation of *Pleurotus ostreatus* and other edible mushrooms. Appl. Microbiol. Biotechnol. 85 (5), 1321–1337.

Smith, J.E., 1998. Biotechnology, third ed. Cambridge University Press, UK, pp. 56–70.

Smith, J.E., Anderson, J.G., Senior, E.K., 1987. Bioprocessing of lignocelluloses. Philos. Trans. Royal Soc. A 321, 507–521.

Songulashvili, G., Elisashvili, V., Penninckx, M., Metreveli, E., Hadar, Y., Aladashvili, N., et al., 2005. Bioconversion of plant raw materials in value-added products by *Lentinus edodes* (Berk.) singer and *Pleurotus* spp. Int. J. Med. Mushrooms 7 (3), 467–468.

Stamets, 2000. Growing Gourmet and Medicinal Mushrooms, third ed., Ten Speed Press, Berkeley, CA, pp. 123–127.

Subramaniyam, R., Vimala, R., 2012. Solid state and submerged fermentation for the production of bioactive substances: a comparative study. Int. J. Sci. Nat. 3 (3), 480–486.

Tanaka, M., Matsuno, R., 1985. Conversion of lignocellulosic materials to single-cell protein (SCP): review developments and problems. Enzyme Microb. Technol. 7, 197–207.

Turlo, J., 2014. The biotechnology of higher fungi—current state and perspectives. Folia Biol. Oecol. 10, 49–65.

Uphoff, N., 2002. Agroecological Innovations: Increasing Food Production with Participatory Development. Earthscan, London, UK, pp. 153–160.

Verstraete, W., Top, E., 1992. Holistic Environmental Biotechnology. Cambridge University Press, UK, pp. 1–18.

Wainwright, M., 1992. An Introduction to Fungal Biotechnology. Wiley, Chichester, West Sussex, pp. 56–73.

Wedde, M., Iacobs, M., Stahl, U., 1999. Fungi: important organisms in history and today. In: Oliver, R.P., Schweizer, M. (Eds.), Molecular Fungal Biology Cambridge University Press, UK, pp. 21–35.

Zarnea, G., 1994. Theoretical bases of microbial ecology Treaty of Microbiology. Romanian Academy Publishing House, Bucharest, vol. 5, pp. 154–163.

CHAPTER

4

VIRTUAL ROBOTIC PROTOTYPE FOR SAFE AND EFFICIENT CULTIVATION OF MUSHROOMS

Florin Adrian Nicolescu[1], Dan Andrei Marinescu[2] and Georgia Cezara Avram[1]

[1]*Faculty for Engineering and Management of Technological Systems, Politehnica University of Bucharest, Bucharest, Romania* [2]*EDAG Engineering GmbH, Wolfsburg–Westhagen, Germany*

4.1 INTRODUCTION

Recent research on the production of organic foods produced from biomass of different species of mushrooms, which have beneficial effects on human health, are well advanced. In-depth studies carried out in China, Japan, the United States, and Russia have proved highly beneficial in the production of foods with high nutritional value from the edible mushroom fruit bodies obtained under standardized conditions using biotechnological systems and continuous-flow processing in terms of food quality and safety (Cho et al., 2002; Reed et al., 2001; Belforte et al., 2006b).

Currently, research in the field of robotic systems with applications in the agro-food industry take place in over 30 prestigious universities in the Netherlands, Germany, France, Spain, UK, Sweden, Israel, the United States, Canada, Japan, China, Taiwan, Malaysia, and Australia. Unfortunately, only some of these studies are involved strictly in the field of robotic systems development for growing and collecting mushrooms, the most advanced of them being Silsoe Research Institute, Robotics & Automation Group; Warwick University, Horticulture Research International, Wellesbourne in the UK; Pennsylvania State University, Department of Agricultural and Biological Engineering in the United States; Pingtung University of Science and Technology, Department of Food Science in Taiwan; Okayama University, Faculty of Agriculture in Japan; and TNO—Netherlands Organization for Applied Scientific Research in the Netherlands (Noble et al., 1997; Heinemann et al., 1994; van Galen et al., 2003).

In conjunction with the specific technology and related equipment for cultivation and harvesting of mushrooms, a clear distinction must be made between the work environment and specific operations performed by human operators in mushroom farms and industrial pilot plants with automatic production processes and harvesting of mushrooms. In this respect, the first group mainly produces mushrooms to supply the fresh market, while the second group has as its main target to perform high production of mushrooms for the food industry (Jarvis, 1997; Che and Ting, 2004; Reed et al., 2001).

Thus, in the Netherlands, the most advanced country in Europe in the field of mushroom cultivation, there are currently more than 300 mechanized farms, with different levels of automation of mushroom production processes (van Galen et al., 2003).

Mushroom Biotechnology. DOI: http://dx.doi.org/10.1016/B978-0-12-802794-3.00004-7

A first main feature of these farms consists of much larger areas of cultivation (usually up to 10,000 square meters) compared to farms where manufacturing operations are carried out by humans (usually limited to about 1200 cultivation areas, up to 2400 square meters).

The second major feature –common to both types of mushroom producing—is the use of growth substrates arranged in horizontal supports (cultivation beds), loaded with organic matter made up of agricultural wastes, horse manure or poultry droppings, gypsum, and other ingredients. The main distinction between these two categories of mushroom production systems consists in specific production processes carried out in each one of them. Thus, in the case of mushroom farms with no automatic equipment, all processes, from the step of filling with nutritive substrates of cultivation beds to mushroom harvesting are performed only by human personnel (Reed and Tillett, 1994; Connolly, 2003).

Mechanized farms have automatic dispensing systems for compost in the beds of cultivation, transportation systems of frames in storage areas for incubation and fruiting body formation, as well as mechanized harvesting systems for mushroom on the cultivation beds. Such automatic collection systems can collect up to 250 kg of mushroom fruit bodies/h, as opposed to collection by human operators, by which can be obtained only about 20–30 kg of mushroom fruit bodies/h. These farms use a cultivation and collection cycle of 5 weeks, which includes mechanized operations of evacuation and emptying of frames with compost, their refill, and relocation in cultivation spaces after these areas were themselves mechanically cleaned.

The beneficial economic effects of the automation processes of cultivation, harvesting, and processing mushrooms in mechanized farms can be summarized as including an increase of cultivated areas by about 5–8 times within the same production cycle, and also an increase of harvesting productivity by 8–12 times, a reduction of 66% of staff costs, 50% decrease of fixed costs, reduction by 60% of the sale costs toward beneficiaries of the products obtained from such farms, and the increase of profits from sales by about 30–35% on average. As a result of these beneficial effects, there is a strong tendency to increase the number of research projects developed in the field of automatic cultivation, harvest, and processing of mushrooms as well as ongoing development of new concepts and specific equipment for greater automation of mushroom biotechnology.

The cultivation and production of mushrooms involves a lot of different operations, each of them requiring careful performance. To design a robotic growing system for mushroom production, it is compulsory to have specialized knowledge from many advanced fields of science and technology (Masoudian and McIsaac, 2013; Belforte et al., 2006a; Che and Ting, 2004).

The unceasing improvement and development of modern cultivation technologies, such as computerized control of the whole production process, new methods applied for spawn preparation and sterilization of substrates needed for cultivation, and automated harvesting of mushroom fruit bodies, will increase the large-scale productivity of mushroom cultivation with lower costs than conventional technologies (Noble et al., 1997; Connolly, 2003).

In the last two decades, a lot of machinery has been designed for picking up the mushroom fruit bodies, especially in the case of button mushrooms (*Agaricus bisporus*), and there have been experiments testing them in mushroom production. In many of these experiments, the locating and picking performance of robotic machineries for mushroom harvesting was assessed (Reed and Tillett, 1994; Belforte et al., 2006a). Some of the recent results regarding the modern technology and research are increasingly used in agriculture, especially in intensive cultures that ensure remunerative returns.

For instance, most organic cultures grown in greenhouses are in a category where, despite the wide use of technology, a lot of human operators still manually perform almost all operations on the crops, although they are often highly repetitive. In this respect, a multipurpose, low-cost robot prototype was designed and built, although more research is needed to improve the productivity of such a prototype (Belforte et al., 2006b). The results proved to be promising and showed some advantages that can be achieved with robotic automation. This fact greatly impacts on the quality of the product, on the production costs, and on collateral issues such as pollution and safety. Knowledge of the state of advanced research in robotic automation with application in agriculture, outlining the characteristics that robots should have to allow their profitable use, is compulsory in order to select the best solutions for robotic automation to put into practice (Belforte et al., 2006b).

However, though the actual results are not significant enough to be applied on a large scale, the newest automation technologies for harvesting, transporting, and grading are in great demand. Thus, a robotic system was developed for harvesting lettuce plants, comprised of a three degrees of freedom manipulator, an end-effector, a feeding conveyor, an air blower, a machine vision device, six photoelectric sensors, and a fuzzy logic controller (Cho et al., 2002).

4.2 CONVENTIONAL TECHNOLOGIES USED IN MUSHROOM CULTIVATION

Current methods for growing mushrooms on various nutritional substrates contain mostly synthetic ingredients, which require nonperforming, energy-intensive, and less productive and efficient machines compared with the structure and functionality of the robotic installation that is the subject of this chapter. There are various ways of growing mushrooms using conventional methods which have a lot of disadvantages, for the following reasons:

1. They require the application of energy-intensive processes, characterized by using a large number of appliances and heating, electrical, and electronic installations;
2. They do not fully ensure fully aseptic conditions imposed by proper cultivation technology for growth and multiplication of biological material, with a permanent risk of contamination of the working environment by human operators, compromising the production of mushrooms;
3. The mushroom production cycle does not work in a continuous flow and requires human operators in the sterile zone, increasing the risk of infection sources;
4. They do not lead to continuous production of mycelium-inoculated compost bags;
5. They do not allow a high and continuous production of fruit bodies and sterilized and inoculated substrate in bags with economical efficiency.

4.3 CONCEPTUAL MODEL OF ROBOTIC CULTIVATION AND INTEGRATED PROCESSING OF MUSHROOMS

The conceptual model of robotic cultivation and integrated processing of mushrooms was designed as a fully automatic installation with a modular structure and multiple workflows, which are compulsory for continuous production of organic fruiting bodies belonging to edible and medicinal mushrooms in fully aseptic conditions, with complete elimination of any human operator along the overall production chain (Petre et al., 2010).

Starting from the thermo-sterilization of heat-resistant plastic bags filled with nutritive substrates, going through robotic inoculation with liquid mycelium, and continuing the cultivation flow through mushroom incubation and fruit body formation inside sterilized and inoculated bags placed in special growth rooms (with fully controlled atmosphere and robotic manipulation devices), the biotechnological process was designed to be finished with automatic crop harvesting of mushroom fruit bodies and aseptic removal of bags containing the exhausted substrates used for cultivation. Then, by preparing and processing the plastic bags filled with sterilized organic substrates and inoculated with liquid mycelium, they could be commercialized to customers interested in the processing and sale of fruiting bodies belonging to edible and medicinal mushrooms produced through continuous controlled cultivation in a fully aseptic environment (Petre et al., 2010; Nicolescu et al., 2009a).

To put into practice this conceptual model of robotic cultivation and integrated processing of mushrooms, two zones were designed as the main important areas of the robotic prototype structure. The first was designed as being the unsterile zone designed to be served by human operators, dedicated to the preliminary processing of substrates, filling the plastic bags with nutritive cultivation substrates, loading bags on the transfer devices, and positioning them in front of automated entry gates in the processing system.

The second zone was organized to contain four specific sections, established as the sterile zones, with specific areas having completely automated functioning. The first section was dedicated to substrate sterilization and robotic inoculation of sterilized substrates in plastic bags with liquid mycelium, and the second was designed as the control unit for the inoculated bags, performed by an automated vision inspection system, and then their corresponding distribution (Nicolescu et al., 2009b).

The bag distribution area was designed to have three distinct functionalities: evacuation from the production system of the transfer devices identified as containing damaged bags; transport of transfer devices containing bags identified as being in good shape out of the sterile section in order to be commercialized as the first deliverable products, dedicated to customers interested in developing their own mushroom fruiting body production, and the transport of the transfer devices containing the appropriate inoculated bags which are to be placed in the third section, reserved for the growth rooms for fully automated incubation and fruiting body formation. Each of these sections may be independently operated, though they are integrated with the same robotic system for uploading and downloading of the transfer devices containing inoculated bags through a set of two storage racks located inside each room for incubation or fruit body formation (Petre et al., 2010; Nicolescu et al., 2010a).

As a result, a fully robotic prototype with a modular structure was designed to have an integrative functionality as scalable variants in single or multiple workflows for continuous production and processing of plastic bags filled with sterilized organic substrates inoculated with liquid mycelium, as well as production of edible and medicinal mushroom fruiting bodies, cultivated on natural substrates in growth rooms equipped with automatic devices for leading and controlling the whole process of mushroom cultivation (as shown in Figures 4.1 and 4.2).

Such a modular robotic prototype is suitable to for use in continuous controlled cultivation of well-known mushroom species such as *Ganoderma lucidum*, *Lentinula edodes*, *Pleurotus ostratus*, and *Pleurotus eryngii*, taking into account that these species do not require composted substrates and casing layers, and their fruiting bodies emerge all over the surface of cultivation substrates, which are packed in plastic bags with a cylindrical shape (Petre et al., 2010).

There are a number of technical problems that the proposed model could solve, such as the prevention and removal of any potential sources of infection on producing substrates for mushrooms cultivation, due

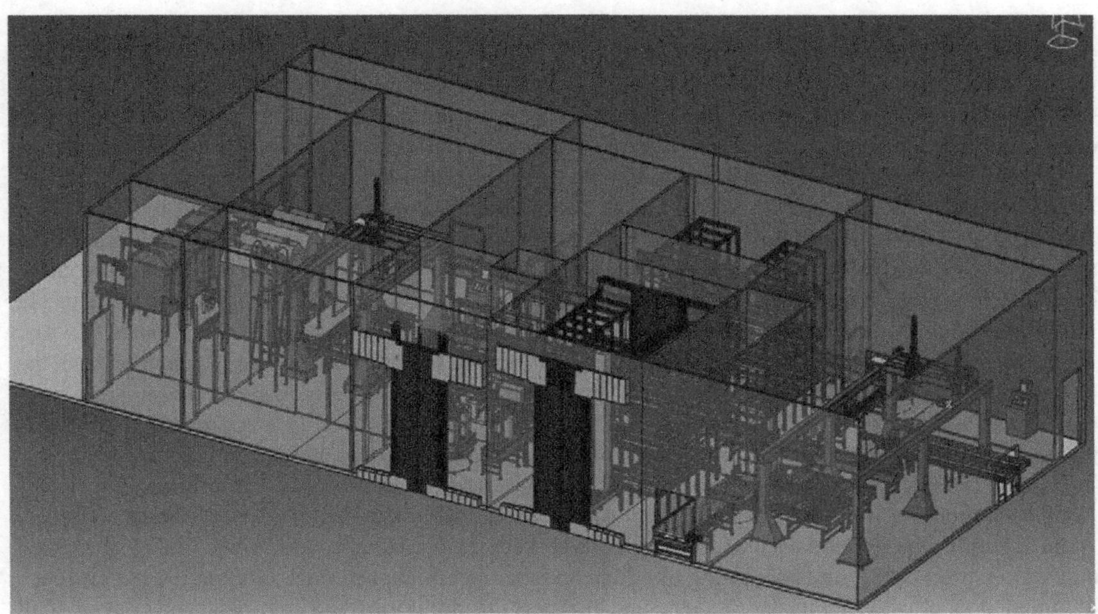

FIGURE 4.1

General overview of modular robotic prototype for mushroom cultivation from lateral right side.

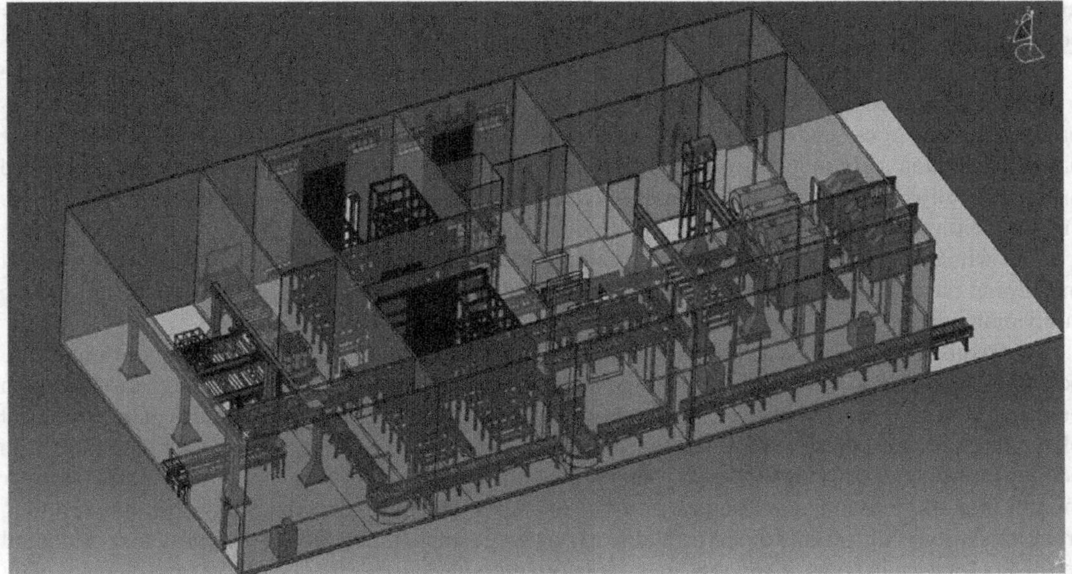

FIGURE 4.2

General overview of modular robotic prototype for mushroom cultivation from lateral left side.

to the obvious fact that in the case of "classical" technology of mushroom production there is always a potential risk of contamination with pathogens, which could compromise the whole production of mushrooms. Also, this modular robotic prototype was designed to ensure high-quality production and food safety of the mushrooms produced in a continuous cycle of production, with an increase of economic efficiency of 70% compared to other processes and equipment used in conventional cultivation systems.

Through the biotechnological process applied by such a robotic prototype, the disadvantages outlined above will be removed, providing for the simultaneous production of bags filled with sterilized and inoculated substrates with mycelium as well as the production of mushroom fruiting bodies. The technological solution to create more areas for sterile cultivation of mushrooms in continuous modular robotic flow was designed so as to allow maximum optimization of production efficiency by ensuring the maximum degree of aseptic conditions for mushroom cultivation (Nicolescu et al., 2009b, 2010a).

At the same time, the modular robotic prototype removes other disadvantages mentioned, as the equipment in sterile areas ensures automated handling, transfer, storage, sterilization, cooling, and automatic inoculation with mycelium, incubation and fruiting body formation in a controlled atmosphere, maintaining optimum and permanent cleaning of air, optimal temperature and relative humidity, harvesting and automatic collection of fruiting bodies, and the washing, disinfection, and collection of wastes from incubation and fruit body formation rooms, as well as final processing of the mushrooms produced.

In addition, the previously mentioned prototype keeps a permanent workflow of fully automated cultivation, production, and processing of edible and therapeutic mushrooms, eliminating the presence of human operators during its deployment, and also permits monitoring and command and control of all production parameters in sterile areas and ensures continuous production of sterilized and inoculated substrates with mycelium from the plastic bags, as well as the production of mushroom fruiting bodies with maximum food safety and security. In this way, significant savings are achieved per finished product unit, and significantly reduced costs of raw materials, energy, and labor, which are reflected in by over 100% higher compared to other facilities used now in the cultivation of mushrooms for consumption.

Concerning the novelty and originality of this conceptual model, which is already patented, the following aspects should be remarked. To avoid any direct human involvement in the production bioprocess, all mushroom cultivation stages, starting from the substrate sterilization and mycelium inoculation, through mycelium incubation and fruiting body formation as well as harvesting, are performed using this fully automatic system, which operates inside a completely controlled and 100% aseptic environment without any human handling.

The whole bioprocess of mushroom growing does not require any use of pesticides to prevent mushroom pathogens and pests. As a result, all harvested mushrooms are produced only by organic cultivation with no use of any kind of pesticide, and they are 100% natural products.

Using this robotic prototype produces two sorts of final products to be traded: packed mushrooms to be commercialized as fresh food, and plastic bags containing sterilized and inoculated substrates with pure liquid mycelium of edible or medicinal mushrooms. At a middle level capacity of an industrial plant, the continuous production of sterilized and inoculated substrates packed in plastic bags (250 bags per day, every day all through the year) and production of fresh mushroom fruit bodies (1500 kilos per month) in a 100% aseptic environment is simultaneously ensured. The exhausted substrates resulting from mushroom cultivation can be used directly as feed supplements (containing proteins) or natural fertilizers for certain types of soil that are poor in organic substances, closing a short food chain of plant waste recycling, making from this point of view a significant contribution to environmental protection.

4.4 MODULAR ROBOTIC PROTOTYPE FOR CONTINUOUS CULTIVATION AND INTEGRATED PROCESSING OF MUSHROOMS

The modular robotic prototype for continuous controlled cultivation and integrated processing of edible and medicinal mushrooms was already patented in 2010 (Petre et al., 2010). This prototype is the first fully automated system for mushrooms cultivation and integrated processing that has been designed so far in mushroom cultivation field.

The use of such a system has a number of technological and economical advantages, such as fully preventing human errors by ensuring biotechnological safety through the application of standardized conditions of workspace purity in the sterile zones, and guaranteed safety and microbial purity of mushroom fruit bodies of cultivated mushroom species as primary organic sources of food and medicines. In addition, this system uses an efficient sterilization method for cultivation substrates by total decontamination with microwaves, which is accomplished in a 70% shorter period of time compared to pasteurization method and with 40% lower energy costs compared to the classic procedures used in existing mushroom farms.

Also, it fully ensures the totally sterile conditions by removal of any hazards regarding human contamination of the workspace, through exclusion of human access to the sterile zones, all the bags being handled, transported, sterilized, cooled, transferred, inoculated, and stored; and finally, the mushroom fruiting bodies are harvested and processed by modular robotic equipment. All functions of the robotic prototype are controlled by a central command system, which also monitors the system parameters as well as controllers for the sterilization system, the cooling area, the robotic inoculation system, the inoculated bag transfer system, storage systems, the air control systems for the incubation areas, the robotic harvesting system, and the final product packing equipment.

The modular robotic prototype for mushroom cultivation was designed to be useful for different production variants, with single or multiple workflows, which could include several nonsterile zones (ZN-A…ZN-K) for nutritive substrate preparation, as well as several sterile zones (ZS-A…ZS-K) for both automatic sterilization and inoculation with liquid mycelium of the solid substrates, followed in order by incubation, fruiting body formation, harvesting, and automatic processing of fruiting bodies belonging to cultivated mushrooms (Figure 4.3).

The programming and monitoring of each component system as well as the whole modular robotic prototype were designed as informatics systems placed in a command room, which is completely isolated from the sterile zones. Human access in the command room was projected to be only through the sealing gates and special equipment having the role of eliminating all pathogenic agents and pests (Petre et al., 2010; Nicolescu et al., 2010b).

4.4.1 GENERAL STRUCTURE OF MODULAR ROBOTIC SYSTEM FOR GROWING MUSHROOMS

The full sets of nonsterile and sterile zones include:

- nonsterile zone A (ZN-A) for solid substrates storage/processing,
- nonsterile zone B (ZN-B) for the distribution of plastic bags and filling with solid substrates by human operators,
- nonsterile zone C (ZN-C) for storage of transfer devices/pallets for bags,

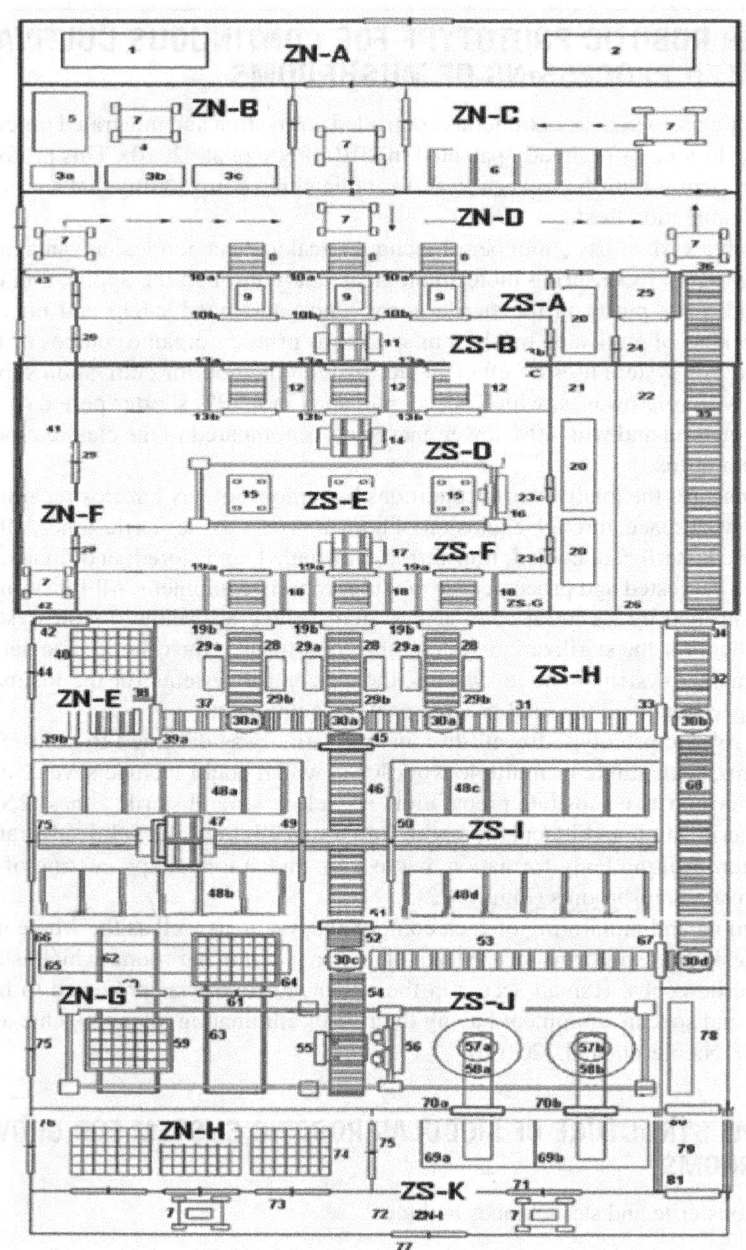

FIGURE 4.3

Representative scheme of modular robotic prototype for mushroom cultivation.

- nonsterile zone D (ZN-D) for transfer devices of bags transported by human operators,
- nonsterile zone E (ZN-E) for the evacuation outside the sterile area of plastic bags filled with sterilized solid substrates already inoculated with liquid mycelium, to be delivered to commercial customers,
- nonsterile zone F (ZN-F) for return of transfer devices/pallets emptied of bags to the input area and their storage,
- nonsterile zone G (ZN-G) for centralized removal of bags filled with exhausted substrates outside the cultivation system,
- nonsterile zone H (ZN-H) for storage of final products in warehouses,
- nonsterile zone I (ZN-I) for human operator access to warehouses of final products,
- sterile zone A (ZS-A) for sterilization of plastic bags filled with nutritive substrates,
- sterile zone B (ZS-B) for automatic moving of transfer devices/pallets with sterilized bags filled with substrates,
- sterile zone C (ZS-C) for cooling of sterilized bags filled with substrates through the temperature regulation in controlled ovens, previous to their inoculation with liquid mycelia,
- sterile zone D (ZS-D) for automatic moving of transfer devices/pallets of bags through multiple inoculation stations,
- sterile zone E (ZS-E) for robotic controlled inoculation with liquid mycelium of the sterilized substrates from plastic bags,
- sterile zone F (ZS-F) for automatic moving of plastic bags filled with inoculated substrates to the automatic inspection and distribution areas,
- sterile zone G (ZS-G) for transfer devices/pallets of bags and transition from sterile zones through a nonsterile zone,
- sterile zone H (ZS-H) for automatic visual inspection of bags' integrity and automatic moving and distribution of transfer devices/pallets of bags filled with sterilized and inoculated substrates with liquid mycelia,
- sterile zone I (ZS-I) for incubation and fruit body formation in fully automatic growth rooms,
- sterile zone J (ZS-J) for automatic harvesting of fruiting bodies of cultivate mushrooms,
- sterile zone K (ZS-K) for automatic processing of harvested mushrooms as fresh final products.

4.4.2 SPECIFIC TECHNOLOGICAL OPERATIONS OF MODULAR ROBOTIC PROTOTYPE

The nonsterile zones, from ZN-A to ZN-I (Figure 4.4), are designed for storage and preliminary processing of raw materials needed for the preparation of substrates for mushroom cultivation. ZN-A zone includes chopping and milling of substrates as well as soaking and homogenizing systems. The next nonsterile zone, B (ZN-B), is served by human operators for filling the bags with cultivation substrates and fixing supporting devices for sustaining bags on transfer devices such as transport pallets. Nonsterile zone C (ZN-C) is designed for temporary storage of transfer devices for bags. The next nonsterile zone, D (ZN-D), is the interface between the last nonsterile zone, ZN-I, and the first sterile one, ZS-A, being dedicated for loading the transfer devices of bags on input conveyors, which will lead them through the access gates into the sterile automatic processing zones. Nonsterile zone E (ZN-E) is placed upstream of the proper cultivation zone. In this area, the bags are collected, packed, and transferred to potential mushroom farmers. The next nonsterile zone, F (ZN-F), is used by human operators to handle the pallets with the supporting devices after the bags have been unloaded.

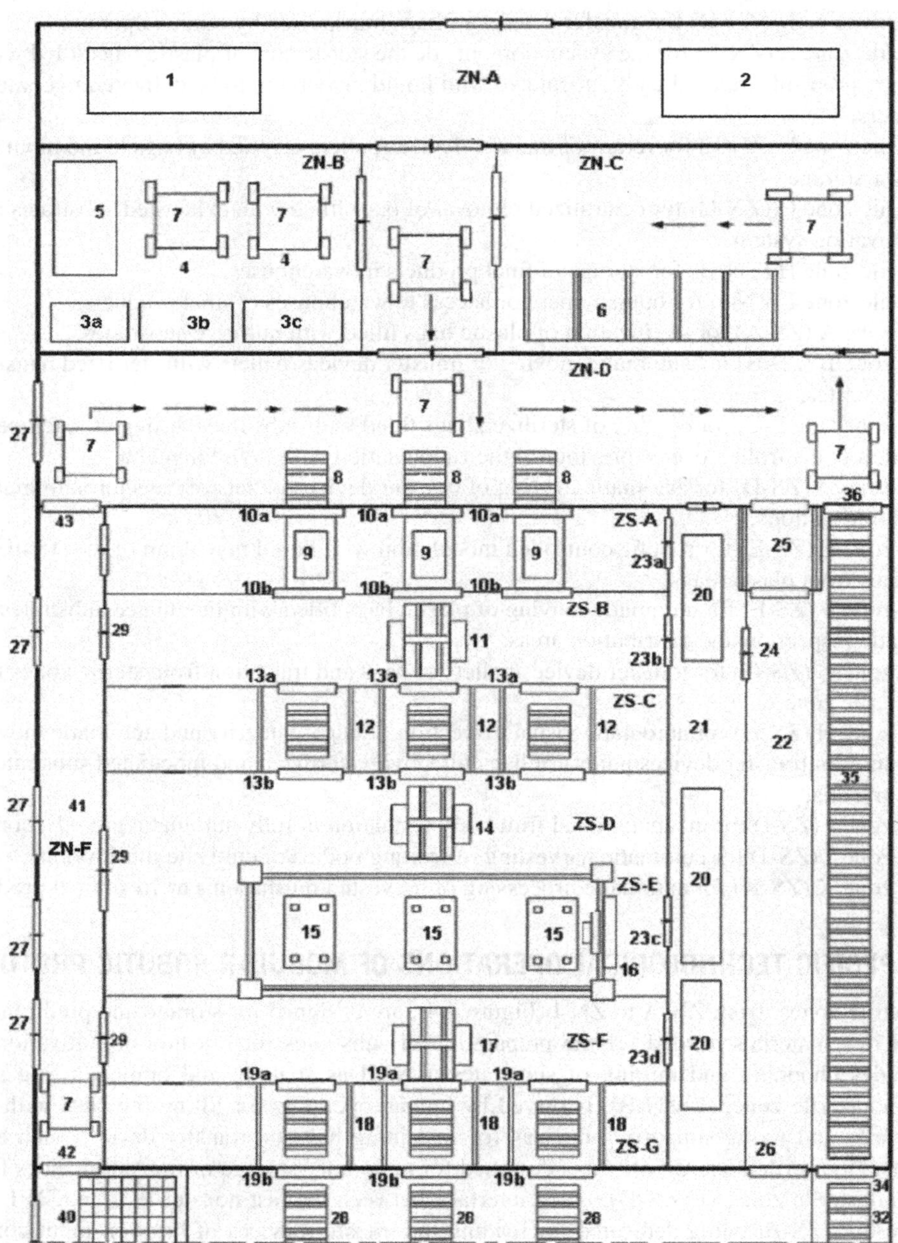

FIGURE 4.4

The nonsterile zones and main devices of modular robotic prototype.

Nonsterile zone G (ZN-G) is designed for centralized collection of bags with exhausted substrates after the mushroom fruiting bodies have been harvested. Nonsterile zone H (ZN-H) is used to store the packed final (post-processed) products, and nonsterile zone I (ZN-I) is designed to allow human access to the final product storage area (Petre et al., 2010).

Regarding the sterile zones of the modular robotic prototype, Figure 4.5 presents all sterile areas in which are performed all the main technological operations involved in the aseptic inoculation, incubation, fruiting body formation, harvesting of mushrooms, and their aseptic processing to get the final fresh products.

The modular robotic prototype includes sterilization enclosures (9), palletizing systems for automatic manipulation of transfer devices for plastic bags (11, 14, 17), microwave sterilization devices, rooms for controlled cooling in aseptic atmosphere of the plastic bags filled with sterilized substrates for cultivation (12, 18, 38), a workstation for temporary storage of sterilized bags during the automatic inoculation procedure (15), a gantry robot for automatic control of inoculation with liquid mycelia (16), video inspection systems for checking the integrity of bags (28a, 29b), and systems for the automatic transfer of inoculated bags (28, 31, 32, 35, 37, 46, 52, 53, 54) and redirecting the bags through transfer devices (30a, 30b, 30c, 30d) to a collecting station for outdoor delivery of plastic bags filled with sterilized compost and inoculated with liquid mycelia (38, 39, 40).

This modular robotic prototype also has incubation and fruiting body formation rooms (49, 50), including racks for storage of the transfer devices of bags (48a, 48b, 48c, 48d) and a Cartesian column-type robot (47) for automatic loading and unloading of the bags through transfer devices, a station for pallets stopping and indexing at a fixed point (55), a gantry robot (56) for automatic harvest of mushroom fruit bodies, four stations for automatic collection and transport of the harvested fruit bodies (57a, 57b, 58a, 58b), the systems (62, 63) for palletizing the rack-type pallets containing bags with the exhausted substrates (59, 64), and the final processing systems (69a, 69b) for the harvested mushroom fruit bodies (Petre et al., 2010).

Sterile zone A (ZS-A) includes a microwave oven area (9) for thermal sterilization of the bags filled with cultivation substrates. Zone B (ZS-B) has as its main structure an automatic transfer system (11), which moves the pallets from zone A (ZS-A) to zone C (ZS-C), including the controlled cooling area (12). The second automatic transfer system moves the pallets from the cooling area (12) to the temporary storage area (15) included in zone D (ZS-D). In zone E (ZS-E), a gantry robot with three degrees of freedom (16) allows simultaneous or successively controlled inoculation with liquid mycelium inside of three bags stored on a transfer device (Figure 4.5). After inoculation, the third automatic transfer system (17), located in sterile zone F (ZS-F), carries the pallets through a transfer gate (18) from zone G (ZS-G) into sterile zone H (ZS-H). The programming and monitoring of the automatic functioning of each system and of the entire robotic prototype (from sterile zone A to the sterile zone G) is done by informatics systems (20) located in the control room (21), which is completely separate from the nonsterile zones previously mentioned (Figure 4.4).

Zone H (ZS-H) includes three visual control systems for automated examination of bags for physical integrity after sterilization and inoculation processes and before their transfer and loading into the incubation and fruit body formation rooms located in I (ZS-I) zone. If the visual control system confirms the full integrity of the bags, they are transported by conveyors (28, 30, 46) through the Cartesian column-type robot unit (47) for automated loading into the storage racks (48a, 48b, 48c, 48d). If inadequate bags are identified on a transfer device, they are evacuated through the redirecting systems (30a, 30b) for extraction from further processing workflow, by means of the special conveyor sections (31, 32), the bag transfer device being leaded to nonsterile zone D (ZN-D).

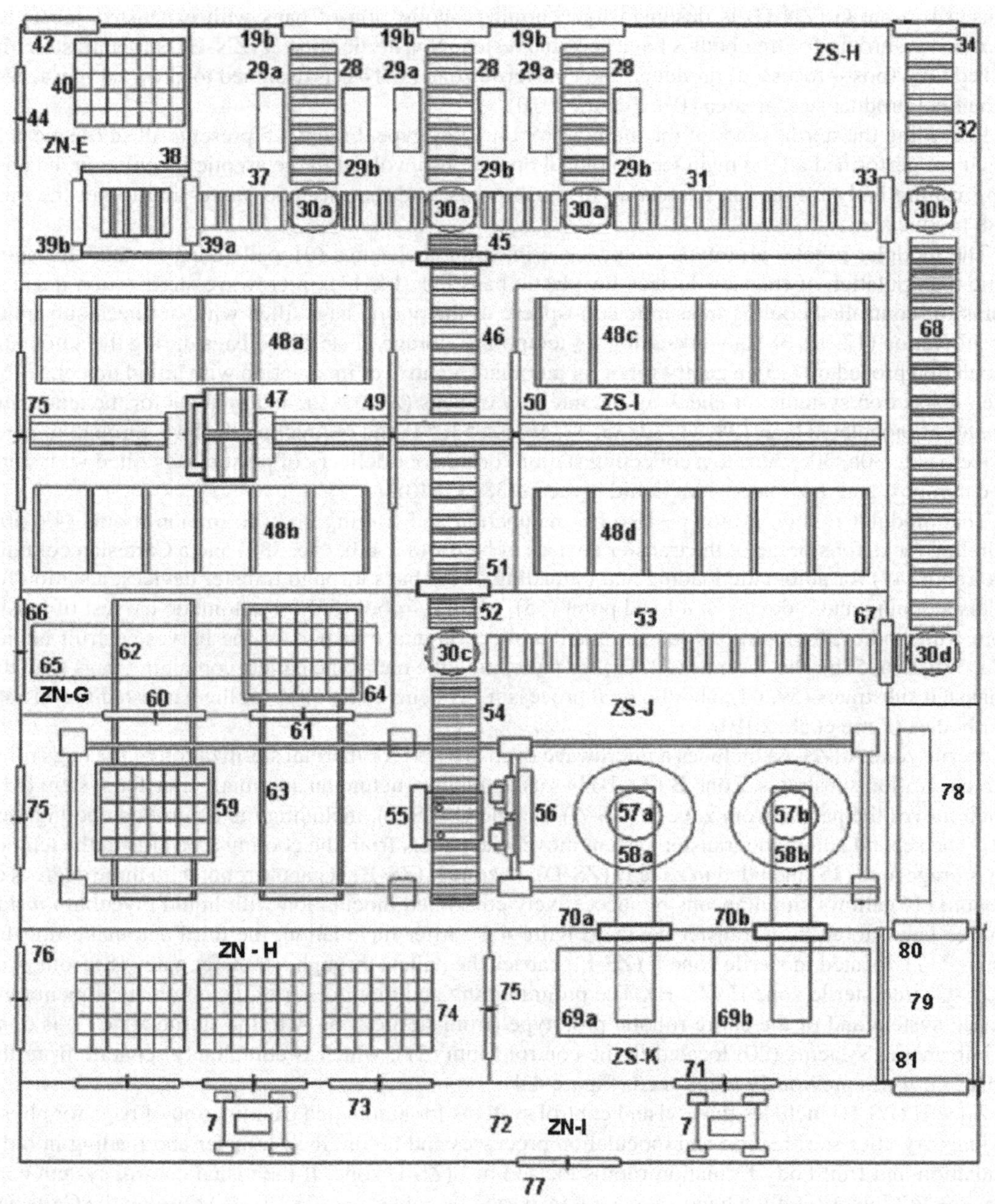

FIGURE 4.5

The sterile zones and main devices of modular robotic prototype.

All fully processed bags, once loaded with sterilized substrates and inoculated with mycelium, are evacuated by a system which contains the transporter (28), the pallet redirecting device (30a), the transporter (37), and the gate (38). The bags are transferred to the incubation rooms in zone I (ZS-I) by the transporters (45, 46). An automated system for transfer and transport (47) loads the four storage systems (48a, 48b, 48c, 48d) and places them in two incubation rooms. After the bags are unloaded from the storage systems, the incubation rooms are cleaned, washed, and sterilized by the system (47). At the same time, the disinfecting liquids and the wastes resulting from the harvest process are removed through gravitational collection systems on the ground.

The inoculation robot (56) includes a rectangular gantry system sustained by eight poles and two mobile trolleys sliding on a cross mounted on the gantry structure; one trolley has a vertical slide with a gripper (in order to manipulate the bags), and the other trolley has a vertical slide with an end-effector that harvests the mushrooms from the surface of bags (3, 5).

The mushrooms are finally collected by the flat band conveyors (58a, 58b). The bags containing exhausted substrates are evacuated by the rack-type pallets (59, 64), the palletizing systems (62, 63), as well as the ground transfer system (65) and some sliding access doors. There is also a system that evacuates the bag-sustaining pallets, which is composed of the transporter (54), the pallet redirecting system (30c), the transporter (53), the pallet redirecting system (30d), the transporter (68), the pallet redirecting system (30b), and the transporters (32, 35) (Figure 4.5).

The entire robotic prototype is controlled by a central command system, which also controls the system parameters and the specific controllers of the sterilizing system, the cooling area, the robotic inoculation system, the evacuation system for inoculated bags, the pallet moving, transfer, transport, and storage systems, the climatic systems for the incubation and fruit body formation rooms, the robotic harvest system, and the packing equipment for final products.

4.4.3 THE ROBOT OF INOCULATION

The gantry robot with three degrees of freedom has the role of inoculating mycelium in the plastic bags loaded with sterilized substrates. For this purpose, this type of robot will gradually take a position above each inoculation post, and the end-effectors will perform the operation of introducing the liquid mycelium inside the content of each group of three bags. The needed amount of mycelium for inoculation is downloaded from a low-capacity collector, which is provided with a device for continuous homogenization of its content, and transferred to the final injection devices by a pump and several injectors. After inoculation, the robot will retract to a "home position," allowing the recharge of the collector from a main storage tank of mycelium through a pipeline with direct connection.

The robot is specially designed for working in a sterile environment and has three degrees of freedom, as follows: *X* longitudinal axis, *Y* transversal axis, and *Z* vertical axis. The drive system is composed of electrical motors, and the programming is done through teach-in techniques (Nicolescu et al., 2010a,b).

The automatic inoculation with liquid mycelium of the bags loaded with sterile compost is done in sterile zone ZS-E, which includes temporary storage posts for pallets loaded with transfer supports of bags and a gantry robot with three degrees of freedom. The robot (Figure 4.6) is specially designed for inoculation, being equipped with a multiple end-effector (Figure 4.7), which may operate by simultaneously injecting a controlled volume of liquid mycelium in all the bags placed on the same support or successively injecting each of them. Before the inoculation, the end-effector is positioned according

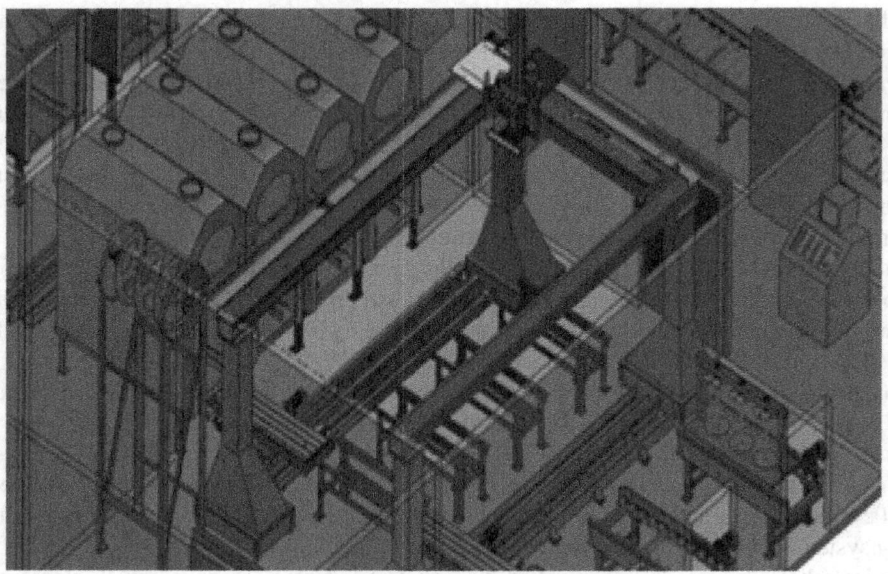

FIGURE 4.6

The gantry robot inside the inoculation room.

FIGURE 4.7

The gantry robot equipped with a multiple end-effector.

to the coordinates of palette location on the temporary storage post. After the inoculation process is completed, the pallets are automatically transferred from zone ZS-E to zone ZS-G by another transfer system placed in zone ZS-F. This transfer system takes the pallets from the temporary storage posts and introduces them into the transfer gate placed in zone ZS-G. The transfer gate is equipped with two vertically sliding doors (Nicolescu et al., 2009b, 2010a).

4.4.4 THE ROBOTIC HARVESTING CELL

Once the pallets with bags containing emerged mushroom fruit bodies reach the robotic harvesting area, a specially designed gantry robot with two end-effectors allows the picking-up of each bag with fruiting bodies from the pallet/support, using a first end-effector, transport of the bags to the collecting tables and evacuation conveyors, mushroom harvesting (using a second specially designed end-effector), and, finally, the transport of bags and their storage in a specially designed support/pallet system for collection and automatic transport of bags outside the sterile area. The central element of the robotic harvesting cell (Figure 4.8) is a specially designed gantry robot equipped with two end-effectors, having a total of 5 NC axes. Two of these are set as independent *Z*-axes, two as independent *Y*-axes, and one as a common *X*-axis. Each *Z*-axis is equipped with a specific end-effector, one designed for manipulation and transport of bags and the second for mushroom harvesting (Nicolescu et al., 2010a,b). To pick up the bags from the pallet/support and transport them to the harvest area, a three-finger gripper was designed,

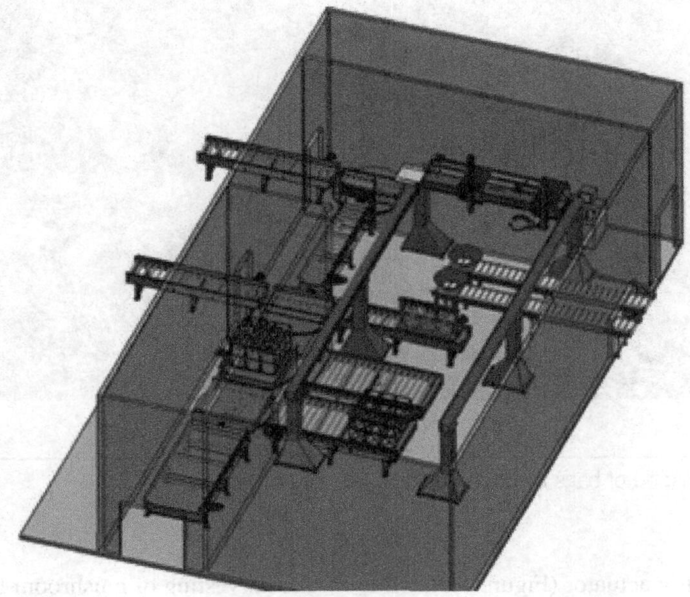

FIGURE 4.8

Automatic harvesting cell of virtual prototype.

FIGURE 4.9

Virtual prototype of specially equipped gantry robot.

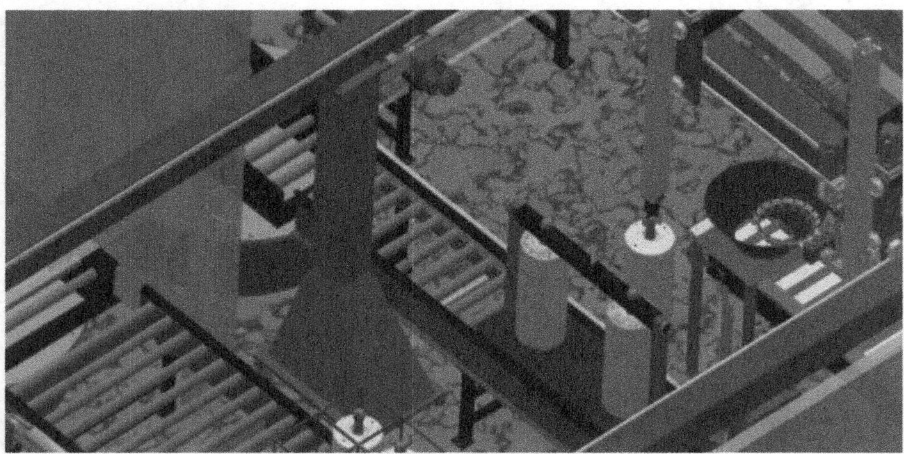

FIGURE 4.10

Manipulation and transport of bags by the end-effector in action.

powered by an electric actuator (Figure 4.9), while for the harvesting of mushroom fruit bodies a special end-effector has been designed, as shown in Figure 4.10, which shows the virtual prototype for the harvesting robot equipped with both end-effectors.

Figure 4.11A and B highlight two sequences captured from the harvesting process simulation and reveals the functional role of each robot's end-effector at the appropriate time.

FIGURE 4.11

Harvesting of mushroom fruiting bodies using a specially designed end-effector: (A) starting-up phase; (B) end phase.

However, the harvesting of mushroom fruit bodies represents a difficult task due to their non-uniform shapes and sizes on the surface of bags. To overcome this difficulty, a completely new technical solution was designed for a "harvesting end-effector with self-adaptive sliding blades," both closing and opening of end-effector being done by a bilateral electric actuator. The self-adaptive sliding blades of the harvesting end-effector have been designed using standardized mechanical components (Nicolescu et al., 2009b, 2010a).

After the harvesting operation is accomplished by the gantry robot, the depleted bags are transported to a storage pallet, which is led outside the sterile area of the facility and replaced with an empty one using the pallet transfer system and roller conveyors.

4.5 CONCLUSIONS

The modular robotic prototype presented, already patent protected, is the first fully automated system for continuous controlled cultivation and integrated processing of mushrooms.

Through its application at the industrial plant level, this system guarantees the nutritional and microbiological safety of final products by ensuring all standardized conditions of workspace purity in the sterile zones and also by preventing human error in the sterile zones, ensuring biotechnological safety. In the workflow put into practice by this prototype, efficient methods of sterilizing the cultivation substrates are used, involving total decontamination with microwaves, which is accomplished in a time 70% shorter and with energy costs lower by 40% compared to the "classical" procedures.

This robotic system fully ensures the total sterile conditions imposed by appropriate biological material cultivation, eliminating any hazards regarding human contamination of the workspace. Also, it prevents human access to the sterile zones, because the bag handling, transport, transfer, storage, inoculation, cooling, sterilizing, harvesting, and other operations are performed by modular robotic equipment.

All functions of installations and devices are effected by a central command system, which also controls the parameters for the sterilizing system, the cooling area, the robotic inoculation system, the evacuation system for inoculated bags, the pallet handling, transfer, transport, and storage systems, the control systems for air and temperature inside the incubation and fruiting body formation areas, the robotic harvesting system, and the packaging equipment for final products.

Concerning the novelty and originality of this conceptual model already patented, the following aspects should be remarked. To avoid any direct human involvement in the production bioprocess, all mushroom cultivation stages, starting from the substrate sterilization and mycelium inoculation, through mycelium incubation and fruiting body formation, as well as mushroom harvesting, are performed using this fully automatic system, which operates inside a completely controlled and 100% aseptic environment without any human handling.

The whole bioprocess of mushroom growing does not require any use of pesticides to prevent mushroom pathogens and pests. As a result, all harvested mushrooms are produced by organic cultivation only, with no use of any kind of pesticides, and they are 100% natural products.

Using this robotic prototype achieves two sorts of final products to be traded: packed mushrooms to be commercialized as fresh food, and plastic bags containing sterilized and inoculated substrates with pure liquid mycelium of edible or medicinal mushrooms.

At a middle level capacity in an industrial plant, the continuous production of sterilized and inoculated substrates packed in plastic bags (250 bags per day, every day all through the year) and production of fresh mushroom fruit bodies (1500 kilos per month) in 100% aseptic environment are simultaneously ensured.

The exhausted substrates resulting from mushroom cultivation can be directly used as feed supplements (containing proteins) or natural fertilizers for certain types of soil that are poor in organic substances, closing a short food chain of plant waste recycling, making from this point of view a significant contribution to environmental protection.

REFERENCES

Belforte, G., Deboli, R., Gay, P., Piccarolo, P., Ricauda Aimonino, D., 2006a. Robot design and testing for greenhouse applications. Biosyst. Eng. 95 (3), 309–321.

Belforte, G., Gay, P., Aimonino, D.R., 2006b. Robotics for improving quality, safety and productivity in intensive agriculture: challenges and opportunities. In: Huat, L.K. (Ed.), Industrial Robotics: Programming, Simulation and Application, vol. 1, pp. 677–690, ISBN 3-86611-286-6.

Che, H.H., Ting, C.H., 2004. The development of a machine vision system for Shiitake grading. J. Food Qual. 5, 120–125.

Cho, S.I., Chang, S.J., Kim, Y.Y., An, K.J., 2002. AE - automation and emerging technologies: development of a three-degrees-of-freedom robot for harvesting lettuce using machine vision and fuzzy logic control. Biosyst. Eng. 82 (2), 143–149.

Connolly, C., 2003. Gripping developments at Silsoe. J. Ind. Robot 30 (4), 322–325.

van Galen, M.A., Hammerstein, J.J.C.M., Stallen, M.P.K., Kamphuis, B.M., 2003. The dutch mushroom sector: small mushrooms, big business. European Aid Cooperation Service, Project Report 1, Big Business, pp. 7–15.

Heinemann, P.H., Hughes, R., Morrow, C.T., Sommer, H.J., Beelman, R.B., Wuest, P.J., 1994. Grading of mushrooms using a machine vision system. Trans. Am. Soc. Agric. Eng. 37, 1671–1681.

Jarvis, R., 1997. Sensor-based robotic automation of mushroom farming - Preliminary considerations Advanced Topics in Artificial Intelligence, Series: Lecture Notes in Computer Science, vol. 1342. Springer, Berlin Heidelberg, pp. 446–455.

Masoudian, A., McIsaac, K.A., 2013. Application of support vector machine to detect microbial spoilage of mushrooms. Proceedings of International Conference on Computer and Robot Vision (CRV), pp. 281–287.

Nicolescu, A., Petre, M., Dobre, M., Enciu, G., Ivan, A., 2009a. Conceptual model of a modular robotic system for mushroom's controlled cultivation and integrated processing. Annals of DAAAM for 2009 & Proceedings of the 20th International DAAAM Symposium "Intelligent Manufacturing & Automation: Focus on Theory, Practice and Education". Vienna, Austria, pp. 687–688.

Nicolescu, A., Enciu, G., Dobrescu, T., Ivan, A., Dobre, M., 2009b. Virtual prototyping a modular robotic system for mushroom controlled cultivation and integrated processing. Annals of DAAAM for 2009 & Proceedings of the 20th International DAAAM Symposium "Intelligent Manufacturing & Automation: Focus on Theory, Practice and Education". Vienna, Austria, pp. 685–686.

Nicolescu, A., Ivan, A., Petre, M., Dobre, M., 2010a. Virtual prototyping robotic cell for mushroom cultivation in controlled atmosphere. Annals of DAAAM for 2010 & Proceedings of the 21st International DAAAM Symposium "Intelligent Manufacturing & Automation: Focus on Interdisciplinary Solutions, pp. 59–60.

Nicolescu, A., Marinescu, D., Dobre, M., Petre, M., 2010b. Virtual prototyping robotic cell for mushrooms crops automatic harvesting. Annals of DAAAM for 2010 & Proceedings of the 21st International DAAAM Symposium "Intelligent Manufacturing & Automation: Focus on Interdisciplinary Solutions", pp. 61–62.

Noble, R., Reed, J.N., Miles, S.J., Jackson, A.F., Butler, J., 1997. Influence of mushroom strains and population density on the performance of a robotic harvester. J. Agric. Eng. Res. 68 (3), 215–222.

Petre, M., Nicolescu, A., Dobre, M., 2010. Process and installation for cultivating eatable and therapeutical mushrooms. Patent number RO123132 B1, International classification A01G1/04, Application number: RO/2008/000610, granted by Romanian Office of Patents and Marks (OSIM) on September 30th 2010, Bucharest, Romania.

Reed, J.N., Tillett, R.D., 1994. Initial experiments in robotic mushroom harvesting. Mechatronics 4 (3), 265–279.

Reed, J.N., Miles, S.J., Butler, J., Baldwin, M., Noble, R., 2001. AE - automation and emerging technologies: automatic mushroom harvester development. J. Agric. Eng. Res. 78 (1), 15–23.

CHAPTER

GROWING *AGARICUS BISPORUS* AS A CONTRIBUTION TO SUSTAINABLE AGRICULTURAL DEVELOPMENT

5

Jean-Michel Savoie[1] and Gerardo Mata[2]

[1]*INRA, UR1264 MycSA, Villenave d'Ornon, France*

[2]*Instituto de Ecología, A.C., Red de Manejo Biotecnólogico de Recursos, Xalapa, Veracruz, Mexico*

5.1 INTRODUCTION

Mushrooms have been consumed by humans since ancient times as a part of the normal diet and as a delicacy due to their desirable taste and aroma. In addition, mycotherapy has been used for a long time and currently is receiving increasing interest in Europe for prevention and treatment purposes. Consequently, humans have developed cultivation methods that yield abundant and constant resources of specific edible and medicinal mushrooms. All over the world, cultivated mushrooms are mainly saprotrophic species, being grown on various agroindustrial wastes. Increasing cultivation of mushrooms may then contribute to the development of a new agriculture and address consumer demand for healthy products. This may satisfy the three interrelated dimensions of sustainable development as proposed by the United Nation Conference on Sustainable Development (Rio+20): environmental, economic, and social. According to the UN Environmental Programme, "Innovative and green technologies can simultaneously increase employment, foster better use of science and traditional knowledge, upscale provision of basic services, and at the same time protect the environment" (UNEP Post 2015 #1: http://ozone.unep.org/Publications/UNEP_Post_2015_Note1.pdf). Progress in mushroom science will help the mushroom industry to address these objectives.

Despite the button mushroom having been cultivated for centuries, recent progress in scientific knowledge of the fungal species *Agaricus bisporus* offers new opportunities to innovate cultivation processes with a contribution to a sustainable agriculture as a target. The history of button mushroom cultivation is uncertain. There is some reason to suspect cultivation of agarics in ancient Egypt and then Greece. An Egyptian wall painting indicated that they cultivated an *Agaricus* sp. strain. In France, during the seventeenth century, the cultivation of what was probably *A. bisporus* had been developed. One story explains that Jean-Baptiste La Quintinie, an expert gardener and agronomist at that time, who created the vegetable garden at Chateau de Versailles, cultivated button mushrooms for the French King Louis XIV. What is undoubtedly known is that at the beginning of the eighteenth century, cultures

Mushroom Biotechnology. DOI: http://dx.doi.org/10.1016/B978-0-12-802794-3.00005-9

were developed in cellars and underground below the buildings of Paris, and then moved into caves and stone quarries around Paris and in the Loire Valley in France. It is now cultivated all over the world in temperate areas. Past and recent research on the ecology, biology, genetics, and cultivation techniques of this mushroom opens new possibilities for the production of this common fungal species under various conditions, having sustainable development as an objective.

The control of pests and pathogens by biological treatments in place of pesticides as well as by the use of resistant cultivars are significant strategies to reach this objective. This is the topic of the following chapter in this volume. The present chapter gives, on the one hand, an overview of the improvement of agro-waste valorization in *A. bisporus* cultivation, and, on the other hand, progress in biodiversity preservation and utilization, as well as developments in genetics are highlighted through the interest in producing strains able to fruit at high temperatures and having health promoting action and low safety risks.

5.2 THE IMPROVEMENT OF AGRO-WASTE VALORIZATION

Edible and medicinal mushrooms are cultivated on lignocellulosic residues, and the economic value of their recycling with these fungal species is considerable. The wastes can be viewed as reusable materials, at least in terms of providing new economic opportunities and positive environmental consequences. Contrary to other cultivated mushrooms, such as *Lentinula edodes* and *Pleurotus* spp., *A. bisporus* is not a lignicolous species, rather, it is a leaf-litter degrading white-rot basidiomycete acting as a secondary decomposer. Forest litter decomposition is usually initiated by generalist primary colonizers involving a diverse community of fungi and bacteria which utilize simple sugars, oligosaccharides, and other low molecular weight compounds. After this initial flush of microbial activity, specialist secondary colonizers, which are less competitive than the microfungi in exploiting labile resources (Frankland, 1992), develop via the decomposition of more recalcitrant plant polymers such as lignocellulose complexes and humic compounds formed during the first phase of decomposition. Taking into account these ecological traits, the cultivation of *A. bisporus* is conventionally performed by inoculating pre-decomposed substrates obtained by composting. Composting is a human action which is designed to accelerate the process of complex organic matter decomposition by the action of microbial communities under high temperature, high humidity, and oxygenation. For the cultivation of *A. bisporus*, the overall goal of composting is to produce a selective nutrient medium for mushroom growing (Miller, 1993). This refers to both chemical and biological aspects that must be managed to obtain the conversion of raw agricultural wastes into compost which is more resistant to degradation and inhabited by a steady microbiota (Savoie et al., 1993). The two controllers that can be activated to reach the composting goals are the quality of raw materials and the physicochemical conditions during the process. Improving the valorization of agro-wastes by cultivation of *A. bisporus* with sustainable development as an objective leads to the use of local resources in composting processes that are environmentally safe and energetically moderate.

5.2.1 THE USE OF LOCAL RESOURCES

Wheat straw and horse manure including straw have been the major components of mushroom compost formulations for a long time in Western countries. Other cereal straws (rice, oat, barley) and sugarcane bagasse are used as lignocellulosic materials in other countries, when these wastes are locally available.

Horse manure is an important ingredient of compost for supplying both microorganisms and nitrogen plus some minerals. In fact, the C/N ratio is an index that has to be taken into account in the choice of raw ingredients. Optimal values in the cultivation substrate are around 20 (Sharma and Kilpatrick, 2000). Due to degradation during composting, this value can be reached when the C/N ratio in the mixture of raw materials is at 40–60. This means that nitrogen sources have to be added to the straw or bagasse (C/N > 100). Due to the limited availability of horse manure in large amounts around the world, composts whose formulations are prepared without the addition of this manure but using other nitrogen-rich materials are proposed. They must provide the compost microbiota with readily available N sources for their growth, and this nitrogen is transformed into more stable forms in the microbial biomass. For instance, ammonium nitrate or urea must be added at the outset of the composting process. Poultry manure, cereal grains and bran, and oilseeds are also commonly used, with soybean being the most often used in countries where available (Zied et al., 2011). Soybean or other seed meals may be treated to delay the immediate availability of nutrients. However, chemical compounds such as formaldehyde have been used for this purpose, which is not in agreement with the objectives of sustainable development due to the toxicity of this product.

As an illustration of the use of local resources for both nitrogen-rich compounds and lignocellulosic materials, Andrade et al. (2008) worked on formulations with soybean mill and urea as a nitrogen source and compared three lignocellulosic materials available in Brazil (two varieties of *Cynodon dactylon* straw and oat straw) mixed with sugarcane bagasse. Mushroom productivity with four *A. bisporus* strains varied from 7 to 18 kg mushroom/100 kg compost. Different types of raw materials can be used depending on their availability, but experiments are needed to define the best formulae for growing a specific strain under local conditions. The quality of materials is known to influence compost ability to support high yields of mushrooms. It may be variable, even with the same ingredient such as wheat straw. It has been shown that the geographical source, the variety, and agronomical practices such as use of growth regulators or fungicides affect straw quality and may have consequences for the composting process and mushroom yields (Savoie et al., 1992, 1993, 1997).

Apart from the raw ingredients of compost, the casing material is an environmental issue. Covering the colonized compost with a casing layer is the way to induce the production of fruiting bodies. In Western countries peat moss is used alone, or in a mixture with calcareous materials as the casing material. Peat is forming very slowly in natural wetlands, which are sources of pure water and ecological reserves for many plants and animals. Mining peat in these ecosystems for the mushroom industry is an environmental problem, and peat availability is a great concern in some regions around the world. It is a great challenge to find a substitute that is available in volume and at low cost to meet the demands of mushroom production (Colauto et al., 2011). The casing layer provides the support and storage of water for developing mushrooms and protects the substrate from desiccation; it adsorbs volatile and other chemical components produced by the vegetative mycelium of *A. bisporus* and provides for certain beneficial bacteria. Alternatives to peat used alone or in mixture for decreasing the quantity of peat have been tested and sometimes practically used, including soil (clay-loam), vermiculite, weathered mushroom compost, activated carbon, bagasse and filter cake, coconut fiber, among others (Bechara et al., 2009b; Siyoum et al., 2010). In the mushroom industry, aerated steam treatment of casing is recommended to reduce the incidence of mushroom pests, but with low intensity and followed by a delay before use, for substrate recolonization by beneficial microorganisms. Such pasteurization might also suppress pathogens of the human body that would contaminate the mushrooms. Soil fumigation might also be used for cleaning the casing material, but there is a lack of environmental friendly fumigants.

Whereas these treatments are mandatory with soil-based and spent-mushroom compost casing, they might be avoided with other materials, saving energy and cost.

Finally, the use of new local resources as raw ingredients for composting and the casing layer is possible and is environmental friendly, but it needs years of experience for the development of mushroom cultivation at a commercial scale, with mushroom production reaching the standards of quality and yields of the places where *A. bisporus* has been cultivated for a long time. The path of commercialization, social issues, and economic balance also contribute to the choice of the use of local raw materials.

5.2.2 FROM OUTDOOR TO INDOOR COMPOSTING

Since the 1950s, conventional composting has been carried out in two steps. Phase I is an initial step carried out outdoors with rather limited process control, while phase II is a conditioning step carried out indoors under generally well-defined process controls (Miller, 1993).

After a pre-wetting period, ingredients are mixed and phase I is driven in long rectangular stacks for 1–2 weeks. Phase I is both biological and chemical, with the breakdown of organic substrates for energy supply and incorporation into microbial biomass being the crucial mechanisms (Figure 5.1). Composting is mainly an aerobic process, but during phase I, anaerobic areas develop in the inner parts of stacks (Miller et al., 1989). In addition, temperature increases due to microbial activity, and in some parts temperatures might reach extreme values greater than 70°C, which, with an atmosphere rich in ammonia, results in partial sterilization (Miller et al., 1989). On the other hand, high temperature and high ammonia levels make the lignocellulosic components of compost more readily available for microbial degradation and stabilize nitrogen resources into steady forms (Savoie et al., 1996b). For management of both aeration and temperature, the stacks are turned frequently. This turning mixes the different parts of the stacks and introduces successive cycles of raw ingredient transformation, which enhances the overall microbial activity and tends to foster better homogenization of the substrate.

Research since the early 1960s has been devoted to techniques for preparing high-quality mushroom compost in the shortest possible time. Meanwhile, large composting facilities were developed in Western countries and residents began to complain about composting odors. At the same time there was an increasing demand from the public for ecologically acceptable production techniques, with a focus on water quality and chemical pollution. As a consequence, several methods of indoor composting were developed and experimented with. They helped to clarify the biological mechanisms involved during composting, mushroom growth, and fruiting. However, empirical approaches allowed the rapid development of indoor composting in tunnels on a large industrial scale in Italy, Austria, Australia, and the Netherlands (Laborde et al., 1993). During indoor composting, the odors can be trapped before air is vented to the outside, and the climatic parameters of the process are controlled. The calories produced can be used for heating the tunnels and other buildings. Nowadays, growers across the world have adopted this environmentally friendly way of making compost in a short time because of the good results obtained by adequately managing the climatic parameters. This reaches the objectives of sustainable development by saving time, energy, raw materials, and space, by improving sanitary control and favoring the stabilization of compost and mushroom quality. However, the main problem is that it means sophisticated facilities with computerized monitoring and management. The technology might be available for small local growers, but it requires a grouping of several growers in cooperatives or other sharing structures.

FIGURE 5.1

Two phases of composting for the cultivation of *Agaricus bisporus*. (A) pre-wetting of raw ingredients; (B) outdoor phase I; (C) tunnel filling; (D) end of phase II compost, in a tunnel.

At the midterm between indoor and outdoor phase I composting, new devices for aerated windrow have been developed for improved rapid composting procedures. These generally use polyvinyl chloride pipes with holes drilled at various distances and positioned in different parts of the windrows, providing either forced or passive aeration (Song et al., 2014; Wakchaure et al., 2014).

5.2.3 REUSE OF THE SAME COMPOST SEVERAL TIMES

Mushrooms are produced in a series of breaks or flushes at one-week intervals. After the second break, each successive flush produces fewer mushrooms due to the depletion of available nutrients in the compost. Compost supplementation with nitrogen-rich nutrients having delayed release and mixed in compost at spawning or casing can limit this phenomenon (Zied et al., 2011), but it is theoretically less efficient than regular injection of nutrients during the culture. This was experimented with in the 1960s (San Antonio, 1966) but not implemented.

More recently, Royse and collaborators, renewing an old idea (Schisler, 1990), developed a methodology to obtain additional yield of mushrooms from compost after one or two flushes, which may help to reduce production costs and reduce the amount of spent mushroom compost requiring disposal: double-cropping (Royse et al., 2008; Royse and Sanchez, 2008a,b; Royse and Chalupa, 2009). This requires removing the casing layer after one, two, or three breaks and incorporating various supplements into the first crop compost (Royse, 2010; Royse et al., 2008). Then the compost is reconditioned and re-cased. In various attempts to improve the double-cropping method, the authors showed that addition of spawn had the effect of supplementation but was not enough, and other supplements were used with higher efficiencies. Addition of 20% of phase II compost for introducing new nutrients and balancing the loss of dry matter during the first crop was experimented with. It proved efficient when supplements were added and thoroughly mixed with de-cased first crop compost (Royse, 2010). Compost fragmentation proved to be efficient in improving the yield of the second crop. Finally, by using de-cased compost collected after two flushes of harvest, fragmented and mixed with commercial supplements, second crops of two flushes may reach the same level of yield per m^2 as the first crops. Double-cropping could help growers become more efficient and competitive, and ensure sustainability of production by increasing the efficiency of raw material use. However, the risks of pathogen transmission from crop to crop may limit the use of the double-cropping, though Royse and Chalupa (2009) did not notice any interference from potential pests and pathogens.

5.2.4 A CULTIVATION SUBSTRATE WITHOUT COMPOSTING?

Composting being a source of some environmental drawbacks and loss of raw materials before inoculation of *A. bisporus*, early in the 1960s, some methods were developed for mushroom cultivation on non-composted substrates. Till (1962) succeeded with sterilized complex substrates by maintaining axenic conditions up to fruiting. This was the first report of *Agaricus* cultivation on a non-composted substrate (NCS). This NCS was the subject of various studies, with some patents registered, showing it is possible to cultivate *A. bisporus* on NCS; however, formulas and techniques need to be optimized and the economic balance needs to be improved before commercial development is achieved. The maintenance of sterile conditions being technically complex and costly, another approach was carried out. This method directly uses grain spawn normally used for inoculating *A. bisporus* in compost and adds a casing layer onto it (San Antonio, 1971; Bechara et al., 2006). Using commercial millet grain spawn to which a delayed-release supplement was added, experiments were conclusive at the laboratory scale (Bechara et al., 2009a).

NCS without sterilization was also experimented with. Sanchez and Royse (2001) reported a procedure for preparing mushroom substrate without composting of ingredients used for the cultivation of shiitake and with a nonsterile phase for fruiting induction. Fermentation processes (Huhnke and Sengbush, 1968), or reinoculation of pasteurized substrates (8 h at 60°C) with specific thermophile microorganisms (Sanchez et al., 2008; Coello-Castillo et al., 2009), were also tested. The objective was focused on inhibiting and preventing the subsequent development of microorganisms which could be compete with the cultivated mushroom. These techniques used various raw materials that were locally available, such as pergola grass or corn cobs in Mexico (Sanchez et al., 2008; Sanchez and Royse, 2009), but require the production of pure inoculum of *Scytalidium thermophilum*, a thermophile fungus known to contribute to the selectivity of composts and to enhance yields of *A. bisporus*. However, supplementation of substrate at casing is required for correct yields (Coello-Castillo et al., 2009).

With NCS developed up to now, yields are rather low compared to traditional compost, and the economic balance of the process is not in favor of development to large-scale production, but this original method of mushroom production is worth studying, as are new composting processes for the improvement of sustainable agricultural waste valorization through *A. bisporus* cultivation.

5.3 THE PRESERVATION AND MANAGEMENT OF BIOLOGICAL DIVERSITY

5.3.1 THE LOSS OF GENETIC DIVERSITY IN CULTIVATED LINES

All cultivated crops and domesticated animals have resulted from the human management of biological diversity, and their continuous evolution through improvement by breeders and farmers constantly responds to new challenges to maintain and increase productivity. In this respect, cultivated mushrooms are no exception. The scientific names given to the button mushroom have changed several times, finally stabilizing 70 years ago (in 1946) as *A. bisporus* (J.E. Lange) Imbach (Table 5.1), but one can speculate that the same biological species has been cultivated since its first appearance in France at the end of the seventeenth century, with improvements to increase productivity.

After the first cultivation, mushroom growers' selection for desirable characteristics, such as faster growth and yield or color, has dramatically changed the cultivated button mushroom compared to its native relatives. Growers used selected mushrooms for inoculating subsequent cultures and accumulated some characteristics over time. With the easy vegetative dissemination of *A. bisporus* and the possibility of storing mycelium alive for months, the diversity in cultivated strains decreased rapidly. In the second part of the twentieth century, all the cultivated strains of *A. bisporus* came from five ancestral lineages defined by phenotypes as "off white," "small white," "white," "brown," "small brown," and "golden white." The lineages, which are genetically different, have a European origin (Xu et al., 1997; Royse and May, 1982; Foulongne-Oriol et al., 2009). The first hybrids, Horst® U1 and Horst® U3, were obtained in the 1980s in the Netherlands by crossing "white" and "off white" cultivars. Except for a hybrid developed during the 1980s in China (Wang et al., 1995), no new white hybrids with genetic backgrounds different from Horst® U1 and Horst® U3 have been developed since (Foulongne-Oriol et al., 2011b; Sonnenberg et al., 2011). White hybrids derived from these original ones are cultivated widely throughout the world. As a result, *A. bisporus* is nearly a monolineage crop.

Today, the lack of diversity in cultivated strains of the button mushroom is considered an important risk for this culture, and efforts have been made over the past 30 years to overcome this problem. The

Table 5.1 The Scientific Names Given to the Button Mushroom Have Changed Several Times

Name	Author	Reference	Year
Agaricus campestris var. *hortensis*	Cooke	Handb. Brit. Fungi 1: 138	1871
Psalliota horensis var. *bispora*	J.E. Lange	Dansk Bot. Ark. 4 (12): 8	1926
Psalliota bispora	(J.E. Lange) F.H. Møller & Jul. Schäff	Annals Mycol. 36 (1): 69	1939
Agaricus bisporus	(J.E. Lange) Imbach	Mitt. Naturf. Ges. Luzern 15: 15	1946

native crop relatives and local varieties are the elements of agricultural biodiversity most likely to contain the high levels of genetic diversity and biological novelty needed to sustain innovations in breeding programs. This implies preservation of the native resources and an evaluation of their genetic and phenotypic diversity, which are significant components of the development of a sustainable agriculture.

5.3.2 THE NATIVE RESERVOIR OF BIODIVERSITY

The native populations of fungi are undoubtedly very important for the productive sector of cultivated mushrooms because individuals physiologically adapted to certain substrates and environmental conditions provide the genetic variation necessary to potentiate desirable characteristics for the commercial sector (Salmones and Mata, 2012). However, in developing countries, and especially in tropical and subtropical areas, which harbor enormous biological diversity, few government agencies have shown interest in supporting regular and organized research on knowledge, conservation, management, and implementation of fungi as a genetic resource. It would be appropriate to avoid the study of native germplasm of an isolated and disintegrated form, and instead encourage efforts to collect, review, analyze, and integrate the different aspects of edible mushrooms in order to obtain a real picture of the strengths, problems, needs, and current priorities. Strain collections are fundamental tools for the development of research currently being carried out in the area of cultivation of edible fungi. Existing collections are dedicated to maintaining mycelia and safeguarding related information. Some of these collections provide deposit services, identification, and authentication of the strains. According to data from the World Federation of Culture Collections (WFCC, 2015), there are 678 collections of microorganisms distributed over 71 countries. In these collections about 2.5 million strains are safeguarded, and approximately 30% of the strains (725,000) correspond to fungi. The American Type Culture Collection (ATCC) holds approximately 56,000 strains of fungi and yeasts, among which are 427 strains of *Agaricus*, many of which do not have data about their origin or are copies of the same strain sheltered in collections of other countries. Other collections not registered in the WFCC also exist.

In the 1980s, there appear to have been fewer than 20 independent lines of *A. bisporus* in mainstream culture collections worldwide, including those of commercial laboratories (Kerrigan, 1996). Mycologists had not reported the species in their native form for a while. Collecting the native strains and their preservation in germplasms appeared necessary to R.W. Kerrigan in North America and P. Callac in Europe (Kerrigan, 1996; Callac et al., 2002), because there was concern about losing their genetic diversity forever. With the contribution of some mycological societies, mycologists, and researchers from many countries, and thanks to specific collaborative programs, more than 600 native specimens have been collected from their natural environments and preserved in two collections: the ARP (*Agaricus* Resource Program) in the United States (Kerrigan, 1996) and the CGAB (Collection du Germoplasme des Agarics à Bordeaux) in France (Callac et al., 2002). Currently, native strains are also preserved in collections of various universities and research institutions. The geographic distribution of the germplasm mainly covers Europe, the Mediterranean region, and North America. However, native strains also have been collected from the Tibetan Plateau (Wang et al., 2008) and other places in Asia, South America, and the Middle East (Malekzadeh et al., 2011). The known geographic range of the species extends from the equatorial climate of Congo (Heinemann, 1956) to the boreal region of Alaska (Geml et al., 2008). Unfortunately, no living specimen from extreme climates is available in free culture collections.

Recently, *A. bisporus* obtained the status of a model fungus for its adaptation, persistence, and growth in the humic-rich leaf-litter environment (Morin et al., 2012). The leaf litter of *Cupressus* spp.

FIGURE 5.2

A native strain of *Agaricus bisporus* from *Cupressus* litter.

was identified as a good habitat for isolating native strains in various countries (Figure 5.2). The needle litter of *Picea*, plant wastes, and manures in fields or parks also furnished numerous native strains. However, *A. bisporus* has been found in various habitats such as in coastal dunes in France or under *Prosopis* and *Tamarix* in arid place of the Sonoran Desert, California, USA (Kerrigan, 1995; Xu et al., 1997; Callac et al., 2002). In light of this, adaptation of geographical populations and/or varieties of the species to different climates and habitats could be expected (Largeteau et al., 2011). This ability is associated with genetic diversity for the use of the native isolates as new sources of breeding material to restore genetic variability in cultivars. Preserving the habitats where *A. bisporus* is frequently found should be a way to maintain the natural reservoir of diversity for this mushroom species. Unfortunately, such habitats are frequently threatened by extreme climatic events and human activity. That is the case of cypress forests in coastal areas in Europe and North America. Development of a durable germplasm is increasingly of concern for *A. bisporus* biodiversity preservation.

5.3.3 GENOTYPIC AND PHENOTYPIC RICHNESS OF GERMPLASMS

Genetic studies on native mushroom populations have revealed a large diversity, and three varieties having different life cycles were identified. *Agaricus bisporus* is an amphitallic species with a homothallic or heterothallic cycle depending on the ploidy level of the spores, which can be heterokaryotic (n + n) or homokaryotic (n) (Figure 5.3). Each dominant life cycle is characteristic of a variety. The cultivated strains and most of the native strains belong to *A. bisporus* var. *bisporus*, which has a predominantly pseudohomothallic life cycle. In this life cycle, most of the basidia are bisporic and produce heterokaryotic spores, giving rise to fertile heterokaryons able to produce fruiting bodies, whereas plasmogamy between two sexually compatible homokaryons is needed to restore a fertile heterokaryon in the amphithallic life cycle of *A. bisporus* var. *burnettii* (Kerrigan et al., 1994) (Figure 5.3). The variety *burnettii*

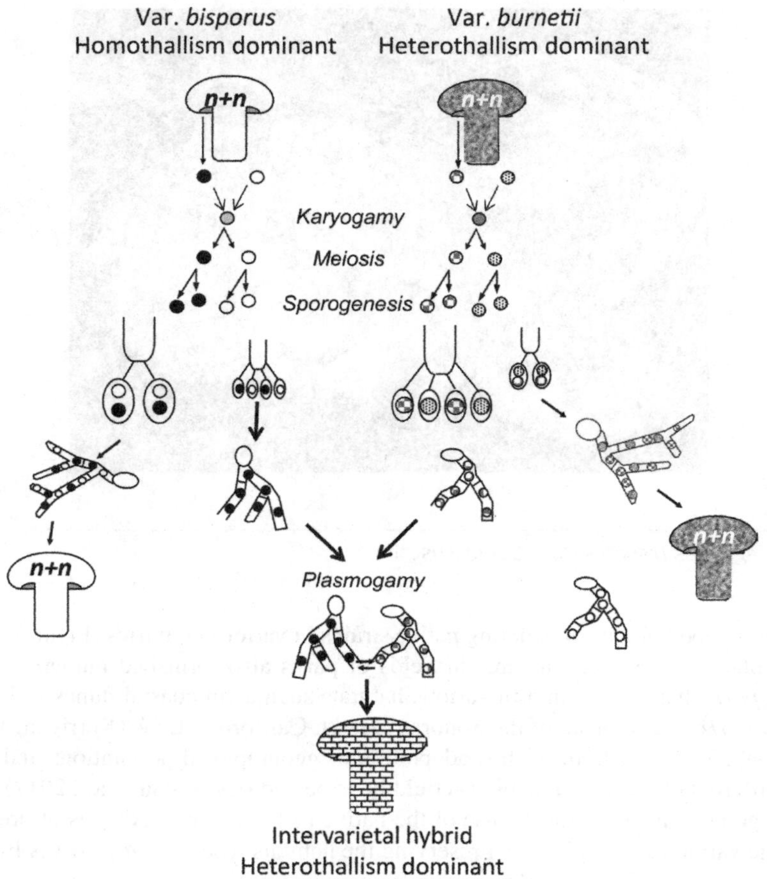

FIGURE 5.3

Life cycles and production of an intervarietal hybrid of *Agaricus bisporus*.

was discovered in the Sonoran Desert of California (Callac et al., 1993), and up to now it has been restricted to this local population. The basidia are mostly tetrasporic and the variety is completely interfertile with the variety *bisporus*. A homothallic life cycle in which homokaryotic sporophores produce homokaryotic spores giving rise to fertile homokaryons has been found in rare tetrasporic specimens in France and in Greece. They belong to the same clonal colony and are the only known members of *A. bisporus* var. *eurotetrasporus* (Callac et al., 2003).

In a survey of genetic diversity in North American populations with molecular markers, 144 different nuclear haplotypes were characterized, showing a wide genetic diversity, but many strains had the same 10 mitochondrial haplotypes as the traditional cultivars originating in Europe. Among such cultivar-like isolates, nuclear genetic divergence between geographical populations was significantly less pronounced than among the strains that do not bear such haplotypes (Kerrigan et al., 1998). This suggested that certain populations, such as those of the California coast, are contaminated by cultivars via outcrossing since nuclear genotypes of cultivars are rarely found in the wild (Kerrigan et al., 1998), whereas other populations such as that of Alberta in Canada have proven to be relatively uncontaminated by cultivar-like genotypes (Xu et al., 1997). From these results the question of the contamination of the native populations has arisen and spurred mycologists to preserve the biodiversity of *A. bisporus* in a germplasm.

Xu et al. (2002) analyzed the genetic diversity of samples from two local populations in the west of France over a two-year period. One site (50 × 70 m) was a field frequently receiving horse manure; the other site, 450 km away, was under Monterey cypress trees, *Cupressus macrocarpa*. High levels of genetic variation were recorded in each population, with the largest potential clonal colony found in no more than 1 m^2. Genetic drift, hybridization between neighboring individuals, and gene flow among populations could contribute to genotypic changes over a period of years. The significant genetic differentiation between the two sites, however, suggested that long-distance gene flow was relatively limited. It might also be due to differences in the ability of each population to develop in the habitat of the other. This was observed by Savoie et al. (1996a) in a study on the ability to colonize and degrade mushroom compost made from horse manure using isolates of these two local populations. The strains from the cypress litter were less efficient at compost colonization, whereas their ability to resist or to adapt to metabolites of *Trichoderma* spp., which is an antagonist of *A. bisporus* in cultures on compost, was higher than that from the open area confronted with horse manure on a field (Savoie et al., 2008b). By comparing various European native strains of *A. bisporus*, the litter of *Cupressus sempervirens* and the climatic conditions associated with the development of this tree in Mediterranean regions appeared to select for strains of *A. bisporus* with low susceptibility to *Trichoderma* spp. metabolites (Savoie et al., 2008b). Resistance to pests and pathogens is searched for in native relatives of cultivars with the expectation of introducing such resistance in cultivated strains and then protecting the culture without using pesticides (Chapter 6).

In addition to adaptation to the cultivation substrate and resistance to pathogen attacks, thermotolerance is an important trait for the promotion of sustainable production of the button mushroom. A large sample of native strains representing the major geographical populations of *A. bisporus* was screened for the ability to produce mature fruiting bodies at a higher temperature (25°C) than the normal temperature (17–19°C) used for the induction of fruiting in commercial cultures (Largeteau et al., 2011). One quarter of the strains of *A. bisporus* var. *bisporus* were able to fruit at high temperature, but with significantly lower yields than at the normal temperature for most. No correlation between either habitat or climate and the frequency of thermo-tolerant strains was found (Largeteau et al., 2011). All the *A. bisporus* var. *burnettii* strains originating from the Sonoran Desert produced high yields of fruiting bodies at 25°C, and some strains retain this ability at 30°C (Navarro and Savoie, 2015). Apparently, this variety is not dependent on or is less susceptible to temperature for fruiting. The mushrooms produced are small, and they rapidly break their veil for dissemination of the spores. All these properties, combined with the amphithallic life cycle, contribute to the adaptation of *A. bisporus* var. *burnettii* to rapidly completing its life cycle in an environment where the favorable period with enough humidity is short.

Recent evidence suggests that *A. bisporus* mushrooms also contain high levels of substances of possible medicinal importance such as tyrosinase, aromatase inhibitors, immunomodulating and antitumor polysaccharides, ergosterol, vitamin D_2, and various antioxidants (Beelman et al., 2003). All these components are interesting for the development of *A. bisporus* as a nutraceutical for the future

(Gopalakrishnan et al., 2005; Lelley and Vetter, 2005). Genetic background is also expected to affect the quality and quantity of molecules with health-promoting activity in *A. bisporus*. Shao et al. (2010) measured higher concentrations of ergosterols and antioxidant activity in a brown cultivated strain than in a white hybrid. Savoie et al. (2008a) reported higher radical scavenging activity correlating with higher phenol content in methanol extracts of two native strains than in a white hybrid cultivated under the same conditions. But there has been no publication on variations in the potentials of health benefits with the genotypes of a large number of strains of *A. bisporus*. The only published work on a large number of strains was a comparison of agaritine contents in 50 *A. bisporus* strains from the native germplasm (Sabourin et al., 2008). Each of the American samples had lower mean agaritine levels than any of the European population samples, and cultivars had intermediate agaritine levels corresponding to the lower-middle range recorded from European strains. The tetrasporic var. *burnettii* strains had significantly lower levels of agaritine than the cultivars. This work highlights a potentially useful resource available for developing *A. bisporus* strains with high or low contents of specific components.

The diversity of interest traits in native isolates and the knowledge of population features is a mine for mushroom breeders who pay attention to improving cultivation characteristics and post-harvest quality reaching consumer expectations and sustainable development.

5.4 GENETIC PROGRESS FOR SUSTAINABLE GROWING OF *AGARICUS BISPORUS*

5.4.1 GENERATING VARIABILITY BY OUTCROSSING

Both availability and characterization of the genetic and phenotypic variability in a large germplasm on the one hand, and possibilities for outcrosses and recombination generating variability are required for genetic selection in breeding programs.

The main method for outcrossing is crosses between compatible homokaryons. In *A. bisporus* var. *bisporus*, the predominantly pseudohomothallic life, equivalent to a pseudoclonal inbreeding system, hampers outcrossing and limits breeding success since only a small proportion of the offspring is homokaryotic. Single spore isolates must be obtained, and 1 to 10% of homokaryotic ones are selected on phenotypic criteria such as mycelial growth rate and their inability to fructify. This is time consuming and results in high levels of error. Fortunately, the basic concept of marker-assisted selection (MAS), which refers to selection on the genotype of molecular marker(s) tightly linked to the trait rather than selection on the phenotype itself, has been developed and applied to homokaryotic spore isolation, giving a quick result with a high level of confidence (Kerrigan, 1992; Callac et al., 1997). On the other hand, the discovery of the heterothallic *A. bisporus* var. *burnettii* among the native genetic resources makes it possible to partly overcome the limitations due to selection of homokaryotic spores as the percentages of bi-, tri-, and tetrasporic basidia was estimated to be 1%, 14%, and 85%, respectively (Callac et al., 1996) (Figure 5.3). The first generation of intervarietal hybrids (var. *bisporus* × var. *burnettii*) predominantly inherit the heterothallic life cycle of *A. bisporus* var. *burnettii*, giving rise to large recombined homokaryotic progenies useful for breeding works and genetic studies.

The second method of outcrossing in *A. bisporus* is a cross between a homokaryon and a heterokaryon via the Buller phenomenon (Buller, 1931). Callac et al. (2006) developed an easy way to get numerous hybrids in a single experiment by inoculating compost simultaneously with homokaryotic mycelium from one parent and predominantly heterokaryotic spores from a second parent of *A. bisporus* var. *bisporus*. Due to the intramictic process, these spores are not equivalent (Kerrigan, 1993). Taking into consideration the rearrangements by crossing over during the karyogamy and meiosis phases in basidia (Figure 5.3), the exchanged fragments are placed in a new genetic configuration because of interactions with the other loci of the chromatids. This leads to phenotypic variability, which could be used directly in breeding programs using single spore cultures to generate small changes in a variety (Moquet et al., 1998). The intramictic process might have been used to generate new cultivars from hybrids Horst® U1 and Horst® U3 exhibiting several phenotypic differences with the same genetic background (see II-1). The enhancement of variability by the "multispore culture" crossing method proposed by Callac et al. (2006) offers additional advantages. When positive traits such as resistance to a pathogen are linked to a locus bearing a lethal recessive allele, they cannot be exploited in breeding programs using conventional crosses between homokaryons since the interesting homokaryon is not viable. Hybrids resulting from crosses between a homokaryon and heterokaryons via the Buller phenomenon can receive recessive deleterious alleles since there is no haploid step (no gametic/haploid selection) for the material coming from the parent of the spores. In this way, alleles unavailable by conventional crossing are introduced. An example is given by Callac et al. (2008) with a strain for which only one of the parental nuclei can be isolated either from homokaryotic spores or by de-dikaryotization techniques. The hybrids which received recessive lethal alleles from the spores of this strain were on average less susceptible to the dry bubble disease caused by a fungal pathogen, *Lecanicillium fungicola*, than those that did not receive these alleles (Callac et al., 2008).

5.4.2 MODERN GENETICS APPLIED TO *A. BISPORUS*

Two haploid isolates of *A. bisporus* have recently been genome sequenced (Morin et al., 2012). Isolate H97 is from a cultivated European strain of *A. bisporus* var. *bisporus* which was a parent of the genuine hybrid Horst® U1, while isolate JB137-s8 is a native strain of *A. bisporus* var. *tetrasporus* isolated in the Sonoran Desert. The genome sequence offers a platform to develop genomic and transcriptomic analyses of *A. bisporus* for optimizations of the cultivation substrate utilization, improving the management of fruiting induction, for identifying and using genetic sources of resistance to pathogens, for improving shelf life, and for controlling metabolite biosynthesis. For instance, Eastwood et al. (2013) analyzed transcriptomic changes during the reproductive phase change and identified three gene clusters with coordinated regulation of transcription in each. Two of them could represent a response to morphogenetic changes occurring during the reproductive phase change, and the third may form part of a genetic response to environmental signals which prepare the hyphae for fruiting. Temperature reduction, CO_2, and 1-octen-3-ol all exert regulatory control on the initiation of fruiting body development, but independently and at different points on the morphogenetic pathway and to different extents (Eastwood et al., 2013). The genes identified are functional targets for helping in the selection of strains with varying regulations by these factors. As another example, Bailey et al. (2013) studied the cap-spotting symptoms that are produced when infection by the pathogen *L. fungicola* occurs late in the development of *A. bisporus*. They identified host-derived genes in a collection of suppression subtractive hybridizations, but only a few differential expressions were observed. Overall, the genes identified could be exploited as markers to assist in breeding programs. In sum, the genome provides for practical advances with an endless source of molecular markers useful for further genetic applications in the button mushroom (Foulongne-Oriol et al., 2013).

Agaricus bisporus is entering the genomics era, with perspectives for integrative studies merging genomics, phenomics, and classical genetics. Genetic linkage mapping, which has been extensively used in plant or animal models, is more recent in fungal research and could become a major asset in the researcher's toolbox (Foulongne-Oriol, 2012). A genetic linkage map is a representation of the genome that shows the relative position and genetic distances between markers or genes along chromosomes. A genetic distance is defined as a function of the crossover frequency during meiosis. The reference linkage of *A. bisporus* (Foulongne-Oriol et al., 2010) was useful for assembling the genome, and it can now feed from the genome for developing new markers and fine mapping. Genetic linkage maps are useful for the localization of genes or quantitative trait loci (QTL) responsible for economically important traits and then for marker-assisted selection in breeding (Foulongne-Oriol et al., 2011a; Foulongne-Oriol, 2012). The genetics of yield-related traits and cap color have been investigated, making it possible to refine our understanding of the inheritance of these traits (Foulongne-Oriol et al., 2012a). The combination of phenotypic data along with genotypic data has permitted the genomic location of either genes or QTLs for resistance to three pathogens (Foulongne-Oriol et al., 2012b; Savoie et al., 2013). In addition to yield and pathogen resistance, two other traits are of concern for the promotion of a sustainable agriculture and may benefit new prospects in the genetics of *A. bisporus*: aptitude to fruit at high temperature, and the composition of fruiting bodies in terms of health promoting actions and low safety risks. These are developed in the following paragraphs.

5.4.3 THE SELECTION OF STRAINS ABLE TO FRUIT AT HIGH TEMPERATURE

In Europe and America, the button mushroom has been grown in caves and cellars for two centuries. This was because of the climatic conditions favorable to the development of fruiting bodies, allowing off-season cultivation: high relative humidity in the air and constant temperatures of 14–16°C. A temperature reduction from 25°C used during spawn running to 18°C induces the differentiation of primordia, which then develop further into fruiting bodies (Eastwood et al., 2013). With the objective of improving the management of fruiting body development and of both yield and quality, climatic chambers have replaced caves since the 1980s. Cleaning and prophylaxis are easier for controlling pests and pathogens, the work environment for the staffs is microbiologically secure, and the organization of mushroom production is highly optimized. The synchronization of flushes may be improved by increasing the temperature to 21–25°C between flushes, and the fruiting temperature may be adapted to each commercial strain in the range of 16–19°C. However, this modern cultivation process necessitates expensive facilities and it consumes large quantities of energy for climatic regulation, especially in hot countries and during hot seasons in temperate areas. Consequently, it is not suitable to sustainable small-scale farming. Therefore, identifying sources of thermo-tolerance, understanding the genetic background of the trait, and breeding thermo-tolerant strains are of practical significance for new opportunities for sustainable cultivation of *A. bisporus*.

Strains of *A. bisporus* tolerant to relatively high temperatures for both mycelial growth rate and fruiting ability have been identified in the germplasm of the species, either in China (Chen et al., 2003) or in the European and North American mushroom populations (Largeteau et al., 2011). In these strains able to fruit at high temperature (FHT+), differentiation of primordia is not controlled by a decrease of temperature to 18°C and may occur at a constant temperature of 25°C. A few strains of *A. bisporus* var. producing high yields of mushrooms at 25°C have been selected (Navarro and Savoie, 2015). Unfortunately, none of them combines all the traits the growers expect: rusticity, thermo-tolerance, resistance to pathogens, high yield, high quality, and long shelf life. Therefore, breeding for combining different agronomical traits in a single cultivar is a way to reach the objective of cultivation of *A. bisporus* at temperatures higher than 20°C in a context of sustainable agriculture together with the economic expansion of emerging countries. Understanding the biological mechanisms that underlie thermo-tolerance during fruiting is helping in breeding.

Genes potentially involved in thermo-tolerance have been identified. The first three thermo-tolerance-related gene fragments were obtained using DD-RT-PCR to analyze gene expression in the vegetative mycelium of a thermo-tolerant Chinese strain under normal and higher-temperature cultivation (Chen et al., 2003). Later the full-length cDNA sequence of one gene (*028-1* GenBank accession number DQ235473) was published, but no obvious homological sequence was found in the databases (Chen et al., 2005). Constructing a binary expression vector of *028-1* and transferring the gene in a non-thermo-tolerant strain by *Agrobacterium* spp. mediated transformation; Chen et al. (2009) observed transformants able to growth at 34°C, whereas the native type did not grow at temperatures higher than 30°C. However, the mycelial growth rate of both the thermo-tolerant parent and the transformants was strongly affected, with mycelial diameters after 22 days at 34°C being 15% to 25% of that at 28°C. The consequences for fruiting ability were not tested on these transformants. Navarro et al. (2013) studied *028-1* in several FHT+ and FHT− strains and finally identified the encoded protein as an activating protein Yap1-like transcription factor participating in yeast thermo-tolerance. The protein was revealed as belonging to a new Agaricomycotina (Basidiomycota) subfamily of YAP1 homologs, and *aap1* was proposed as a new name for *028-1*. The authors analyzed the sequence polymorphism and alignment of the deduced polypeptide sequences in 24 strains of *A. bisporus* having different abilities for mycelial growth at temperatures above 30°C and for producing mature fruiting bodies at 25°C. No specific allele was observed for either the mycelium surviving at 33°C or the ability to fruit at 25°C (Navarro et al., 2013). On the other hand, measuring the relative expression levels of *aap1* in primordia or fruiting bodies produced at 17°C and 25°C in nine FHT+ strains showed low levels of regulation and absence of a general pattern for the different strains. *Aap1* is not a dominant contributor to the thermo-tolerance of *A. bisporus*, but the protein it encodes may be involved as an overall stress resistance transcription factor (Navarro et al., 2013).

Heat shock proteins (HSPs) constitute a large group of chaperone proteins found in virtually all organisms. They are a class of ubiquitous and highly conserved proteins which show increased expression in response to elevated temperature or other forms of environmental stress including increased concentrations of reactive oxygen species (ROS) and interactions with pathogens. The *hspA* gene (Genbank X98508) was identified in *A. bisporus* thanks to differential expressions in primordia and fruiting bodies of a commercial strain (Ospina-Giraldo et al., 2000). Regulation of *hspA*, a gene for a heat shock protein of the hsp70 family, was studied by qPCR in primordia and sporophores of 10 strains cultivated at 17°C or 25°C. Higher levels of expression were observed in the critical stages of primordia than in sporophores (unpublished data). The *hspA* gene might help prepare the primordia physiologically to rapidly progress in differentiation, leading to the production of mature sporophores, but the levels of expression in primordia produced at 25°C were lower than at 17°C, or not significantly different (unpublished data). On the other hand, measuring the time course of expression of heat shock proteins of *A. bisporus*, Lu et al. (2014) observed faster and more robust accumulation of these defense-related proteins in a thermo-tolerant strain than in a sensitive strain during high-temperature treatment of vegetative mycelium. Meanwhile, heat stress induced fast accumulation of H_2O_2 in the thermo-sensitive strain, but had a more moderate effect in the thermo-tolerant strain, where H_2O_2 accumulation might be limited by higher activities of catalase and superoxide dismutases (Lu et al., 2014). Para-amino-benzoic acid (PABA) has been identified as a component involved in the thermo-tolerance of the mycelium, by combining proteomic analysis, gene expression, quantification of the final product (PABA), and thermo-tolerance analysis after addition of PABA or inhibitors of its synthesis. PABA was shown to alleviate oxidative damage caused by heat stress, by increasing *hsp* genes and catalase and SOD activities (Lu et al., 2014). Finally, while mycelial thermo-tolerance involves general stress resistance processes, the ability to produce mature sporophores at 25°C is not likely linked to a stress response mechanism. This might explain why Foulongne-Oriol et al. (2014) did not observe significant differences in transcript profiling of *A. bisporus* var. *burnettii* primordia under two fruit-producing temperature conditions (17°C and 25°C) and why the numbers of fruiting bodies produced at both temperatures were correlated in an analysis of hybrids from an intervarietal *A. bisporus* var. *bisporus* × *A. bisporus* var. *burnettii* cross. A hypothesis for the mechanisms leading to the FHT+ trait is a lack of susceptibility to the temperature signal, which is necessary for induction of sporophores development in FHT− strains.

By phenotyping the progeny of the intervarietal *A. bisporus* var. *bisporus* × *A. bisporus* var. *burnettii* hybrid previously used for building the reference linkage map of the species (Foulongne-Oriol et al., 2010), the genetic inheritance of the FHT+ trait brought by the *burnettii* parental strain was demonstrated (Foulongne-Oriol et al., 2014). The phenotypic distribution of the number of fruiting bodies produced at both 17°C and 25°C and of the earliness of production showed continuous variation. The *A. bisporus* var. *bisporus* parent did not fructify at 25°C. Depending on the experiment, 65–87% of the hybrids from the offspring fructified at 25°C, but most of them were less productive than at 17°C and none reached the yield of the *A. bisporus* var. *burnettii* parent. A significant delay of first picking day was observed at 25°C, whereas the *A. bisporus* var. *burnettii* parent fructified earlier at 25°C than at 17°C. Broad-sense heritability analyses have suggested that most of the phenotypic variation was genetically determined (Foulongne-Oriol et al., 2014).

Therefore, one can plan to breed strains of *A. bisporus* combining thermo-tolerance, high yields, and pathogen resistance (see Chapter 6) of the variety *burnettii* and other quality traits (firmness, low susceptibility to discoloration, color, etc.) found in strains of the variety *bisporus*. However, like the quality traits which might be affected under higher fruiting temperature conditions (Gao et al., 2013), FHT+ is a complex trait, with polygenic inheritance and sensitive to the environmental changes; thus, MAS based on QTLs might be useful for breeders. QTLs have been detected for FHT+ traits (Foulongne-Oriol et al., 2014). Only two genomic regions, on two different linkage groups, are involved in determination of the ability to fructify at 25°C. Using the genome sequence, colocations between candidate genes and QTL related to FHT trait are possible. Two genes proved to be involved in thermo-tolerance in mycelium by Lu et al. (2014), an *Hsp70* gene and *Pabs* gene encoding for PABA, are colocated with the most significant QTL of the FHT+ trait and might be used as markers for selection (Foulongne-Oriol et al., 2014). MAS applied to either pyramiding or backcross breeding schemes is a promising strategy to develop original cultivars of *A. bisporus*.

The studies reported above illustrate how the use of modern genetics and the toolbox now available may contribute to generating markers and efficiently assisting breeding work for research groups or private companies willing to develop *A. bisporus* strains attaining some of the requirements of sustainable agriculture.

5.4.4 SELECTION OF STRAINS WITH HEALTH-PROMOTING COMPOUNDS AND LOW SAFETY RISK

Among natural products, the use of mushrooms with therapeutic properties is growing day by day because they are readily obtained in relatively large quantities and are inexpensive. It is commonly accepted by the scientific community that this resource remains largely underexploited in terms of molecules (metabolites, proteins, enzymes, polymers), the existence of which we may even not suspect. Some cultivated mushrooms are famous for their medicinal properties (*L. edodes*, *Agaricus subrufescens*, *Ganoderma lucidum*, and other species), but *A. bisporus* also contains high levels of substances of possible medicinal importance (Beelman et al., 2003). In addition to the effects of cultivation conditions and of mushroom maturation, one can expect to improve the mushrooms' quality and their natural ability to produce selected health-promoting compounds by breeding programs, as for other traits.

Agaritine is a phenylhydrazine derivative naturally present in *A. bisporus* and other *Agaricus* spp. There is controversy about its status as a potential carcinogen, depending on the doses used in *in vitro* tests or in feeding tests (Roupas et al., 2010). Moreover, in contrast to the carcinogenic activity previously ascribed to this compound, Endo et al. (2010) reported antitumor activity against leukemic tumor cells *in vitro* for agaritine from *A. suburefescens*. Agaritine might be converted to a stronger carcinogenic derivative when it is metabolized (Walton et al., 1997), which could explain its toxicity in certain experiments. Though a review of feeding studies using mushroom carpophores and mushroom extracts suggests that agaritine from consumption of cultivated *A. bisporus* mushrooms poses no known toxicological risk to healthy humans (Roupas et al., 2010), it is a challenge for the mushroom industry to guarantee low levels of agaritine in mushrooms. The fact that the tetrasporic var. *burnettii* strains and some American native strains of *A. bisporus* var. *bisporus* had significantly lower levels of agaritine than the cultivars of *A. bisporus* var. *bisporus* (used in all the feeding tests) and some European native strains (Sabourin et al., 2008) opens up prospects for the development of improved intervarietal hybrids with low agaritine content. As for the FHT trait discussed above, phenotyping the progeny of the intervarietal *A. bisporus* var. *bisporus* × *A. bisporus* var. *burnettii* hybrid used for building the reference linkage map of the species would allow deciphering the genetic heritability of the agaritine content, and molecular markers would be identified for MAS application in breeding of low agaritine strains. This work still has to be done.

5.4.5 VALORIZATION OF GENETIC PROGRESS FOR SUSTAINABLE GROWING OF *AGARICUS BISPORUS*

A biological translation of the practical needs for developing sustainable growth of *A. bisporus* throughout the world is the development of homogeneous, vigorous, and rustic strains able to produce under various conditions large quantities of fruiting bodies rich in active ingredients. The research performed on this mushroom clearly finds an available biodiversity within the species that is a source of interesting traits for the development of improved cultivars. The examples developed above show that intervarietal hybrids of *A. bisporus* var. *burnettii* × *A. bisporus* var. *bisporus* might be a primary source for the development of cultivars combining various traits in favor of sustainable methods of cultivation: high mycelial growth rates in compost, thermo-tolerance, ease of fruiting induction, high yields, resistance to pests and pathogens, richness in active ingredients, safety, but also firmness, low susceptibility to discoloration, and long shelf life.

Breeding usually requires considerable investment in time and resources, even taking advantage of the best of all the genome-based tools. This is a challenge now for mushroom breeders, but they might avoid the development of new varieties if they know it will be difficult to cover their investment due to a lack of protection of the new variety or strain. On another hand, growers and consumers are waiting for genetic progress allowing the development of sustainable mushroom production. Therefore, there is a need for worldwide regulation for the protection of new mushroom varieties which both encourages breeders to develop new varieties expecting a return on investment and favors the global qualitative and quantitative improvement of mushroom production as one of the common public goods that should benefit the welfare of mankind.

The massive development of mycelium under sterile conditions is one of the main stages in the cultivation of edible fungi (Salmones et al., 2007). In the best case, it should be performed in well-equipped laboratories with highly trained staff, which is a complex activity (Royse, 1997) and one of the main "bottleneck" processes for obtaining mushrooms with high commercial quality. The quality of the inoculum is related to the type of grain used, strain, and method and conditions of preparation (Mata and Savoie, 2007). Although the technology for spawn production has been properly established and considered a standard methodology, much of the information has been disseminated by the academic sector based on the experimental and control "hot spots" process, both from the point of biotechnology as microbiologically through numerous scientific publications and outreach. In Western countries, specialized companies have developed and improved their own technologies without dissemination of information, and the links between mushroom cultivators and research centers have not been kept sufficiently robust as should be expected from a developing industry. In developing countries, some large mushroom growers have specialized laboratories to produce their own spawn, while others, usually small businesses or cooperatives, buy their spawn from private laboratories, but also from public research centers and universities (Salmones et al., 2007) working in the field of edible mushrooms.

5.5 CONCLUSIONS

Agaricus bisporus mushroom growth is an ancient activity which has developed in recent years with the introduction of more sophisticated cultivation techniques such as indoor composting, cultivation in climatic rooms, and use of hybrids. This progress leading to build large production facilities has mostly been motivated by market and economic constraints, but there still are several challenges in terms of environmental issues, management of fruiting induction, control of pests and pathogens, and food quality which growers and mushroom industry are faced with. In Western countries, there is a new demand for a re-localization of food production under environmentally friendly conditions, whereas the globalization of food leads to a demand among consumers for the same products all over the world. The production of *A. bisporus* is an interesting way to respond to this demand and contribute to promoting sustainable production, both with the development of a small-scale mushroom farming and improvement of the efficiency of production in large mushroom farms. They both will benefit from new understanding of biological processes now facilitated through the availability of modern biological tools and data as well as old and recent research results produced by universities and research institutes around the world. Current and future research thus concerns the improvement of agro-waste valorization during *A. bisporus* cultivation and the use of natural biodiversity along with genetic knowledge.

REFERENCES

Andrade, M.C.N.D., Zied, D.C., Minhoni, M.T.D.A., Kopytowski Filho, J., 2008. Yield of four *Agaricus bisporus* strains in three compost formulations and chemical composition analyses of the mushrooms. Braz. J. Microbiol. 39, 593–598.

Bailey, A.M., Collopy, P.D., Thomas, D.J., Sergeant, M.R., Costa, A.M.S.B., Barker, G.L.A., et al., 2013. Transcriptomic analysis of the interactions between *Agaricus bisporus* and *Lecanicillium fungicola*. Fungal Genet. Biol. 55, 67–76.

Bechara, M.A., Heinemann, P., Walker, P.N., Romaine, C.P., 2006. Non-composted grain-based substrates for mushroom production (*Agaricus bisporus*). Trans. ASABE 49, 819–824.

Bechara, M.A., Heinemann, P.H., Walker, P.N., Romaine, C.P., 2009a. Effect of delayed-release supplements in grain-based substrate on yield of the mushroom *Agaricus bisporus*. Trans. ASABE 51, 1501–1505.

Bechara, M.A., Heinemann, P.H., Walker, P.N., Demirci, A., Romaine, C.P., 2009b. Evaluating the addition of activated carbon to heat-treated mushroom casing for grain-based and compost-based substrates. Bioresour. Technol. 100, 4441–4416.

Beelman, R.B., Royse, D., Chikthimmah, N., 2003. Bioactive components in button mushroom *Agaricus bisporus* (J. Lge) Imbach (Agaricomycetideae) of nutritional, medicinal or biological importance. Int. J. Med. Mushrooms 5, 321–337.

Buller, A.H.R., 1931. Researches on Fungi, vol. IV. Longmans, Green and Co, London.

Callac, P., Billette, C., Imbernon, M., Kerrigan, R.W., 1993. Morphological, genetic, and interfertility analyses reveal a novel, tetrasporic variety of *Agaricus bisporus* from the Sonoran Desert of California. Mycologia 85, 335–351.

Callac, P., Imbernon, M., Kerrigan, R.W., Olivier, J.M., 1996. The two life cycles of *Agaricus bisporus*. In: Royse, D.J. (Ed.), Proceedings of the Second International Conference on Mushroom Biology and Mushroom Products Pennsylvania State University, University Park, PA, pp. 57–66.

Callac, P., Desmerger, C., Kerrigan, R.W., Imbernon, M., 1997. Conservation of genetic linkage with map expansion in distantly related crosses of *Agaricus bisporus*. FEMS Microbiol. Lett. 146, 235–240.

Callac, P., Moquet, F., Imbernon, M., Ramos Guedes-Lafargue, M., Mamoun, M., Olivier, J.M., 1998. Evidence for PPC1 a determinant of the pilei-pellis color of *Agaricus bisporus* fruiting bodies. Fungal Genet. Biol. 23, 181–188.

Callac, P., Theochari, I., Kerrigan, R.W., 2002. The germplasm of *Agaricus bisporus*: main results after ten years of collection in France, in Greece, and in North America. Acta Hortic. 579, 49–55.

Callac, P., Jacobé de Haut, I., Imbernon, M., Guinberteau, J., Desmerger, C., Theochari, I., 2003. A novel homothallic variety of *Agaricus bisporus* comprises rare tetrasporic isolates from Europe. Mycologia 95, 222–223.

Callac, P., Spataro, C., Caillé, A., Imbernon, M., 2006. Evidence for outcrossing via Buller phenomenon in substrate simultaneously inoculated with spores and mycelium of *Agaricus bisporus*. Appl. Environ. Microbiol. 72, 2366–2372.

Callac, P., Imbernon, M., Savoie, J.M., 2008. Outcrossing via the Buller phenomenon in a substrate simultaneously inoculated with spores and mycelium of *Agaricus bisporus* creates variability for agronomic traits. In: Lelley, J.L., Buswell, J.A. (Eds.), Proceedings of the Sixth International Conference on Mushroom Biology and Mushroom Products GAMU, Krefeld, pp. 113–119.

Chen, M.Y., Wang, Z.S., Liao, J.H., Lu, Z.D., Guo, Z.G., Li, H.G., 2005. Full-length cDNA sequence of a gene related to the thermo-tolerance of *Agaricus bisporus*. Acta Edulis Fungi 12, 85–88.

Chen, M.Y., Liao, J.H., Guo, Z.J., Li, H.R., Lu, Z.H., Cai, D.F., et al., 2009. The expression vector construction and transformation of thermo-tolerance-related gene of *Agaricus bisporus*. Mycosystema 28, 797–801.

Chen, R., Chen, L.F., Song, S.Y., 2003. Identification of two thermotolerance-related genes in *Agaricus bisporus*. Food Technol. Biotechnol. 41, 339–344.

Coello-Castillo, M.M., Sánchez, J.E., Royse, D.J., 2009. Production of *Agaricus bisporus* on substrates pre-colonized by *Scytalidium thermophilum* and supplemented at casing with protein-rich supplements. Bioresour. Technol. 100, 4488–4492.

Colauto, N.B., da SilveiraII, A.R., da EiraII, A.F., LindeI, G.A., 2011. Production flush of *Agaricus blazei* on Brazilian casing layers. Braz. J. Microbiol. 42, 616–623.

Eastwood, D.C., Herman, B., Noble, R., Dobrovin-Pennington, A., Sreenivasaprasad, S., Burton, K.S., 2013. Environmental regulation of reproductive phase change in *Agaricus bisporus* by 1-octen-3-ol, temperature and CO_2. Fungal Genet. Biol. 55, 54–66.

Endo, M., Beppu, H., Akiyama, H., Wakamatsu, K., Ito, S., Kawamoto, Y., et al., 2010. Agaritine purified from *Agaricus blazei* Murrill exerts anti-tumor activity against leukemic cells. Biochim. Biophys. Acta 1800, 669–673.

Foulongne-Oriol, M., 2012. Genetic linkage mapping in fungi: current state, applications, and future trends. Appl. Microbiol. Biotechnol. 95, 891–904.

Foulongne-Oriol, M., Spataro, C., Savoie, J.M., 2009. Novel microsatellite markers suitable for genetic studies in the white button mushroom *Agaricus bisporus*. Appl. Microbiol. Biotechnol. 1184, 1125–1135.

Foulongne-Oriol, M., Spataro, C., Cathalot, V., Monllor, S., Savoie, J.M., 2010. An expanded genetic linkage map of an intervarietal *Agaricus bisporus* var. *bisporus* × *A. bisporus* var. *burnettii* hybrid based on AFLP, SSR and CAPS markers sheds light on the recombination behaviour of the species. Fungal Genet. Biol. 47, 226–236.

Foulongne-Oriol, M., Dufourcq, R., Spataro, C., Devesse, C., Rodier, A., Savoie, J.M., 2011a. Comparative linkage mapping in the white button mushroom *Agaricus bisporus* provides foundation for breeding management. Curr. Genet. 57, 39–50.

Foulongne-Oriol, M., Rodier, A., Caumont, P., Spataro, C., Savoie, J.M., 2011b. *Agaricus bisporus* cultivars: hidden diversity beyond apparent uniformity ?Savoie, J.M. Foulongne-Oriol, M. Largeteau, M. Barroso, G. (Eds.), Proceedings of the Seventh International Conference on Mushroom Biology and Mushroom Products, vol. 2 INRA, France, pp. 9–16. http://wsmbmp.org/Previous_Conference_7.html.

Foulongne-Oriol, M., Rodier, A., Rousseau, T., Savoie, J.M., 2012a. Quantitative trait locus mapping of yield-related components and oligogenic control of the cap color of the button mushroom, *Agaricus bisporus*. Appl. Environ. Microbiol. 78, 2422–2434.

Foulongne-Oriol, M., Rodier, A., Savoie, J.M., 2012b. Relationship between yield components and partial resistance to *Lecanicillium fungicola* in the button mushroom, *Agaricus bisporus*, assessed by quantitative trait locus mapping. Appl. Environ. Microbiol. 78, 2435–2442.

Foulongne-Oriol, M., Murat, C., Castanera, R., Ramírez, L., Sonnenberg, S.M., 2013. Genome-wide survey of repetitive DNA elements in the button mushroom *Agaricus bisporus*. Fungal Genet. Biol. 55, 6–21.

Foulongne-Oriol, M., Navarro, P., Spataro, C., Ferrer, N., Savoie, J.M., 2014. Deciphering the ability of *Agaricus bisporus* var. *burnettii* to produce mushrooms at high temperature (25°C). Fungal Genet. Biol. 73, 1–11.

Frankland, J.C., 1992. Mechanisms in fungal succession. In: Carroll, G.C., Wicklow, D.T. (Eds.), The Fungal Community Dekker, New York, NY. pp. 383–340.

Gao, W., Baars, J.J.P., Dolstra, O., Visser, R.G.F., Sonnenberg, A.S.M., 2013. Genetic variation and combining ability analysis of bruising sensitivity in *Agaricus bisporus*. PLoS One 8, e76826. http://dx.doi.org/10.1371/journal.pone.0076826.

Geml, J., Laursen, G.A., Taylor, D.L., 2008. Molecular diversity assessment of arctic and boreal *Agaricus* taxa. Mycologia 100, 577–589.

Gopalakrishnan, C., Pawar, R.S., Bhutani, K.K., 2005. Development of *Agaricus bisporus* as a nutraceutical of tomorrow. Acta Hortic. (ISHS) 680, 45–47.

Heinemann, P., 1956. Champignons récoltés au Congo Belge par Madame M. Goosens-Fontana. II Agaricus Fries s.s. Bulletin du Jardin Botanique de l'Etat a Bruxelles 26, 1–127.

Huhnke, W., Sengbush, R.V., 1968. Champignonanbau auf nicht kompostiertem Nährsubstrat. Mushroom Sci. 7, 405–419.

Kerrigan, R.W., 1992. Strategies for the efficient recovery of *Agaricus bisporus* homokaryons. Mycologia 84, 575–579.

Kerrigan, R.W., 1993. New prospects for *Agaricus bisporus* strain improvement. Rept. Tottori Mycol. Inst. 31, 188–200.

Kerrigan, R.W., 1995. Global genetic resources for *Agaricus* breeding and cultivation. Can. J. Bot. 73, S973–S979.

Kerrigan, R.W., 1996. Characteristics of a Large Collection of Wild Edible Mushroom Germ Plasm: The *Agaricus* Resource Program. Centraalbureau voor Schimmelcultures and the World Federation for Culture Collection, Veldhoven, pp. 302–308.

Kerrigan, R.W., Imbernon, M., Callac, P., Billette, C., Olivier, J.M., 1994. The heterothallic life cycle of *Agaricus bisporus* var. *burnettii*, and the inheritance of its tetrasporic trait. Exp. Mycol. 18, 193–210.

Kerrigan, R.W., Carvalho, D.B., Horgen, P.A., Anderson, J.B., 1998. The indigenous coastal Californian population of the mushroom *Agaricus bisporus*, a cultivated species, may be at risk of extinction. Mol. Ecol. 7, 35–45.

Kerrigan, R.W., Challen, M.P., Burton, K.S., 2013. *Agaricus bisporus* genome sequence: a commentary. Fungal Genet. Biol. 55, 2–5.

Laborde, J., Lanzi, G., Francescutti, B., Giordani, E., 1993. Indoor composting: general principles and large scale development in Italy. In: Chang, S.T., Buswell, J.A., Chiu, S.W. (Eds.), Mushroom Biology and Mushroom Products The Chinese University Press, Hong Kong, pp. 93–113.

Largeteau, M.L., Callac, P., Navarro-Rodriguez, A.M., Savoie, J.M., 2011. Diversity in the ability of *Agaricus bisporus* wild isolates to fruit at high temperature (25°C). Fungal Biol. 115, 1186–1195.

Lelley, J.I., Vetter, J., 2005. The possible role of mushrooms in maintaining good health and preventing diseases. Acta Edulis Fungi 12 (Suppl.), 412–419.

Loftus, M., Bouchti, K.L., Robles, C., Van Griensven, L.J.L.D., 2000. Use of SCAR marker for cap color in *Agaricus bisporus* breeding programs. Mushroom Sci. 15, 201–205.

Lu, Z., Kong, X., Xiao, M., Chen, M., Zhu, L., Shen, Y., et al., 2014. Para-aminobenzoic acid (PABA) synthase enhances thermotolerance of mushroom *Agaricus bisporus*. PLoS One 9, e91298.

Malekzadeh, K., Shahri, B.J.M., Mohsenifard, E., 2011. Use of ISSR markers for strain identification in the button mushroom, *Agaricus bisporus*. In: Savoie, J.M., Foulongne-Oriol, M., Largeteau, M., Barroso, G. (Eds.), Proceedings of the Seventh International Conference on Mushroom Biology and Mushroom Products, vol. 1 INRA, France, pp. 30–34. http://wsmbmp.org/Previous_Conference_7.html.

Mata, G., Savoie, J.M., 2007. Producción de semilla y conservación de cepas de *Agaricus bisporus*. In: Sánchez, J.E., Royse, D.J., Leal Lara, H. (Eds.), Cultivo, mercadotecnia e inocuidad alimenticia de *Agaricus bisporus* ECOSUR, Tapachula, Chiapas, pp. 37–48.

Miller, F.C., 1993. Conventional composting system. In: Tan Nair, N.G. (Ed.), Proceedings of the second AMGA/ISMS International Workshop-Seminar on *Agaricus* Compost, Sydney, pp. 1–18.

Miller, F.C., Harper, E.R., Macauley, B.J., 1989. Field examination of temperature and oxygen relationships in mushroom composting stacks – a consideration of stack oxygenation based on utilization and supply. Aust. J. Exp. Agric. 29, 741–749.

Moquet, F., Guedes-Lafargue, M.R., Mamoun, M., Olivier, J.M., 1998. Selfreproduction induced variability in agronomic traits for a wild *Agaricus bisporus*. Mycologia 90, 806–812.

Morin, E., Kohler, A., Baker, A.R., Foulongne-Oriol, M., Lombard, V., Nagy, L.G., et al., 2012. Genome sequence of the button mushroom *Agaricus bisporus* reveals mechanisms governing adaptation to a humic-rich ecological niche. Proc. Natl. Acad. Sci. USA 109, 17501–17506.

Navarro, P., Savoie, J.M., 2015. Selected wild strains of *Agaricus bisporus* produce high yields of mushrooms at 25°C. Rev. Iberoam. Micol. 32, 54–58.

Navarro, P., Billette, C., Ferrer, N., Savoie, J.M., 2013. Characterization of the *aap1* gene of *Agaricus bisporus*, a homolog of the yeast *YAP1*. C. R. Biol. 337, 29–43.

Ospina-Giraldo, M.D., Collopy, P.D., Romaine, C.P., Royse, D.J., 2000. Classification of sequences expressed during the primordial and basidiome stages of the cultivated mushroom *Agaricus bisporus*. Fungal Genet. Biol. 29, 81–94.

Roupas, P., Keogh, J., Noakes, M., Margetts, C., Taylor, P., 2010. Mushrooms and agaritine: a mini-review. J. Funct. Food 2, 91–98.

Royse, D.J., 1997. Specialty mushrooms and their cultivation. Hortic. Rev. 19, 19–97.

Royse, D.J., 2010. Effects of fragmentation, supplementation and the addition of phase II compost to 2nd break compost on mushroom (*Agaricus bisporus*) yield. Bioresour. Technol. 101, 188–192.

Royse, D.J., May, B., 1982. Genetic relatedness and its application in selective breeding of *Agaricus brunnescens*. Mycologia 74, 569–575.

Royse, D.J., Sanchez, J.E., 2008a. Supplementation of first break mushroom compost with hydrolyzed protein, commercial supplements and crystalline amino acids. World J. Microbiol. Biotechnol. 24, 1333–1339.

Royse, D.J., Sanchez, J.E., 2008b. Supplementation of 2nd break mushroom compost with isoleucine, leucine, valine, pheylalanine, Fermenten and SoyPlus. World J. Microbiol. Biotechnol. 24, 2011–2017.

Royse, D.J., Chalupa, W., 2009. Effects of spawn, supplement, and phase II compost additions and time of re-casing second break compost on mushroom (*Agaricus bisporus*) yield and biological efficiency. Bioresour. Technol. 100, 5277–5282.

Royse, D.J., Sanchez, J.E., Beelman, R.B., Davidson, J., 2008. Re-supplementing and recasing mushroom (*Agaricus bisporus*) compost for a second crop. World J. Microbiol. Biotechnol. 24, 319–325.

Sabourin, R.E., Brant, B.L., Wach, M.P., Kerrigan, R.W., 2008. Variation in agaritine levels among individuals in natural populations of *Agaricus bisporus*. Mushroom Sci. 17, 184–190.

Salmones, D., Mata, G., 2012. Ceparios de hongos en México. In: Sánchez Vázquez, J.E., Mata, G. (Eds.), Hongos comestibles y medicinales en Iberoamérica: investigación y desarrollo en un entorno multicultural El Colegio de la Frontera Sur – Instituto de Ecología, A.C., Tapachula, Chiapas, pp. 69–77.

Salmones, D., Mata, G., Gaitán Hernández, R., 2007. Aportaciones del sector académico en la producción de inóculo de *Pleurotus* spp. In: Sánchez Vázquez, J.E., Martínez Carrera, D., Mata, G., Leal Lara, H. (Eds.), El cultivo de setas Pleurotus spp. en México ECOSUR, Tapachula, Chiapas, pp. 41–44.

San Antonio, J.P., 1966. Effects of injection of nutrient solutions into compost on the yield of mushrooms (*Agaricus bisporus*). Proc. Am. Hortic. Soc. 89, 415–422.

San Antonio, J.P., 1971. A laboratory method to obtain fruit from cased grain spawn of the cultivated mushroom, *Agaricus bisporus*. Mycologia 63, 16–21.

Sanchez, J.E., Royse, D.J., 2001. Adapting substrate formulas used for shiitake for production of brown *Agaricus bisporus*. Bioresour. Technol. 77, 65–69.

Sanchez, J.E., Royse, D.J., 2009. *Scytalidium thermophilum*-colonized grain, corncobs and chopped wheat straw substrates for the production of *Agaricus bisporus*. Bioresour. Technol. 100, 1670–1674.

Sanchez, J.E., Mejia, L., Royse, D.J., 2008. Colonization of pangola grass with *Scytalidium thermophilum* for production of *Agaricus bisporus*. Bioresour. Technol. 99, 655–662.

Savoie, J.M., Chalaux, N., Olivier, J.M., 1992. Variability in straw quality and mushroom production: importance of fungicide schedules on chemical composition and potential degradability of wheat straw. Bioresour. Technol. 41, 161–166.

Savoie, J.M., Chalaux, N., Libmond, S., Laborde, J., Olivier, J.M., 1993. French research progress: quality of raw ingredients and of composts, improvement of microbial activities. In: Tan Nair, N.G. (Ed.), Proceedings of the Second AMGA/ISMS International Workshop-Seminar on *Agaricus* Compost ISMS, Sydney, NSW, pp. 109–126.

Savoie, J.M., Bruneau, D., Mamoun, M., 1996a. Resource allocation ability of wild isolates of *Agaricus bisporus* on conventional mushroom compost. FEMS Microbiol. Ecol. 21, 285–292.

Savoie, J.M., Olivier, J.M., Laborde, J., 1996b. Changes in nitrogen resources with increases in temperature during production of mushroom compost. World J. Microbiol. Biotechnol. 12, 379–384.

Savoie, J.M., Chalaux, N., Libmond, S., 1997. Solid state fermentation of wheat straw: methods for detecting straw quality and improving biodegradability of poor quality straw. In: Roussos, S., Lonsane, B.K., Raimbault, M., Viniegra-Gonzalez, G. (Eds.), Advances in Solid State Fermentation Kluwer Academic Publishers, Dordrecht, pp. 299–309.

Savoie, J.M., Minvielle, N., Largeteau, M., 2008a. Radical scavenging properties of extracts from the white button mushroom, *Agaricus bisporus*. J. Sci. Food Agric. 88, 970–975.

Savoie, J.M., Reixtouex, S., Minvielle, N., Callac, P., 2008b. Susceptibility of European populations of *Agaricus bisporus* to *Trichoderma metabolites*. In: Lelley, J.I., Buswell, J.A. (Eds.), Proceedings of the Sixth International Conference on Mushroom Biology and Mushroom Products GAMU, Krefeld, pp. 305–311.

Savoie, J.M., Foulongne-Oriol, M., Barroso, G., Callac, P., 2013. Genetics and genomics of cultivated mushrooms, application to breeding of Agaric. In: Kempken, F. (Ed.), The Mycota, Vol. 11: Agricultural Applications Springer-Verlag, Berlin, pp. 3–33.

Schisler, L.C., 1990. Why mushroom production declines with each successive break, and the production of a second crop of *Agaricu*s mushrooms on "spent" compost. Appl. Agric. Res. 5, 44–47.

Shao, S., Hernandez, M., Kramer, J.K.G., Rinker, D.L., Tsao, R., 2010. Ergosterol profiles, fatty acid composition, and antioxidant activities of button mushrooms as affected by tissue part and developmental stage. J. Agric. Food Chem. 58, 11616–11625.

Sharma, H.S., Kilpatrick, M., 2000. Mushroom (*Agaricus bisporus*) compost quality factors for predicting potential yield of fruiting bodies. Can. J. Microbiol. 46, 515–519.

Siyoum, N.A., Surridge, K., Korsten, L., 2010. Bacterial profiling of casing materials for white button mushrooms (*Agaricus bisporus*) using denaturing gradient gel electrophoresis. South Afr. J. Sci. 106, 1–6.

Song, T.T., Cai, W.M., Jin, Q.L., Feng, W.L., Fan, L.J., Shen, Y.Y., et al., 2014. Comparison of microbial communities and histological changes in Phase I rice straw-based *Agaricus bisporus* compost prepared using two composting methods. Sci. Hortic. 174, 96–104.

Sonnenberg, A.S.M., Baars, J.P., Hendrickx P.M., Lavrijssen B., Gao W., Weijn A., et al., 2011. Breeding and strain protection in the button mushroom *Agaricus bisporus*. In: Savoie, J.M., Foulongne-Oriol, M., Largeteau, M., Barroso, G. (Eds.), Proceedings of the 7th international conference on mushroom biology and mushroom products, vol. 1, pp. 7–15. <http://wsmbmp.org/Previous_Conference_7.html>.

Till, O., 1962. Cultivation of mushroom on sterile substrates and reutilization of spent compost. Mushroom Sci. 5, 127–133.

Wakchaure, G.C., Meena, K.K., Choudhary, R.L., Singh, M., Yandigeri, M.S., 2014. An improved rapid composting procedure enhance the substrate quality and yield of *Agaricus bisporus*. Afr. J. Agric. Res. 8, 4523–4536.

Walton, K., Coombs, M.M., Walker, R., Ioannides, C., 1997. Bioactivation of mushroom hydrazines to mutagenic products by mammalian and fungal enzymes. Mutat. Res. 381, 131–139.

Wang, Z.S., Liao, J.H., Li, F.G., 1995. Studies on breeding hybrid strain As2796 of *Agaricus bisporus* for canning in China. Mushroom Sci. 14, 71–72.

Wang, Z.S., Liao, J., Li, H., Wang, B., Chen, M., Lu, Z., et al., 2008. Study on the biological characteristics of wild *Agaricus bisporus* strains from China. Mushroom Science 17. Proceedings of the 17th Congress of the International Society for Mushroom Science. South African Mushroom Farmers Association, Pretoria, South Africa (CD-ROM), pp. 149–158.

Xu, J., Kerrigan, R.W., Callac, P., Horgen, P.A., Anderson, J.B., 1997. Genetic structure of natural populations of *Agaricus bisporus*, the commercial button mushroom. J. Hered. 88, 482–488.

Xu, J., Desmerger, C., Callac, P., 2002. Fine-scale genetic analyses reveal unexpected spatial-temporal heterogeneity in two natural populations of the commercial mushroom *Agaricus bisporus*. Microbiology 148, 1253–1262.

Zied, C.D., Savoie, J.M., Pardo-Giménez, A., 2011. Soybean the main nitrogen source in cultivation substrates of edible and medicinal mushrooms. In: El-Shemy, H. (Ed.), Soybean and Nutrition InTech Open Access Publisher, Rijeka, pp. 433–452.

CHAPTER

NEW PROSPECTS IN PATHOGEN CONTROL OF BUTTON MUSHROOM CULTURES

6

Jean-Michel Savoie[1], Gerardo Mata[2] and Michèle Largeteau[1]

[1]INRA, UR1264 MycSA, Villenave d'Ornon, France [2]Instituto de Ecología, A.C., Red de Manejo Biotechnológico de Recursos, Xalapa, Veracruz, Mexico

6.1 INTRODUCTION

The mushroom industry is under pressure to comply with changing legislation regarding environmental issues, human health, and food safety; it also faces economic constraints to reduce production costs. As other crops are, cultivated mushrooms are susceptible to a variety of pests and pathogens including viruses, bacteria, and fungi, which affect the yield and quality of mushroom harvested by growers. Few chemicals might be used to control mushroom pests and pathogens because of the susceptibility of the cultivated fungi to many molecules, the short duration of the cultivation cycles with risks of pesticide residues in harvested mushrooms, the decision that many molecules are going to be banned in Western countries, and resistance developed by pathogens to the commonly used pesticides. Consequently, mushroom growers have to change their practices for pest and pathogen suppression in their cultures. Four kinds of alternative actions might be used separately or in combination in integrated pest and pathogen management: (i) good hygiene practices; (ii) using mushroom varieties resistant to the main pests and pathogens; (iii) introducing biocontrol agents acting both directly as microbial antagonists and indirectly by stimulating the mushrooms' defense mechanisms; and (iv) applying environmentally friendly biomolecules from agricultural wastes as substitutes for pesticides.

The most widely grown edible mushroom in many countries is the button mushroom *Agaricus bisporus* (J.E. Lange) Imbach. Because it is a humicolous fungus growing in the natural habitat in manures and leaf litters, its cultivation technique is quite different from that of most of the other cultivated species [*Pleurotus* spp., *Lentinula edodes* (Berk.) Pegler, *Flammulina velutipes* (Curtis) Singer, etc.], which are lignicolous fungi growing in the natural environment on dead wood. This chapter is focused on *A. bisporus* and presents an outline of past and recent developments in alternatives to pesticides for disease suppression during cultivation (Figure 6.1).

Mushroom Biotechnology. DOI: http://dx.doi.org/10.1016/B978-0-12-802794-3.00006-0

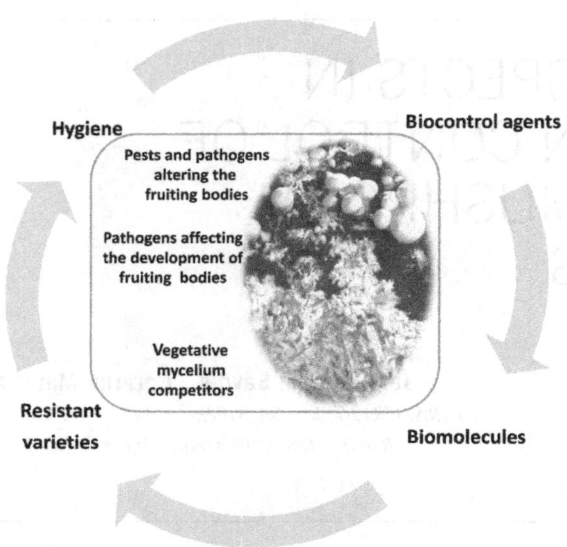

FIGURE 6.1

Integrated pest and pathogen management in button mushroom cultures.

6.2 MAJOR PATHOGENS AFFECTING *AGARICUS BISPORUS* AND THEIR PROPHYLAXIS

The principal mushroom pests and pathogens known to significantly affect cultures of *A. bisporus* are fungal antagonists and fungal pathogens [*Trichoderma aggressivum* Samuels & W. Gam (green mold); *Lecanicillium fungicola* (Preuss) Zare & W. Gam (dry bubble); *Mycogone perniciosa* (Magnus) Delacr. (wet bubble); *Hypomyces rosellus* (Alb. & Schwein.) Tul. & C. Tul. (cobweb); *Diehliomyces microsporus* (Diehl & E.B. Lamb.) Gilkey (false truffle)], bacterial pathogens [*Burkholderia gladioli* Severini (soft rot, cavity disease), *Pseudomonas tolaasii* Paine (bacterial blotch or brown blotch); *Pseudomonas* sp. (mummy disease)], viruses, insects (phorid and sciarid flies, cecidomyiid midges, mites), and nematodes (Fletcher and Gaze, 2008). Three pathogens are specifically presented in this chapter for discussion of new prospects in pathogen control in button mushroom cultures: (i) green mold as the representative of competitors or antagonists of the vegetative mycelium in compost, (ii) dry bubble as a significant disease caused by a fungus affecting the development of fruiting bodies after interaction on the casing layer during the early stage of differentiation, and (iii) brown blotch as a disease caused by bacteria altering the fruiting bodies.

6.2.1 ANTAGONISTS OF *A. BISPORUS*: WEED MOLDS AND *TRICHODERMA* SPP.

Fungi which act as competitors or antagonists of the vegetative mycelium of *A. bisporus* are frequently responsible for crop loss. These weed molds generally develop in bad-quality composts. One can often

FIGURE 6.2

Trichoderma aggressivum f. *europaeum* affecting cultures of *Agaricus bisporus*: (A) and (B) compost surface colonized by *T. aggressivum*; (C) area without *A. bisporus* development due to inoculation with *T. aggressivum* in the center; (D) casing surface colonized by *T. aggressivum*.

Photos (C) & (D) are from T. Rousseau, CTC.

find small patches of the lipstick mold (*Sporendonema purpurescens*) at the surface of the casing layer. The white and brown plaster molds (*Scopulariopsis fimicola* and *Papulaspora byssina*) are present in wet composts which remain sticky and dark colored in infected areas. Later, the surface of the casing may be densely covered with spores looking like flour or plaster. When high ammonia content persists in compost, *Chaetomium* spp. (*C. olivaceum*, *C. globosum*) might colonize it as the olive green mold. Green molds were caused by *Trichoderma* species and had no serious consequences, except during the summer months in Western countries, until a new pathogen appeared in cultures in Ireland in the spring of 1985 and spread across Europe in the following years. In parallel, the new green mold disease was recorded in North America 10 years later. The causal agent is *T. aggressivum*, with a European biotype (*T. aggressivum* f. *europaeum*, formerly Th2; Figure 6.2) and an American biotype (*T. aggressivum* f. *aggressivum*, formerly Th4). The ability of these biotypes to grow in compost and outcompete *A. bisporus* depends on several genetic characteristics. *Trichoderma aggressivum* can flourish in compost inoculated or uninoculated with *A. bisporus*, while less aggressive species only poorly colonize these substrates. Metabolites produced by *A. bisporus* and stimulating the growth of *T. aggressivum*, but inhibiting other *Trichoderma* spp., contribute to the aggressiveness (Mumpuni et al., 1998). On the other hand, *T. aggressivum* does not present a better ability to degrade compost components than the other species, but it exhibits a higher tolerance toward the inhibitory effect of bacteria and fungi present in compost (Mamoun et al., 2000a; Savoie et al., 2001). During direct interaction, *T. aggressivum* produces a coumarin that limits the growth of *A. bisporus* (Krupke et al., 2003). Specific isoforms of chitinases (Guthrie and Castle, 2006) and a β-glucanase are induced as a part of its aggression capacity (Abubaker et al., 2013).

Management of the composting process using indoor composting, or at least a carefully performed pasteurization phase with filtered air, is recommended for producing high-quality composts cleaned of most of the spores of the weed molds. During the cultures, the spread of spores from the infected area to others can be limited by good hygiene and adequate organization of the cultivation areas and circulation of workers. Red pepper mites are often associated with the presence of *Trichoderma* spp. and may contribute to the spread of the spores. Benzimidazole fungicides have been used in America as a preventive treatment for mushroom spawn, but benzimidazole resistance has appeared in *T. aggressivum* (Romaine et al., 2005) and it is not authorized in Europe. There is no specific fungicide treatment for controlling weed molds, though several molecules have been tested (Kredics et al., 2010).

6.2.2 DRY BUBBLE DISEASE

A profile of *Lecanicillium fungicola*, the fungal agent of dry bubble diseases, has been published by Berendsen et al. (2010). When *L. fungicola* spores germinate on developing fruiting bodies, superficial cinnamon-brown lesions are formed (spotty cap). When the sporophores initially are attacked before the mushroom tissues are differentiated into stipes and caps, the mycoparasites affect the morphogenesis of *A. bisporus* fruiting bodies, leading to the production of deformed tissue with no sign of differentiation, the dry bubbles (Figure 6.3). *Lecanicillium fungicola* infects fruiting mycelia but rarely attacks vegetative ones. This might be due to identification of the host via lectins on the surface of hyphae that are developmentally regulated (Bernado et al., 2004). A lectin was actually shown to be downregulated during infection of *A. bisporus* sporocarps by *L. fungicola*. In this way, the mushroom might alter its mycelial surface composition and therefore interfere with recognition by the pathogen, limiting the infection (Bailey et al., 2013). 1-Octen-3-ol is a volatile compound produced by *A. bisporus*; it has been implicated in the self-inhibition of fruiting body formation, while it has been shown to inhibit spore germination of *L. fungicola*. The decrease of 1-octen-3-ol concomitant with the induction of fruiting might also be favorable to the germination of *L. fungicola* and consequently infection at the susceptible stage (Berendsen et al., 2013a). After attachment, the mycoparasite penetrates the host cell walls and modifies the genetic program of differentiation into fruiting bodies.

By studying the expression of three selected genes, Largeteau et al. (2010) observed levels in dry bubbles similar to the first developmental stage of non-infected cultures. This illustrates morphogenesis disruption occurring after infection. Unfortunately, there is no published work on an overall survey of differentially expressed genes in healthy and affected pins. In a transcriptomic analysis of the interactions between *A. bisporus* and *L. fungicola* forming spotty caps, it was impossible to determine whether the differential expression observed was a function of the host defense pathways, or whether it was a result of *L. fungicola* interfering with normal developmental processes (Bailey et al., 2013).

FIGURE 6.3

A culture *of Agaricus bisporus* affected by *Lecanicillium fungicola* and details of various deformed mushrooms and dry bubbles.

The necrotic lesions on the cap surface (spotty cap symptom) may reflect a response of *A. bisporus* similar to the hypersensitive response in plants where cells in contact with the pathogen die and encapsulate it in a necrotic tissue. In the bubbles, the browning was shown actually not to be an efficient mechanism of defense but a tissue alteration resulting from high levels of infection by *L. fungicola* (Largeteau et al., 2007). The oxidative stress reaction may occur during the host-pathogen interaction with involvement of hydrogen peroxide. A negative correlation was found between the susceptibility of *A. bisporus* strains to dry bubble disease and hydrogen peroxide levels in infected tissues of these strains, but not in healthy sporocarps (Savoie and Largeteau, 2004). A gene encoding for a heat shock protein of the HSP70 family in *A. bisporus* involved in oxidative stress reaction (*hspA*) was shown to be overexpressed in primordia of strains partially resistant to *L. fungicola* when compared to susceptible strains (Largeteau et al., 2010). This may help *A. bisporus* to protect itself against the high level of reactive oxygen species which it produces to counterattack *L. fungicola*.

Currently, control of *L. fungicola* relies mainly on prevention and hygiene measures in mushroom farms during picking, on control of mites and flies disseminating spores, and on efficient removal of finished crops. Fungicides may be sprayed on the developing crop if the pathogen appears, but active control is difficult because the chemicals used have become less effective as the pathogen has developed resistance, and legislation is restricting their use. In Europe, chlorothalonil and prochloraz manganese are recommended and authorized, while carbendazim is inefficient because of *L. fungicola* resistance (Anonymous, 2000).

6.2.3 THE BACTERIAL BROWN BLOTCH PATHOGENS

Pseudomonas is the major spoilage genus associated with blotch in fresh mushrooms. A number of diverse species have the ability to cause blotch diseases with various degrees of discolorations and have a certain role in the expression of such symptoms. *Pseudomonas tolaasii* is the main causative agent of the classical bacterial blotch disease in cultivated mushrooms (Figure 6.4). It induces dark-brown, often wet and sunken lesions on button mushroom caps and stalks, which render the crop unmarketable.

The mechanism of *A. bisporus* infection by *P. tolaasii* is well documented (Largeteau and Savoie, 2010). Briefly, the pathogenic type of bacteria produces the extracellular toxin tolaasin, composed of seven lipodepsipeptides. The effect of tolaasin on mushroom caps is attributed to its capacity to disrupt the plasma and vacuole membrane in *A. bisporus* cells by its surfactant properties and its ion channel forming activity. As a consequence, the phenol oxidases (tyrosinases) in the mushroom cells are in contact with phenolic compounds released from vacuoles, and these substrates are transformed into dark pigmented melanin by the enzymes. Mamoun et al. (1999) proposed a model in which the intensity of the symptoms is dependent on the GHB-melanin biosynthesis pathway and the concentration in γ-L(+)-glutamyl-4-hydroxybenzene (GHB). By studying other strains of the *Pseudomonas fluorescens* Migula group, Henkels et al. (2014) suggested that certain antifungal metabolites, long associated with

FIGURE 6.4

Bacterial brown blotch disease symptoms on *Agaricus bisporus* after inoculation of *Pseudomonas tolaasii*.

the biocontrol capabilities of these bacteria against fungal plant pathogens, can be toxic to *A. bisporus*, manifesting as discoloration or pitting of mushroom caps.

Bacterial blotch is strongly influenced by environmental and surface-moisture conditions. Once the disease occurs, blotch-causing bacteria are spread by splash-dispersal during watering. Consequently, pathogen control requires inhibiting its reproduction on the mushroom surface and further dissemination. Temperature and relative humidity should be controlled to ensure that water droplets or moisture films on mushrooms dry up within 2–3 h. Good ventilation should be ensured after watering. Sodium hypochlorite at 150 ppm chlorine, applied every time the casing is watered, helps to keep the incidence down (Anonymous, 2000). Alternatively, an acetic acid solution at 70–80 mM might be used as it has proved to be efficient in *Pleurotus eryngii* (DC.) Quél. cultures (Bruno et al., 2014).

6.3 STRAINS OF *AGARICUS BISPORUS* RESISTANT TO PATHOGENS

6.3.1 GENETIC RESOURCES FOR RESISTANCE TO MUSHROOM PATHOGENS

Local varieties and native relatives of a cultivated mushroom might be a source of biodiversity likely to contain high levels of genetic diversity with potential resistance or low susceptibility to pathogens. Unfortunately, many studies show a narrow genetic variability among the cultivars of *A. bisporus* (Savoie et al., 2013), but hundreds of native specimens are currently preserved in the germplasm of *A. bisporus* in the collection of the *Agaricus* Resource Program (ARP) or Collection of Germplasm of *Agaricus* at Bordeaux (INRA-CGAB), as well as various laboratory collections. Attempts to find *A. bisporus* strains resistant to various pathogens have been undertaken to different extents depending on the pathogen, but not all the germplasm potential has been investigated yet.

6.3.1.1 *Resistance to* Trichoderma aggressivum

To test the susceptibility of *A. bisporus* strains to green mold, the best method is based on crop loss analysis in experimental facilities, in trials where calibrated spore suspensions of *T. aggressivum* have been either mixed with the spawn just before compost spawning, or inoculated at the surface of the compost just after spawning. Using this method, commercially available *A. bisporus* strains have exhibited different susceptibility to *T. aggressivum* f. *aggressivum* (Anderson et al., 2001). Hybrid white strains were extremely susceptible, while hybrid off white strains exhibited intermediate susceptibility and brown strains were less susceptible. However, using 25 native strains ranging from cream to dark brown color, and 12 commercial strains, Mamoun et al. (2000b) did not observe significant relationships between cap color and susceptibility to *T. aggressivum* f. *europaeum*.

Alternative methods have been searched for in the laboratory for screening the mushroom germplasm for pathogen resistance. Chen et al. (2003) proposed to measure the ability of the strains to protect the spawn substrate from colonization by the pathogen, though this grain test does not offer the level of resolution afforded by a crop loss assessment. An *in vitro* test using agar media containing cell wall lytic enzymes and undefined metabolites of *Trichoderma* from a commercial product (Lysing Enzyme®) was used to establish whether improvement of mushroom resistance to *T. aggressivum* could be obtained by inducing reaction mechanisms before contact with the pathogen and whether this ability was strain dependent (Savoie and Mata, 2003). A large variability in response to the treatment was observed, and this method was further used in laboratory screening of germplasms on this component of resistance (Savoie et al., 2008). Recent results have indicated that laccase activity, in particular that encoded by *lcc 2*, serves as a defense response of *A. bisporus* to *T. aggressivum* toxins and contributes to green mold resistance in commercial brown strains (Sjaarda et al., 2015).

Using the different methods, a few strains have appeared slightly susceptible and could be used in breeding programs, but absolute resistance was not observed.

6.3.1.2 *Resistance to* Lecanicillium fungicola

For testing the resistance to a pathogen, management of the contamination by the pathogen is necessary. With *L. fungicola*, it is easy to evaluate the resistance of an *A. bisporus* strain to the development of spotty caps by placing droplets containing spore suspensions of the pathogen on fruiting bodies. However, this gives no information on the susceptibility to the production of dry bubbles. Inoculation of the casing surface before the development of the fruiting bodies is more representative of the natural contamination leading to dry bubbles. Both the yield of healthy mushrooms and the percentage of diseased mushrooms can be measured as criteria for dry bubble susceptibility. However, the date of inoculation is an important parameter. In a casing inoculation experiment, a significant correlation between the time needed by *A. bisporus* strains to form their first fruiting bodies and susceptibility to *L. fungicola* was observed (Largeteau et al., 2004). The earlier fruiting strains were significantly less diseased.

Using such tests, wild *A. bisporus* strains highly tolerant to *L. fungicola* were identified in the INRA-CGAB collection after testing 450 strains (Savoie et al., 2013). In this work, as in others, total resistance was not found (Berendsen et al., 2010; Largeteau and Savoie, 2010), showing that resistance is a multifactorial trait. However, interesting strains have been identified as candidate for introgression of their partial resistance into other strains in breeding programs.

6.3.1.3 *Resistance to* Pseudomonas tolaasii

A standardized method to assess mushroom blotch resistance in cultivated and native *A. bisporus* strains has been developed and used on a wide range of strains. Immediately after harvesting, mushrooms are arranged in moist chambers. Then 20-μL droplets of bacterial suspension are placed at the top of the caps and symptoms are evaluated after incubation for 48 h at 16°C. Alternatively, a toxin (tolaasin) suspension may be used. Three commercial strains and 26 native strains were tested by Moquet et al. (1996). A great diversity of symptom intensity was observed. The susceptibility of white and brown strains was different, and due to a strong genetic linkage with the color allele *PPC1*, two different models for the susceptibility had to be taken into account, depending on the type of strain (Mamoun et al., 1999; Moquet et al., 1999).

As with the two other pathogens, no absolute resistance was found. In addition, there is no correlation between the susceptibility of different *A. bisporus* strains to *P. tolaasii*, to *L. fungicola*, and to *T. aggressivum* (Largeteau et al., 2004), showing that different biological mechanisms are involved.

6.3.2 BREEDING FOR RESISTANCE TO PATHOGENS

As with other economically important production traits such as yield and quality, resistance to pathogens of edible mushroom cultures is under polygenic inheritance. This was suggested by the absence of true resistance in the germplasm of *A. bisporus* and had been clearly stated for resistance to *L. fungicola* (Foulongne-Oriol et al., 2011a). Partial resistances identified in native strains have yet to be introduced into suitable commercial varieties of mushroom. Selecting for such quantitative traits is possible, but not easy and could take a long time due to the difficulties in selective breeding of *A. bisporus*. This species has a complex life cycle involving dikaryotic cells, infrequent meiotic recombination, and usually the production of dikaryotic basidiospores which effectively are already mated and can produce new fruiting bodies without crossing (Savoie et al., 2013).

That is why great attention was paid on the development of molecular tools allowing the dissection of these quantitative traits in individualized loci through quantitative trait loci (QTL) mapping. These approaches may greatly facilitate the effective manipulation of levels of resistance to pathogens in subsequent breeding programs. They have been extensively proven to be successful in plant and animal breeding, and the use of molecular markers in mushroom varietal selection is now on the rails with studies developed with *A. bisporus* as a model (Foulongne-Oriol et al., 2012a,b).

The construction of a genetic linkage map was the first step toward understanding the genetic basis of resistance to pathogens. This was achieved with the construction of the first comprehensive linkage map for *A. bisporus* obtained in intervarietal hybrids between *A. bisporus* var. *bisporus* and *A. bisporus* var. *burnettii* (Foulongne-Oriol et al., 2010). In parallel, numerous traits of interest were assessed on the intervarietal derived materials, and a phenotypic database was established. The combination of phenotypic data along with genotypic data permitted the genomic location of either genes or QTLs for resistance to the three pathogens explored in this chapter (Savoie et al., 2013). A majority of the QTLs was found specific to one pathogen, suggesting that distinct mechanisms are involved, in agreement with the absence of strain combining resistances to the three pathogens in the explored germplasm.

QTL analysis may help to understand the biological bases of the pathogen actions. A QTL on the ability to resist or adapt to *Trichoderma* metabolites was found in the vicinity of a QTL related to mycelium growth in controlled condition (Foulongne-Oriol et al., 2011b), in agreement with the role of the fitness of *A. bisporus* strains for colonizing the compost before strong development of *T. aggressivum*. A QTL controlling the dry bubble symptom caused by *L. fungicola* mapped in the same genomic interval as the fruiting earliness trait and may be related to fitness, in agreement with previous observations on susceptibility of the native strains (Largeteau et al., 2004) presented in Section 2.1.2. A major QTL for resistance to *P. tolaasii* explained about 30% of phenotypic variation, and was closely linked to the cap color locus *PPC1* (Moquet et al., 1999). Interestingly, this region was also found to be involved in resistance against *L. fungicola* during secondary infection leading to spotty caps (Foulongne-Oriol et al., 2012b; Savoie et al., 2013). The resistance alleles to the both symptoms were associated with the brown allele at *PPC1*, suggesting a shared mechanism of resistance based on melanin biosynthesis.

QTL analysis also allows the highlighting of the difficulties and limitations of mushroom breeding for resistance to pathogen. The hybrids most resistant to *L. fungicola* tend to early production of numerous small mushrooms that could not be expected by mushroom growers searching for large mushrooms. This is in agreement with QTL collocations, but such unfavorable linkage drags highlight the difficulty for introgressing desirable pathogen resistance traits from native strains in commercial hybrids maintaining expected yield and quality levels (Savoie et al., 2013). On the other hand, a common genomic region was involved in both spotted cap symptom caused by *L. fungicola* and adaptation to *Trichoderma* metabolites, but the parental origin was different for each trait. Combining multiple pathogen resistances in one genotype will necessitate multi-path breeding schemes.

The whole genome of *A. bisporus* has been publicly available since 2011 (http://genome.jgi.doe.gov/Agabi_varbisH97_2/Agabi_varbisH97_2.home.html), and this is going to contribute hugely to our understanding of interactions with its pathogens and to development of resistant strains. The tight relationship between the linkage map and genome sequence also makes possible custom-made molecular markers tightly linked to target loci for further marker-assisted selection (MAS). MAS is a promising tool that is already used in *A. bisporus* breeding for genetically simple traits and is going to be used for pathogen resistance, based on the studies of QTLs presented above. In combination with the genetic resources available, the whole genome sequence and the new generation sequencing technologies offer the opportunity to perform genome-wide association studies to identify pathogen resistance traits.

6.4 BIOLOGICAL CONTROL AGENTS

Biological control is the practice or process by which an undesirable organism is controlled by means of another (beneficial) organism. Biological control agents (BCAs) act by direct and sometimes indirect effects of microbial antagonism: parasitism, antibiosis, or competition. There are a number of BCAs commercially available for use in plant protection. As is done for fungicides, one can imagine testing their efficiency for the protection of mushroom cultures. Another way of developing biocontrol of mushroom pathogens is to identify new BCAs and define their optimal use by using knowledge on the biology of the host–pathogen interactions.

6.4.1 BIOCONTROL OF *TRICHODERMA AGGRESSIVUM* WITH BACTERIA

A relevant illustration of the extensive use of a BCA is given by the biofungicide Serenade®, tested and registered as a biocontrol agent of *T. aggressivum* in Europe since 2007. On the one hand, Savoie et al. (2001) observed that *Bacillus* species had inhibitory effects on *T. aggressivum* growth in compost, and on the other hand, Krupke et al. (2003) suggested that lactonase-producing bacteria should be involved in disease prevention by degrading methyl-isocoumarins produced by *T. aggressivum* as an antifungal compound active on *A. bisporus*. Based on these publications, several groups have tried to isolate lactonase-producing bacteria such as *Bacillus subtilis* from mushroom cultures and tested their efficiency as BCAs toward *T. aggressivum*. Another method was followed at the French Mushroom Technical Center. Védie and Rousseau (2008) examined commercial formulae of registered BCAs and found Serenade® based on *B. subtilis* QST 713 an efficient biocontrol product for protecting *A. bisporus* when mixed with the spawn just before inoculation into the compost. This strain was already registered as a biopesticide and it was easy to obtain a registration for its use in mushroom cultures. It is supposed to work in three different ways to fight disease-causing pathogens: it stops harmful spores from germinating, it disrupts cell membrane growth, and it inhibits attachment of the pathogen to its host. Up to now, the role of lactonase in its efficiency against green mold disease has not been proved. As a fungus, *A. bisporus* might also be affected by *B. subtilis* QST 713. It is thus obvious that indirect mushroom-mediated mechanisms also play a role in disease suppression. In the case of another mushroom species, *Pleurotus ostreatus*, it was shown that *Paenobacillus polymyxa* and other bacterial isolates inhibiting *Trichoderma* spp. in substrate did not stop the growth of the cultivated fungus, but partially affected mycelium growth and stimulated a defense reaction in which induction of higher levels of laccase activities occurred (Velázquez-Cedeño et al., 2008). Similarly, *B. subtilis* QST 713 could stimulate the defenses of *A. bisporus* in which laccase activity contributes to resistance and is a likely candidate for the enzymatic breakdown of methyl-isocoumarin produced by *T. aggressivum* (Sjaarda et al., 2015). This BCA is currently used by mushroom growers in France, but when tested under different cultivation conditions, the efficiency of Serenade® has not always been confirmed (Kosanović et al., 2013).

Research into defining these bioprotective mechanisms and for selecting BCAs adapted to the compost microbiota is needed. There have been some results registered leading to patents. That is the case for the use of a strain belonging to *Pseudomonas aeruginosa* (Bolkan and Larsen, 1998) and of another strain of *B. subtilis* (Gheshlaghi and Verdellen, 2009). However, to our knowledge, the development of commercial products used by growers is not documented.

6.4.2 BIOCONTROL OF *PSEUDOMONAS TOLAASII* WITH PHAGES AND ANTAGONISTIC BACTERIA

Various viruses affect *A. bisporus*, and some of these may cause diseases, but this issue is not discussed here [see Largeteau and Savoie (2010) for a review]. We paid attention to bacteriophages, which are bacterial viruses that invade bacterial cells, being ubiquitous in the environment. They play a key role in controlling bacterial numbers in various habitats and are involved in gene transfer between bacteria. There are many advantages to the use of bacteriophages. They have a high specificity to certain host bacteria, avoiding their detrimental effect on other *Pseudomonas* species in the casing layer that can be involved in fruiting induction, and they have a rapid sterilization effect on their hosts.

The first report on the isolation of bacteriophages from fruiting bodies of *A. bisporus* and their potential for practical use as BCAs in the mushroom industry was published by Munsch and Olivier (1995). The phage TO.1 was selected by these scientists. The best application time of phage suspension as a spray on the casing layer was at the beginning of the bacterial multiplication, i.e., 4 days after casing. Over 10 years of experiments with artificial inoculation of the same strain of *P. tolaasii*, the efficiency of bioprotection with TO.1 varied from 30% to 80%, depending on uncontrolled variations of cultivation conditions. In another experiment, it was shown that the efficiency of using this phage was dependent on the bacterial strain and there was a risk of selection regarding preexisting resistant strains in mushroom farms. To our knowledge, this bacteriophage has not been commercialized and used in mushroom farms, but the concept of bacteriophage use still is a good strategy to decrease the level of bacterial blotch disease in mushroom cultures. Recently, bacteriophages isolated from various sewage samples were shown to have lysing activities in *P. tolaasii* and to efficiently protect *P. ostreatus* against bacterial blotch (Kim et al., 2011). The sequence of the whole genome of the phage φPto-bp6g was then characterized and proposed to be potentially applicable as a safe biological control reagent against brown blotch disease in mushroom cultivation (Nguyen et al., 2012). Sajben-Nagy et al. (2012) isolated phages from necrotic caps of *P. ostreatus* and characterized one of them.

The search for antagonistic bacteria has received more attention than the bacteriophage method. It was initiated in the 1970s with the use of nonpathogenic *P. tolaasii*, but control levels were low and inconsistent (Nair and Fahy, 1972, 1976; Olivier et al., 1978). After an intensive isolation and screening program for bacteria antagonistic to *P. tolaasii*, Fermor et al. (1991) developed a biocontrol system for bacterial blotch in which the antagonist was applied on the mushroom bed in three treatments during the culture: at casing, pre-pinning, and after first flush. A consistent reduction in bacterial blotch disease levels in mushroom crops, of at least 50%, was achieved in assays with selected antagonists. Recently, antagonist strains have been isolated from compost, soil cover, and button caps with or without visible symptoms. A strain of *Pseudomonas putida* was selected as a potential BCA (Tajalipour et al., 2014). A patent has been registered on an active biocontrol agent (*Pseitdonionas synxantha* PS54) against *Pseudomonas* species causing rotting diseases in mushroom production (Manczinger et al., 2013). This followed an older one on an isolated strain of *P. fluorescens* (Fahy, 1987). Overall results of the different studies suggest that bacterial antagonists may be potential biocontrol agents for biological promotion of the health and growth of button mushroom, but their real use in mushroom farms still needs to be developed.

6.4.3 NO BIOCONTROL OF *LECANICILLIUM FUNGICOLA*

In the specific case of fungal pathogens of cultivated mushrooms, both the host and the pathogen are fungi, making it necessary to test the effect of the biocontrols on the host. Due to the biology of the host-pathogen interaction, the selection of possible biocontrol agents for dry bubble disease should focus on microorganisms that occur naturally in the casing soil in high densities, and preferably associated with *A. bisporus* (Berendsen et al., 2012). There have been attempts to identify BCAs effective against *L. fungicola* reported in the 1970s (de Trogoff and Ricard, 1976) and since 2000 (Bhatt and Singh, 2000; Bhat et al., 2010), but as for the other pathogens presented above, there is no publication on follow-up.

In a recent work, 160 bacterial strains were isolated from colonized casing and screened for *in vitro* antagonism to *L. fungicola* (Berendsen et al., 2012). It appeared *in vitro* that *L. fungicola* was sensitive to siderophore-mediated competition for iron and to antibiosis. Consequently, it was affected by many fluorescent pseudomonads. However, the more promising bacteria did not control the pathogen in bioassays simulating the commercial culture conditions. The authors attributed this ineffectiveness to characteristics of both *L. fungicola* and *A. bisporus*. The survival of the inoculum and the rate of bacteria necessary for biocontrol activity are also questioned. On the other hand, Berendsen et al. (2013b) observed that the white button mushroom did not develop an induced resistance to *L. fungicola* after a first infection. The conclusion of these investigations is that biological control of the dry bubble pathogen with antagonistic bacteria will be very difficult. Future efforts to control the dry bubble pathogen should therefore focus on other methods such as the development of resistant *A. bisporus* strains or the use of environmentally friendly biomolecules.

6.5 USE OF ENVIRONMENTALLY FRIENDLY BIOMOLECULES

Environmentally friendly biomolecules from specific plants and agricultural wastes might be used as substitutes for pesticides for mushroom pathogens control.

6.5.1 ESSENTIAL OILS

Since antiquity, essential oils have been widely used for bactericidal, virucidal, fungicidal, and insecticidal applications. Most of them are extracted by distillation from aromatic plants, and they contain a variety of volatile and nonvolatile molecules. Depending on type and concentration, they exhibit cytotoxic effects on living cells but are usually non-genotoxic. Many of these natural compounds have been tested as alternative agents against pathogens of edible fungi, and their strong fungistatic effect has been demonstrated. For instance, Soković and Van Griensven (2006) tested *in vitro* effects of essential oils extracted from various plant species on the pathogens causing the three major diseases of *A. bisporus*. The essential oils from *Origanum vulgare*, *Thymus vulgaris*, and *Mentha spicata* showed the best *in vitro* effects. Other studies were focused on the fungal pathogens affecting the fruiting bodies (Tanovic et al., 2009) or on green mold (Gorski et al., 2010). All the authors observed strong *in vitro* effects after using certain essential oils.

In vitro activity is a condition that is insufficient for identifying a biopesticide. Tests must be done under growing conditions. In a study on wet bubble disease, only a limited number of essential oils from a wide range of available oils screened were found to have the ability to effectively inhibit the pathogen *M. perniciosa* while exhibiting a minimal effect on the growth of *A. bisporus* (Regnier and Combrinck, 2010). In a preventive application of thyme and lemon verbena essential oils to casings inoculated with the pathogen, healthy mushrooms with no visible signs of *M. perniciosa* infection were harvested, whereas a curative use of these essential oils was ineffective in controlling the occurrence of the wet bubble pathogen on the mushrooms and severe infections (Regnier and Combrinck, 2010). Tea tree oil did not exhibit significant antifungal activity on *Trichoderma* isolates collected on *A. bisporus* from Serbian farms (Kosanović et al., 2013). Finally, essential oil vapors might serve to control proliferation of mold and bacteria in mushroom growing chambers that are now treated with other sanitizing agents, but more studies are required to develop specific methods for practical application of commercial products, which should have a cost acceptable to the mushroom growers.

6.5.2 COMPOST TEA

Apart from essential oils, the application of a variety of water-based extracts from composts made of agricultural wastes was tested for suppressing plant pathogens. Interesting studies on the use of spent mushroom substrate (SMS) tea were performed by Gea et al. (2014). SMS tea was obtained after steam treatment of SMS, including casings composed mainly of peat moss, at 70°C for 12h, followed by recomposting (or maturation) for 57 days, and mixing this treated SMS with water for 1 day under aerobic conditions. SMS teas provided *in vitro* inhibition of mycelial growth of *L. fungicola*. In mushroom crops inoculated with *L. fungicola*, the application of SMS teas decreased the incidence of dry bubble disease. These results and the absence of any fungitoxic effect on *A. bisporus* suggest that the dry bubble pathogen can be controlled by the use of teas made from SMS (Gea et al., 2014). When SMS tea was added to the casing layer, its microbial community might have affected the final composition of the casing layer microbiome, which was less favorable to the development of *L. fungicola*. SMS tea may be considered a biological alternative to fungicides for the integrated control of *L. fungicola* because it has high efficiency against the pathogen without significant detrimental effects on the yield of mushroom, and without the risk of fungicide residues in mushrooms (Navarro et al., 2011). This is an easy method of biocontrol using a homemade product to be included in integrated pest management strategies. For large development of this biocontrol method, variations in microbial and chemical characteristics of the SMS teas must be managed by a standardized method of production for a guarantee of efficacy.

6.5.3 WHITE LINE-INDUCING PRINCIPLE

Bioactive compounds extracted from BCA and various microorganisms from soils and plants have been examined and promoted as replacements for synthetic pesticides to suppress diseases in plant cultures. For mushrooms, this method of biocontrol has not been developed. This might be due to the fact that for fungal diseases, both the host and the pathogen are fungi. In the case of bacterial blotch, a cell-free crude extract of *Pseudomonas reactans* containing a lipodepsipeptide was proposed as a bioactive compound to be used as a preventive treatment to protect *A. bisporus* against *P. tolaasii* (Soler-Rivas

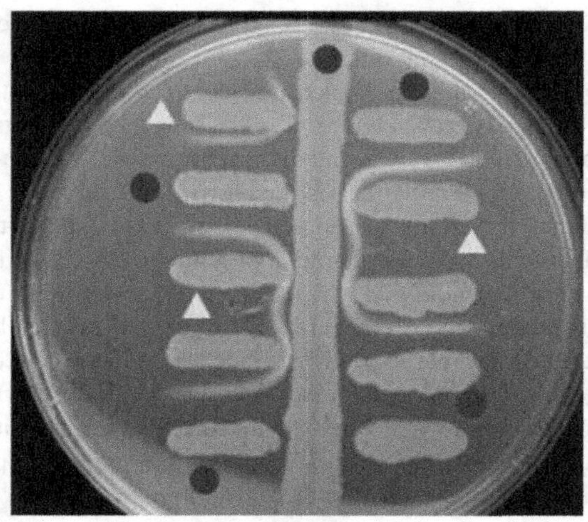

FIGURE 6.5

White line precipitate which is formed by the complexing reaction of tolaasin produced by *Pseudomonas tolaasii* and WLIP produced by *Pseudomonas reactans.* The circles represent *P. reactans,* and the triangles represent *P. tolaasii.*

et al., 1999). This compound, named the white line-inducing principle (WLIP), is a component of *P. reactans* virulence. As tolaasin does, WLIP is able to damage mushroom membranes by forming transmembrane pores, but it doesn't pass through the entire membrane because of its molecular size. It may have a detergent-like activity (Coraiola et al., 2006) with limited damage to the mushrooms.

WLIP was previously used in the identification of pathogenic forms of *P. tolaasii* due to the white precipitate formed by a complex between WLIP and tolaasin (Figure 6.5). Such a complex might neutralize the toxin on pretreated mushrooms, which would explain the reduction of symptoms of brown blotch disease. Both crude extracts and purified WLIP had inhibitory effects on brown discoloration induced by inoculation of *P. tolaasii* on mushroom caps (Soler-Rivas et al., 1999), but to our knowledge, no commercial product has been developed to date.

6.6 CONCLUSIONS

The development of a microbial pesticide or of a product isolated from plants or microorganisms for controlling pathogens requires several steps, beginning with their isolation in pure culture followed by screening of their efficacy by efficient bioassays performed *in vitro*, *ex vivo*, or *in vivo*, up to pilot trials under real conditions of application. Similar steps are needed for the development of a product isolated from plants or microorganisms for controlling pathogens. Whereas efficiencies were shown in various cases, bacterial control of *T. aggressivum* and other green molds is almost the only case of development of a BCA for the cultivation of *A. bisporus* mushrooms. Knowledge of the various mechanisms of interaction between *A. bisporus* and its major pathogens is increasing and will help to develop new BCAs and environmentally friendly biomolecules for controlling the pathogens.

On the other hand, the development of a mushroom strain resistant to pathogens requires other steps addressed to the identification of genetic sources of resistance in native relatives or local strains, the understanding of the heritability of this genetic resistance, the production of hybrids and screening of their qualities, and finally making tests under real conditions of mushroom culture. The resistance mechanisms of *A. bisporus* to its major pathogens are complex and under polygenic control, but recent publications and works in progress on genetics and genomics open new opportunities to rapidly select new hybrids of *A. bisporus* resistant to at least one of its major pathogens. There is now a need for materialization of these prospects in mushroom farms.

REFERENCES

Abubaker, K.S., Sjaarda, C., Castle, A.J., 2013. Regulation of three genes encoding cell-wall-degrading enzymes of *Trichoderma aggressivum* during interaction with *Agaricus bisporus*. Can. J. Microbiol. 59, 417–424.

Anderson, M.G., Beyer, D.M., Wuest, P.J., 2001. Yield comparison of hybrid *Agaricus* mushroom strains as a measure of resistance to *Trichoderma* green mold. Plant Dis. 85, 731–734.

Anonymous, 2000. Guidelines on good plant protection practice: mushrooms. EPPO Standard PP 2/20(1).

Bailey, A.M., Collopy, P.D., Thomas, D.J., Sergeant, M.R., Costa, A.M.S.B., Barker, G.L.A., et al., 2013. Transcriptomic analysis of the interactions between *Agaricus bisporus* and *Lecanicillium fungicola*. Fungal Genet. Biol. 55, 67–76.

Berendsen, R.L., Baars, J.P., Kalkhove, S.I.C., Lugones, L.G., Wosten, H.A.B., Bakker, P.A.H.M., 2010. *Lecanicillium fungicola*: causal agent of dry bubble disease in white-button mushroom. Mol. Plant Pathol. 11, 585–595.

Berendsen, R.L., Kalkhove, S.I.C., Lugones, L.G., Baars, J.J.P., Wosten, H.A.B., Bakker, P.A.H.M., 2012. Effects of fluorescent *Pseudomonas* spp. isolated from mushroom cultures on *Lecanicillium fungicola*. Biol. Control 63, 210–221.

Berendsen, R.L., Kalkhove, S.I.C., Lugones, L.G., Baars, J.J.P., Wosten, H.A.B., Bakker, P.A.H.M., 2013a. Effects of the mushroom-volatile 1-octen-3-ol on dry bubble disease. Appl. Microbiol. Biotechnol. 97, 5535–5543.

Berendsen, R.L., Schrier, N., Kalkhove, S.I.C., Lugones, L.G., Baars, J.J.P., Zijlstra, C., et al., 2013b. Absence of induced resistance in *Agaricus bisporus* against *Lecanicillium fungicola*. Antonie Van Leeuwenhoek Int. J. General Mol. Microbiol. 103, 539–550.

Bernado, D., Cabo, A., Novaes-Ledieu, M., Mendoza, C.G., 2004. *Verticillium* disease or "dry bubble" of cultivated mushrooms: the *Agaricus bisporus* lectin recognises and binds the *Verticillium fungicola* cell wall galactomannan. Can. J. Microbiol. 50, 729–735.

Bhat, M., Simon, S., Munshi, N.A., Bhat, Z.A., 2010. *In vitro* efficacy of casing and compost isolated bacterial inoculants against *Verticillium fungicola* (Preuss) Hassebrauk and *Agaricus bisporus* (Lange) Imbach. Indian J. Biol. Control 24, 137–141.

Bhatt, N., Singh, R.P., 2000. Chemical and biological management of major fungal pathogens of *Agaricus bisporus* (Lange) Imbach. Mushroom Sci. 15, 587–593.

Bolkan, H., Larsen, D.J., 1998. Administering *Pseudomonas aeruginosa*. Patent US 5762928 A.

Bruno, G.L., Ranab, G.L., Sermania, S., Scarolaa, L., Cariddi, C., 2014. Control of bacterial yellowing of cardoncello mushroom *Pleurotus eryngii* using acetic or hydrochloric acid solutions. Crop Protect. 50, 24–29.

Chen, X., Ospina-Giraldo, M.D., Wilkinson, V., Royse, D.J., Romaine, C.P., 2003. Resistance of pre- and post-epidemic strains of *Agaricus bisporus* to *Trichoderma aggressivum* f. *aggressivum*. Plant Dis. 87, 1457–1461.

Coraiola, M., Lo Cantore, P., Lazzaroni, S., Evodente, A., Iacobellis, N.S., Dalla Serra, M., 2006. WLIP and tolaasin I, lipodepside from *Pseudonomas reactans* and *Pseudomonas tolaasii* permealise model membranes. Biochimica Biophysica Acta 1758, 1713–1722.

De Trogoff, H., Ricard, J.L., 1976. Biological control of *Verticillium malthousei* by *Trichoderma viride* spray on casing soil in commercial mushroom production. Plant Dis. Rep. 60, 677–680.

Fahy, P.C., 1987. Mushroom blotch control agent. Patent EP 0210734 A1.

Fermor, T.R., Henry, M.B., Fenlon, J.S., Glenister, M.J., Lincoln, S.P., Lynch, J.M., 1991. Development and application of a biocontrol system for bacterial blotch of the cultivated mushroom. Crop Protect. 10, 271–278.

Fletcher, J.T., Gaze, R.H., 2008. Mushrooms Pest and Disease Control: A Color Handbook. Academic Press, London, 195 p.

Foulongne-Oriol, M., Spataro, C., Cathalot, V., Monllor, S., Savoie, J.M., 2010. An expanded genetic linkage map of an intervarietal *Agaricus bisporus* var. *bisporus* × *A. bisporus* var. *burnettii* hybrid based on AFLP, SSR and CAPS markers sheds light on the recombination behaviour of the species. Fungal Genet. Biol. 47, 226–236.

Foulongne-Oriol, M., Rodier, A., Rousseau, T., Largeteau, M., Savoie, J.M., 2011a. Quantitative genetics to dissect the fungal-fungal interaction between *Lecanicillium fungicola* and the white button mushroom *Agaricus bisporus*. Fungal Biol. 115, 421–431.

Foulongne-Oriol, M., Minvielle, N., Savoie, J.M., 2011b. QTL for resistance to *Trichoderma* lytic enzymes and metabolites in *Agaricus bisporus*Savoie, J.M. Foulongne-Oriol, M. Largeteau, M. Barroso, G. (Eds.), Proceedings of the Seventh International Conference on Mushroom Biology and Mushroom Products, vol. 2. INRA, France, pp. 17–25. <http://www.wsmbmp.org/Previous_Conference_7.html>.

Foulongne-Oriol, M., Rodier, A., Rousseau, T., Savoie, J.M., 2012a. Quantitative trait locus mapping of yield-related components and oligogenic control of the cap color of the button mushroom, *Agaricus bisporus*. Appl. Environ. Microbiol. 78, 2422–2434.

Foulongne-Oriol, M., Rodier, A., Savoie, J.M., 2012b. Relationship between yield components and partial resistance to *Lecanicillium fungicola* in the button mushroom, *Agaricus bisporus*, assessed by quantitative trait locus mapping. Appl. Environ. Microbi. 78, 2435–2442.

Gea, F.J., Carrasco, J., Diánez, F., Santos, M., Navarro, M.J., 2014. Control of dry bubble disease (*Lecanicillium fungicola*) in button mushroom (*Agaricus bisporus*) by spent mushroom substrate tea. Eur. J. Plant Pathol. 138, 711–720.

Gheshlaghi, N., Verdellen, J., 2009. *Bacillus subtilis* and use thereof as a green mold inhibitor. Patent WO 2009105878 A1.

Gorski, R., Sobieralski, K., Siwulski, M., Gora, K., 2010. Effect of selected natural essential oils on *in vitro* development of fungus *Trichoderma harzianum* found in common mushroom (*Agaricus bisporus*) cultivation. Ecol. Chem. Eng. S-Chemia I Inzynieria Ekologiczna S17, 69–77.

Guthrie, J.L., Castle, A.J., 2006. Chitinase production during interaction of *Trichoderma aggressivum* and *Agaricus bisporus*. Can. J. Microbiol. 52, 961–967.

Henkels, M.D., Kidarsa, T.A., Shaffer, B.T., Goebel, N.C., Burlinson, P., Mavrodi, D.V., et al., 2014. *Pseudomonas protegens* Pf-5 causes discoloration and pitting of mushroom caps due to the production of antifungal metabolites. Mol. Plant-Microbe Interact. 27, 733–746.

Kim, M.H., Park, S.W., Kim, Y.K., 2011. Bacteriophages of *Pseudomonas tolaasii* for the biological control of brown blotch disease. J. Korean Soc. Appl. Biol. Chem. 54, 99–104.

Kosanović, D., Potočnik, I., Duduk, B., Vukojevi, J., Staji, M., Rekanović, E., et al., 2013. *Trichoderma* species on *Agaricus bisporus* farms in Serbia and their biocontrol. Ann. Appl. Biol. 163, 218–230.

Kredics, L., García Jimenez, L., Naeimi, S., Czifra, D., Urbán, P., Manczinger, L., et al., 2010. A challenge to mushroom growers: the green mould disease of cultivated champignons. In: Mendez-Vilaz, A. (Ed.), Current Research, Technology and Education Topics in Applied Microbiology and Biotechnological Microbiology Formatex, Bada-joz, pp. 295–305.

Krupke, O.A., Castle, A.J., Rinker, D.L., 2003. The North American mushroom competitor, *Trichoderma aggressivum* f. *aggressivum*, produces antifungal compounds in mushroom compost that inhibits mycelial growth of the commercial mushroom *Agaricus bisporus*. Mycol. Res. 107, 1467–1475.

Largeteau, M.L., Savoie, J.M., 2010. Microbially-induced diseases of *Agaricus bisporus*: biochemical mechanisms and impact on commercial mushroom production. Appl. Microbiol. Biotechnol. 86, 63–73.

Largeteau, M.L., Rodier, A., Rousseau, T., Juarez del Carmen, S., Védie, R., Savoie, J.M., 2004. *Agaricus* susceptibility to *Verticillium fungicola*. In: Romaine, C.P., Keil, C.B., Rinker, D.L., Royse, D.J. (Eds.), Science and Cultivation of Edible and Medicinal Fungi. PennState University, University Park, pp. 515–523.

Largeteau, M.L., Regnault-Roger, C., Savoie, J.M., 2007. *Verticillium* disease of *Agaricus bisporus*: variations in host contribution to the total fungal DNA in relation to symptom heterogeneity. Eur. J. Plant Pathol. 118, 155–164.

Largeteau, M.L., Latapy, C., Minvielle, N., Regnault-Roger, C., Savoie, J.M., 2010. Expression of phenol oxidase and heat-shock genes during the development of *Agaricus bisporus* fruiting bodies, healthy and infected by *Lecanicillium fungicola*. Appl. Microbiol. Biotechnol. 85, 1499–1507.

Mamoun, M., Moquet, F., Savoie, J.M., Devesse, C., Ramos Guedes Lafargue, M., Olivier, J.M., et al., 1999. *Agaricus bisporus* susceptibility to bacterial blotch in relation to environment: biochemical studies. FEMS Microbiol. Lett. 181, 131–136.

Mamoun, M., Savoie, J.M., Olivier, J.M., 2000a. Interaction between the pathogen *Trichoderma harzianum* Th2 and *Agaricus bisporus* in mushroom compost. Mycologia 92, 233–240.

Mamoun, M., Iapicco, R., Savoie, J.M., Olivier, J.M., 2000b. Green mould disease in France: *Trichoderma harzianum* Th2 and other species causing damages on mushroom farms. In: Van Griensven, L.J.L.D. (Ed.), Science and Cultivation of Edible Fungi Balkema, Rotterdam, pp. 625–632.

Manczinger, L., Vágvölgyi, C., Sajben, E., Nagy, Á., Szöke-Kis, Z., Nagy, A., et al., 2013. Active agents against *Pseudomonas* species causing rotting diseases in mushroom production, their use and compositions containing them. Patent WO 2013034939 A2.

Moquet, F., Mamoun, M., Olivier, J.M., 1996. *Pseudomonas tolaasii* and tolaasin: comparison of symptom induction on a wide range of *Agaricus bisporus* strains. FEMS Microbiol. Lett. 142, 99–103.

Moquet, F., Desmerger, C., Mamoun, M., Ramos Guedes-Lafargue, M., 1999. A quantitative trait locus of *Agaricus bisporus* resistance to *Pseudomonas tolaasii* is closely linked to natural cap color. Fungal Genet. Biol. 28, 34–42.

Mumpuni, A., Sharma, H.S.S., Brown, A.E., 1998. Effect of metabolites produced by *Trichoderma harzianum* biotypes and *Agaricus bisporus* on their respective growth radii in culture. Appl. Environ. Microbiol. 12, 5053–5056.

Nair, N.G., Fahy, P.C., 1972. Bacteria antagonistic to *Pseudomonas tolaasii* and their control of the brown blotch of the cultivated mushroom *Agaricus bisporus*. J. Appl. Bacteriol. 35, 439–442.

Nair, N.G., Fahy, P.C., 1976. Commercial application of biological control of mushroom bacterial blotch. Aust. J. Agri. Res. 27, 415–422.

Navarro, M.J., Santos, M., Diánez, F., Tello, J.C., Gea, F.J., 2011. Toxicity of compost tea from spent mushroom substrate and several fungicides towards *Agaricus bisporus*Savoie, J.M. Foulongne-Oriol, M. Largeteau, M. Barroso, G. (Eds.), Proceedings of the Seventh International Conference on Mushroom Biology and Mushroom Products, vol. 2 INRA, Arcachon, pp. 196–201. <http://wsmbmp.org/proceedings/7th%20international%20conference/2/P26.pdf>.

Nguyen, H.T., Yoon, S., Kim, M.H., Kim, Y.K., Yoon, M.Y., Cho, Y.H., et al., 2012. Characterization of bacteriophage φPto-bp6g, a novel phage that lyses *Pseudomonas tolaasii* causing brown blotch disease in mushrooms. J. Microbiol. Methods 91, 514–519.

Olivier, J.M., Guillaumes, J., Martin, G., 1978. Study of a bacterial disease of mushroom caps Proceedings 4th International Conference on Plant Pathogenic Bacteria. INRA, Angers, Part 2, pp. 903–916.

Regnier, T., Combrinck, S., 2010. *In vitro* and *in vivo* screening of essential oils for the control of wet bubble disease of *Agaricus bisporus*. South Afr. J. Bot. 76, 681–685.

Romaine, C.P., Royse, D.J., Schlagnhaufer, B., 2005. Superpathogenic *Trichoderma* resistant to TopsinM found in Pennsylvania and Delaware. Mushroom News 53, 6–9.

Sajben-Nagy, E., Maróti, G., Kredics, L., Horváth, B., Párducz, A., Vágvölgyi, C., et al., 2012. Isolation of new *Pseudomonas tolaasii* bacteriophages and genomic investigation of the lytic phage BF7. FEMS Microbiol. Lett. 332, 162–169.

Savoie, J.M., Largeteau, M.L., 2004. Hydrogen peroxide concentrations detected in *Agaricus bisporus* sporocarps and relation with their susceptibility to the pathogen *Verticillium fungicola*. FEMS Microbiol. Lett. 237, 311–315.

Savoie, J.M., Mata, G., 2003. *Trichoderma harzianum* metabolites pre-adapt mushrooms to *Trichoderma aggressivum* antagonism. Mycologia 95, 191–199.

Savoie, J.M., Iapicco, R., Largeteau-Mamoun, M., 2001. Factors influencing the competitive saprophytic ability of *Trichoderma harzianum* Th2 in mushroom (*Agaricus bisporus*) compost. Mycol. Res. 105, 1348–1356.

Savoie, J.M., Reixtouex, S., Minvielle, N., Callac, P., 2008. Susceptibility of European populations of *Agaricus bisporus* to *Trichoderma* metabolites. In: Lelley, J.I., Buswell, J.A. (Eds.), Proceedings of the Sixth International Conference on Mushroom Biology and Mushroom Products GAMU, Krefeld, pp. 305–311. <http://www.wsmbmp.org/Previous_Conference_6.html>.

Savoie, J.M., Foulongne-Oriol, M., Barroso, G., Callac, P., 2013. Genetics and genomics of cultivated mushrooms, application to breeding of Agarics. In: Kempken, F. (Ed.), The Mycota, Vol.11: Agricultural Applications Springer, Berlin, pp. 3–33.

Sjaarda, C., Abubaker, K.S., Castle, A.J., 2015. Induction of lcc2 expression and activity by *Agaricus bisporus* provides defense against *Trichoderma aggressivum* toxins. Microbial. Biotechnol. Article first published online. DOI: http://dx.doi.org/10.1111/1751-7915.12277.

Soković, M., van Griensven, L.J.L.D., 2006. Antimicrobial activity of essential oils and their components against the three major pathogens of the cultivated button mushroom, *Agaricus bisporus*. Eur. J. Plant Pathol. 116, 211–224.

Soler-Rivas, C., Arpin, N., Olivier, J.M., Wishers, H.J., 1999. WLIP, a lipodepsipeptide of *Pseudomonas 'reactans'*, as inhibitor of the symptoms of the brown blotch disease of *Agaricus bisporus*. J. Appl. Microbiol. 86, 635–641.

Tajalipour, S., Hassanzadeh, N., Jolfaee, H.K., Heydari, A., Ghasemi, A., 2014. Biological control of mushroom brown blotch disease using antagonistic bacteria. Biocontrol Sci. Technol. 24, 473–484.

Tanovic, B., Potocnik, I., Delibasic, G., Ristic, M., Kostic, M., Markovic, M., 2009. *In vitro* effect of essential oils from aromatic and medicinal plants on mushroom pathogens: *Verticillium fungicola* var. *fungicola*, *Mycogone perniciosa*, and *Cladobotryum* sp. Arch. Biol. Sci. 61, 231–237.

Védie, R., Rousseau, T., 2008. Serenade biofungicide: une innovation majeure dans les champignonnières françaises pour lutter contre *Trichoderma aggressivum*, agent de la moisissure verte du compost. La Lettre du CTC 21, 1–2.

Velázquez-Cedeño, M., Farnet, A.M., Mata, G., Savoie, J.M., 2008. Role of *Bacillus* spp. in antagonism between *Pleurotus ostreatus* and *Trichoderma harzianum* in heat-treated wheat-straw substrates. Bioresour. Technol. 99, 6966–6973.

CHAPTER

SCLEROTIUM-FORMING MUSHROOMS AS AN EMERGING SOURCE OF MEDICINALS: CURRENT PERSPECTIVES

7

Beng Fye Lau and Noorlidah Abdullah
Mushroom Research Centre, Institute of Biological Sciences, Faculty of Science, University of Malaya, Kuala Lumpur, Malaysia

7.1 INTRODUCTION

Sclerotia, structures formed by the dense aggregation of mycelia, are important for fungal survival during adverse conditions, such as desiccation, microbial attack, or the long-term absence of a host (Townsend and Willetts, 1954; Coley-Smith and Cooke, 1971). The nutrients contained in the sclerotia allow the mushroom to enter a dormant stage when conditions are not favorable. The formation and development of sclerotia can be divided into three overlapping stages: initiation, development, and maturation; these are accompanied by morphological and biochemical differentiations under tight genetic control. Sclerotial initiation is usually induced by the onset of starvation or depletion of nutrients and other conditions not favoring mycelial growth. Moreover, several endogenous and exogenous factors have been reported to affect the initiation process (Willets and Bullock, 1992). When conditions improve, growth resumes, often with the formation of fruiting bodies.

A recent discovery by Smith et al. (2015) revealed that sclerotium-forming fungi are phylogenetically distributed among 85 genera in 20 orders of Ascomycota and Basidiomycota. Of interest is the small number of Basidiomycetes (mushrooms) that are known to form sclerotia as part of their life cycle, though their sizes and morphology vary depending on the taxonomic position of each species. Some of the sclerotium-forming mushrooms (SFM), e.g., *Lentinus tuber-regium*, are known for their culinary and medicinal properties, whereas others are more commonly consumed for their medicinal benefits, such as *Polyporus umbellatus*, *Wolfiporia cocos*, *Inonotus obliquus*, and *Lignosus* spp. From the taxonomic point of view, these mushrooms belong to the family Polyporaceae.

Background information of selected SFM that will be discussed:

Lentinus tuber-regium (Fr.) Fr. (synonym: *Pleurotus tuber-regium* (Fr.) Singer), also known as the king tuber oyster mushroom, is native to tropical and subtropical regions, including Africa, Asia, and Australasia. In China, *L. tuber-regium* is called *hunai* (literally, tiger's milk) and is found mainly in

Mushroom Biotechnology. DOI: http://dx.doi.org/10.1016/B978-0-12-802794-3.00007-2

southern China (Deng et al., 2000). The sclerotium is round with outer dark brown skin and inner white flesh. Both the sclerotium and fruiting body are edible. Its medicinal properties, especially antitumor and immunopotentiation activity, have been studied extensively (Zhang et al., 2001, 2004, 2006). Despite the growing interest in its medicinal properties and the sclerotial dietary fibers, industrial production of *L. tuber-regium* has not been reported; nevertheless, there are numerous studies on its cultivation using various agroresidues as substrates (Fasidi and Ekuere, 1993; Okhuoya and Etugo, 1993; Fasidi and Olorunmaiye, 1994; Isikhuemhen et al., 2000; Kuforiji and Fasidi, 2009).

Polyporus umbellatus (Pers.) Fr. (synonym, *Grifola umbellata* (Pers.) Pilat; *Boletus umbellatus* Pers.), sometimes referred to as "umbrella polypore" or "lumpy bracket," is usually found growing in the form of fungal rosettes on the roots of deciduous hardwood trees, most often on birch, maple, beech, and oak trees (Choi et al., 2003). The mushroom is distributed in China, Japan, and temperate regions of the northern hemisphere. The fruiting body is deeply umbilicate, light brown, and composed of numerous caps with narrow white pores. The flesh of the sclerotium is white and rather soft when young but hardens with age. The sclerotium of *P. umbellatus*, known as *zhuling* (in Chinese) or *chuling* (in Japanese) has been used in traditional Chinese medicine for a wide range of ailments related to edema, scanty urine, vaginal discharge, and urinary dysfunction, as well as jaundice and diarrhea (Zhao, 2013). It is well known for its diuretic effect, which has been actively studied lately (Zhao et al., 2009a; Zhang et al., 2010). Regarding acclimatization, outdoor cultivation has been reported (Choi et al., 2003). Moreover, there have also been attempts to understand the mechanism of sclerotial formation under artificial cultivation (Xing et al., 2013).

Wolfiporia cocos (F.A. Wolf) Ryvarden & Gilb (synonym: *Poria cocos* F.A. Wolf; *Wolfiporia extensa* (Peck) Ginns) is an example of a wood-decay fungus that is distributed in the southern provinces of China. It is notable for the development of a large underground sclerotium, used in traditional Chinese medicine for inducing diuresis, excreting dampness, invigorating the spleen, and tranquilizing the mind. The inner part of the sclerotium is referred to as fu-ling (Wang et al., 2013). Fu-ling-pi, the epidermis of the sclerotia of *W. cocos*, which is removed in the preparation of fu-ling, is used as a diuretic, and this has been validated recently (Zhao et al., 2012; Feng et al., 2013). Other medicinal properties of *W. cocos*, notably antitumor (Chen and Chang, 2004), anti-inflammatory, and immunopotentiation (Lee and Jeon, 2003), have been investigated. Research on the cultivation of *W. cocos* is still limited, but previous findings have indicated that indoor cultivation using bottles containing pine logs was successful (Kubo et al., 2006).

Inonotus obliquus (Ach. ex Pers.) Pilát (synonym: *Boletus obliquus* Ach. ex Pers), also known as the chaga mushroom, is a parasitic fungus that grows on birch and other trees, and produces a massive black, crusty conk, i.e., the sclerotium. It is also known as the clinker polypore, cinder conk, black mass, and birch canker polypore. Interestingly, it is sometimes called the tinder fungus due to its use as tinder for primitive fire starting techniques. The black appearance of the sclerotium is due to the presence of melanin pigments. The chaga mushroom is considered a medicinal species in Russian and Eastern European folk medicine to treat cancer, cardiovascular diseases, diabetes, digestive disorders, liver ailments, and tuberculosis (Patel, 2015). Research on its medicinal properties, including its anticancer effects, is ongoing (Youn et al., 2008, 2009; Song et al., 2013; Ning et al., 2014). Artificial cultivation of *I. obliquus* has been described by Sun et al. (2011).

Lignosus spp. consist of members with centrally stipitate pilei arising from sclerotia buried in the ground. Of interest are several species that are found in Southeast Asia and China, known in Malaysia as *cendawan susu rimau* (literally, tiger's milk mushroom in Malay) or *hurulingzhi* (literally, tiger milk

Ganoderma, in Chinese). Tiger's milk mushrooms are claimed to have numerous medicinal properties by the local Malays and indigenous people of Peninsular Malaysia. The sclerotium is the only part having medicinal properties (Lee et al., 2009). Overall, research on *Lignosus* spp. is still limited, and most reports have been on *Lignosus rhinocerotis* (Cooke) Ryvarden (synonym: *Polyporus rhinocerus* Cooke), the most commonly encountered *Lignosus* sp. in Malaysia (Choong et al., 2014). To date, its antitumor and immunomodulatory effects have been studied, and some positive results were obtained from *in vitro* and *in vivo* investigations (Lai et al., 2008; Wong et al., 2011; Lee et al., 2012; Lau et al., 2013b; Yap et al., 2013). Successful artificial cultivation of *L. rhinocerotis* was reported by Abdullah et al. (2013).

An earlier work by Wong and Cheung (2008b) discussed the cultivation and biochemical, nutritional, and biopharmacological properties of the sclerotia of *L. tuber-regium*, *W. cocos*, and *L. rhinocerotis*. There are also review articles dedicated solely to each species, including *L. tuber-regium* (Wong and Cheung, 2008a), *W. cocos* (Rios, 2011; Wang et al., 2013), *I. obliquus* (Shashkina et al., 2006; Zhong et al., 2009), and *P. umbellatus* (Zhao, 2013). In view of the expansion of research on various aspects of the SFM, this chapter serves to provide more up-to-date findings on medicinal properties apart from the widely studied antitumor and immunomodulatory effects. Considering the limited supply of native sclerotia and the difficulties associated with the solid-substrate cultivation of mushroom sclerotia, the potential use of mycelia (derived from liquid fermentation) as substitutes for the sclerotia will be highlighted.

7.2 THE IMPORTANCE OF MUSHROOM SCLEROTIA

7.2.1 FOOD

Some mushroom sclerotia are important as food components, especially for indigenous peoples or rural communities; for instance, in Nigeria, the sclerotium of *L. tuber-regium* is considered a delicacy (Oso, 1977). The sclerotium is frequently used in cooking and making various dishes. The sclerotium can be peeled and ground for use in a vegetable soup. Sometimes, the inner tissue is milled into a paste, which can be used to substitute in part or whole for melon seeds in the preparation of "egusi" soup, or mixed with corn flour and fried (Isikhuemhen and Okhuoya, 1995; Nwokolo, 1987). The nutritional attributes of *L. tuber-regium* have been extensively documented (Nwokolo, 1987; Fasidi and Ekuere, 1993; Akindahunsi and Oyetayo, 2006). In addition, the physicochemical and functional properties of the sclerotial dietary fibers have been thoroughly studied (Wong et al., 2003, 2005, 2006; Wong and Cheung, 2005a,b).

7.2.2 FOLK MEDICINE

All of the aforementioned SFM are used as folk remedies. Notably, the sclerotium of *P. tuber-regium* is used in some combinations that are intended to cure headache, stomach ailments, colds, fever, asthma, smallpox, and high blood pressure (Oso, 1977; Fasidi and Olorunmaiye, 1994). Both *P. umbellatus* and *W. cocos* are part of traditional Chinese medicine, in which sclerotia are appreciated for their diuretic activity (Zhao, 2013). The chaga mushroom, on the other hand, has been used as a folk remedy in Russia and Siberia since the sixteenth century. In Russian folk medicine, the chaga mushroom is used to treat cancers, e.g., stomach and lung cancer, and it is likewise considered effective for other common stomach and intestinal ailments such as gastritis, ulcers, colitis, and general pain (Patel, 2015). In Southeast Asia, *Lignosus* spp. are important components of the traditional medicine of the local communities. Information from early literature and recent ethnomycological surveys has indicated that the sclerotia have been claimed to be effective in treating cancer, cough, asthma, fever, wound healing, and other ailments (Lee et al., 2009). In China, *L. rhinocerotis* is used as a medicine by physicians for treating gastric ulcer, liver cancer, and chronic hepatitis (Wong and Cheung, 2008b).

7.2.3 BIOACTIVE COMPONENTS FROM SFM

7.2.3.1 Low-molecular-weight compounds

Chemical investigations have resulted in the isolation and characterization of low-molecular-weight (LMW) compounds, mainly secondary metabolites, from the sclerotia and fruiting bodies of SFM. The compounds are often isolated from crude extracts by a combination of techniques involving liquid–liquid extractions, column chromatography, thin layer chromatography, and high-performance liquid chromatography (HPLC), and further characterized by various hyphenated techniques, e.g., liquid chromatography-mass spectrometry (LC-MS), gas chromatography-mass spectrometry (GC-MS), and spectroscopic methods, e.g., nuclear magnetic resonance for structure elucidation. Most of the compounds isolated from the sclerotia of *P. umbellatus*, *I. obliquus*, and *W. cocos* belong to the classes of steroids. Complete lists of compounds isolated from the mushrooms have been published earlier (Shashkina et al., 2006; Zhong et al., 2009; Rios, 2011; Wang et al., 2013; Zhao, 2013); hence, only some recent findings in chemical investigations of SFM will be highlighted here:

According to Zheng et al. (2010), *I. obliquus* has been documented to produce a diverse range of secondary metabolites, including phenolic compounds, melanins, and lanostane-type triterpenoids. Handa et al. (2010) isolated an unusual lanostane-type triterpenoid, i.e., spiroinonotsuoxodiol, and two lanostane-type triterpenoids identified as inonotsudiol A and inonotsuoxodiol A. A more recent study by Zhao et al. (2015) reported the isolation of three new lanostane-type triterpenes, namely inonotusanes A–C. Bioassay-guided fractionation of ethyl acetate extract from the sclerotium of *P. umbellatus* resulted in the isolation of three ergostane-type ecdysteroids, i.e., polyporoid A–C (Sun and Yasukawa, 2008). In the course of searching for marker components, two new polyporusterones were isolated from the sclerotia of *P. umbellatus*, together with another three known analogs (Zhou et al., 2007). The structures of the new ones were elucidated as (20*S*,22*R*,24*R*)-16,22-epoxy-3β,14α,23β,25-tetrahydroxyergost-7-en-6-one and (23*R*,24*R*,25*R*)-23,26-epoxy-3β,14α,21α,22α-tetrahydroxyergost-7-en-6-one. A new pentacylic triterpene, 1β-hydroxylfriedelin, has been isolated and characterized (Zhao et al., 2009b). Working with the sclerotium of *W. cocos*, Zheng and Yang (2008) reported the isolated of two new lanostane triterpenoids, 29-hydroxypolyporenic acid C and 25-hydroxypachymic acid, together with known compounds, including ergosta-7,22-dien-3β-ol, polyporenic acid C, and pachymic acid.

On the other hand, chemical investigations of the remaining SFM are limited, and in many cases chemical profiling was carried out without isolation and characterization by spectroscopic methods, and compound identification relied on comparison with authentic samples, databases, characteristic fragmentation patterns, and UV absorption profiles. According to Afieroho and Ugoeze (2014), GC-MS analysis of the hexane extract of *L. tuber-regium* resulted in the identification of several fatty acids, such as linoleic acid and oleic acid, as well as steroids. Lin et al. (2014) identified typical phenolic compounds including protocatechuic, chlorogenic, syringic, ferulic, and folic acids in the ethanol extract of *L. tuber-regium* using HPLC. Working with the extracts of *L. rhinocerotis*, Lau et al. (2014) identified several LMV compounds, including sugars, fatty acids, methyl esters, sterols, amides, amino acids, phenolics, and triterpenoids from the aqueous methanol extract of *L. rhinocerotis* using GC-MS and UHPLC-ESI-MS/MS.

7.2.3.2 High-molecular-weight compounds

High-molecular-weight (HMW) components, mostly hydrophilic polysaccharides, proteins, and polysaccharide–protein complexes, have received a lot of attention over the last decade. In general, polysaccharides are commonly extracted using water at different temperatures (e.g., boiling is a common method) or alkaline aqueous solutions. The crude polysaccharides will then be subjected to a series of purification steps which usually involve gel filtration (separation by sizes) and ion-exchange (separation by charge) chromatography to yield pure glucans. Characterization of polysaccharides includes determination of their molecular weight and the composition of monosaccharides by GC-MS following acid-hydrolysis of the polysaccharides. There are some excellent review papers regarding the isolation, characterization, and biological activities of mushroom polysaccharides, in which some of the sclerotial polysaccharides are described (Zhang et al., 2007; Cheung, 2013; Giavasis, 2014; Ruthes et al., 2015).

7.3 SCIENTIFIC VALIDATION OF THE MEDICINAL PROPERTIES OF SFM

The purported medicinal benefits of SFM, based on traditional practices, have been the subject of intense research in the past decade. Among the multitude of medicinal benefits of SFM reported in the literature, it was obvious that their antitumor and immunomodulatory effects received far more attention than other biological activities. In many cases, these effects are more often attributed to the hydrophilic HMW components, mainly in the form of polysaccharides, proteins, and/or polysaccharide–protein complexes that are typically abundant in aqueous preparations of culinary and medicinal mushrooms. These have been subjects of intense research and review by previous workers, although the focus was not only on SFM (Wasser and Weis, 1999; Wasser, 2002; Moradali et al., 2007; Zhang et al., 2007; Patel and Goyal, 2012; Ren et al., 2012). Therefore, this time around, the focus is shifted to other biological properties of SFM that have received less attention and have not been systematically compiled, such as the antioxidative, anti-inflammatory, antimicrobial, antidiabetic, and neuritogenic effects, and their effects on the cardiovascular systems. Major findings pertaining to the aforementioned biological activities, especially information on potentially active compounds and their possible mode of action, will be highlighted.

7.3.1 ANTITUMOR ACTIVITY

Chemical components of the sclerotia, including the polysaccharides, proteins, and polysaccharide–protein complexes, are best known for their antitumor activities. The extracts and fractions of SFM have been demonstrated to exhibit *in vitro* and *in vivo* antitumor activity; some are selective against cancer cells with lesser damage to the normal cells. Findings from *in vitro* studies have also revealed that, in many cases, cell death is attributed to apoptosis involving either the intrinsic or extrinsic pathways.

Among the LMW compounds, secondary metabolites belonging to the classes of triterpenoids (including steroids) and phenolic acids have been identified as active compounds that target multiple pathways related to carcinogenesis, such as induction of apoptosis, prevention of metastasis, and inhibition of angiogenesis. Numerous LMW compounds with cytotoxic effects against various cancer cells have been isolated from alcoholic extracts of SFM, but in-depth studies on potential mechanisms of action have been carried out for only a small number of antitumor metabolites. The mode of action of cytotoxic chemical constituents from the sclerotia of *W. cocos* and *P. umbellatus* has been actively studied. Chemical constituents from the sclerotium of *W. cocos*, such as dehydropachymic acid, pachymic acid, and tumulosic acid, were reported to exhibit moderate cytotoxicity against human colon carcinoma cells (Li et al., 2004). Further, pachymic acid inhibited the proliferation of human pancreatic cancer cells (Panc-1, MiaPaca-2, AsPc-1, BxPc-3) without affecting the normal pancreatic duct epithelial cells (HPDE-6), causing cell cycle arrest at the G1 phase as well as downregulating the expression of KRAS and matrix metalloproteinase-7 (MMP-7) in BxPc-3 cells. At the same time, pachymic acid also suppressed the invasive behavior of BxPc-3 cells, and this was associated with a reduction of MMP-7 at the protein level (Kikuchi et al., 2011). Previously isolated cytotoxic constituents from the sclerotium of *P. umbellatus* are mainly steroids (Zhao et al., 2010). One of the compounds, ergone, displayed remarkable cytotoxicity against Hep G2, Hep-2, and Hela cells without affecting normal cells (HUVEC).

Based on the work of Zhao et al. (2011), the cytotoxicity of ergone against Hep G2 was attributed to apoptosis, which was evidenced by cell cycle arrest at the G2/M stage, chromatin condensation, nuclear fragmentation, and externalization of phosphatidylserine. Both intrinsic and extrinsic apoptotic pathways were involved based on the occurrence of PARP-cleavage, activation of caspase-3, -8, and -9, and upregulation of Bax and downregulation of Bcl-2. Inotodiol isolated from *I. obliquus* was shown to be a potent antitumor agent in a two-stage carcinogenesis test on mouse skin using 7,12-dimethylbenz[a]anthracene as an initiator and tetradecanoyl phorbol acetate (TPA) as a promoter (Nakata et al., 2007). Some of the compounds were able to inhibit DNA topoisomerases; for instance, Mizushina et al. (2004) reported that dehydroebriconic and dehydrotrametenoic acids inhibited DNA topoisomerase II activity (IC_{50}: 4.6 μM), while both compounds moderately inhibited the activities of DNA polymerases α, β, γ, δ, ε, η, ι, κ, and λ in mammals to similar extents. Both compounds also suppressed the growth of human gastric cancer cells and induced cell cycle arrest. Chemical investigation on the sclerotial LMW of *L. tuber-regium* is limited, but a recent study by Lin et al. (2014) demonstrated that a 60% (v/v) ethanol extract of *P. tuber-regium* sclerotium inhibited the vascular endothelial growth factor-induced human umbilical vein endothelial cell (HUVEC) migration and tube formation. Moreover, the phenolics-rich extract also inhibited the formation of subintestinal vessel plexus in zebrafish embryos *in vivo*.

HMW components are known to exhibit antitumor activity via modulation of the immune system. Their activities were affected by a number of factors, namely molecular weight, degree of branching, solubility, type of side linkages, and other physicochemical characteristics, as reviewed by Zhang et al. (2007). The antiproliferative effects of *L. tuber-regium* are largely attributed to the sclerotial water-soluble components, mainly polysaccharides. In a study by Tao et al. (2006), a water-soluble β-glucan extracted from the sclerotia was fractionated into eight fractions, five of which were sulfated. In the *in vitro* cytotoxicity test, the sulfated derivatives were relatively more cytotoxic than the native glucan fractions. Wong et al. (2007) reported that the nonstarch polysaccharides exerted strong cytotoxicity against human acute promyelocytic leukemia cells (HL-60) and induced apoptosis. Zhang et al. (2006) explored the antiproliferative effect of a water-soluble carboxymethylated β-glucan partially synthesized from an insoluble native glucan isolated from *L. tuber-regium* sclerotium. It inhibited the growth of MCF7 (IC_{50}: 204 μg/mL) and induced cell cycle arrest at the G1 phase, which was also associated with the downregulation of cyclin D1 and cyclin E expressions and increased expression of the Bax/Bcl-2 ratio. Zhang et al. (2011) demonstrated that the aqueous extract and polysaccharides of *P. umbellatus* sclerotium inhibited bladder carcinogenesis in rats, which may be associated with upregulation of glutathione *S*-transferase π (GSTPi) and NAD(P)H dehydrogenase (quinone) 1 (NQO1) in the bladder. Chen and Chang (2004) isolated a neutral polysaccharide fraction from *W. cocos* (designated as PC-PS, 160 kDa) by a series of chromatographic separations, and its antiproliferative activity against human leukemic cells U937 and HL-60 was investigated *in vitro*. They found that the conditioned medium prepared with PC-PS (15 μg/mL) stimulated human blood mononuclear cells for 5 days suppressed the proliferation of U937 and HL-60 cells by 87.3% and 74.7%, respectively.

According to Youn et al. (2008), water extract of *I. obliquus* inhibited the growth of Hep G2 cells in a dose-dependent manner, accompanied with G0/G1-phase arrest and apoptotic cell death. At the molecular level, cell cycle arrest was associated with downregulation of p53, pRb, p27, cyclins D1, D2, and E, and cyclin-dependent kinase (Cdk) 2, Cdk4, and Cdk6 expression. Working on the melanoma cell lines B16-F10, Youn et al. (2009) found that the water extract of *I. obliquus* also suppressed the growth of cells and induced apoptosis, and these were associated with the downregulation of pRb, p53, and p27 expression levels. In addition, intraperitoneal administration of *I. obliquus* extract, at a dose of 20 mg/kg for 10 days, inhibited tumor progression in B16-F10 cells implanted mice—lear evidence of *in vivo* antitumor effects. Among the SFM, studies on the antiproliferative activity of the sclerotial aqueous extracts of *L. rhinocerotis* are limited. Sclerotial hot and cold aqueous extract was reported to moderately inhibit the growth of leukemic and solid-tumor cell lines (Lai et al., 2008; Lee et al., 2012; Lau et al., 2013b; Yap et al., 2013). The cytotoxic action of a cold aqueous extract against MCF7 was found to be mediated by apoptosis, based on DNA fragmentation studies (Lee et al., 2012). The nature of cytotoxic components in the sclerotium of *L. rhinocerotis* has yet to be identified but is suspected to involve some HMV components, either proteins or protein–carbohydrate complex (Lee et al., 2012), and these are heat-sensitive (Lau et al., 2013b).

It has been established that chemical modification might enhance the antitumor properties of mushroom polysaccharides; for instance, Tao et al. (2009) extracted two water-soluble polysaccharide–protein complexes from the sclerotium of *P. tuber-regium* and chemically modified them to obtain their sulfated and carboxymethylated derivatives, which both showed good cytotoxicity against cancer cells. In another study by Wang et al. (2004), a water-insoluble (1→3)-β-glucan isolated from the sclerotium of *W. cocos* was sulfated, carboxymethylated, methylated, hydroxyethylated, and hydroxypropylated, respectively, to afford five water-soluble derivatives. The authors reported that the native β-glucan did not show antitumor activity, whereas the sulfated and carboxymethylated derivatives exhibited significant antitumor activities against S-180 and gastric carcinoma tumor cells. As reported by Chen et al. (2009), (1→3)-β-glucan isolated from *W. cocos* was phosphorylated to obtain a series of derivatives that exhibited relatively strong inhibition against S-180 tumor cells (Chen et al., 2009).

7.3.2 IMMUNOMODULATORY ACTIVITY

The HMW from SFM are sometimes called biological response modifiers, as they are known to modulate the immune system. Various polysaccharides, proteins, and polysaccharide–protein complexes from SFM have been reported to exert an immunomodulatory effect. According to Chang et al.

(2009), an immunomodulatory protein (designated PCP, 35.6 kDa) from the sclerotium of *W. cocos* stimulated RAW 264.7 macrophages through the induction of tumor necrosis factor-alpha (TNF-α) and interleukin-1β (IL-1β), and regulation of nuclear factor-kappa B (NF-κB)-related gene expression. Further, *in vivo* work demonstrated that PCP activated peritoneal cavity macrophages to induce Toll-like receptor 4 (TLR4)-mediated myeloid differentiation factor 88-dependent signaling. Ma et al. (2010) evaluated the immunomodulatory effect of a polysaccharide-rich fraction (designated PRF) of the sclerotium of *W. cocos*. Oral administration of PRF at 200 mg/kg body weight to immunized mice increased the T-cell percentage among splenocytes but reduced anti-OVA immunoglobulin G and M levels significantly, indicating that PRF could modulate the specific immune response of Balb/c mice and that the modulation occurs via the activation of T cells. Kim (2005) evaluated the immunomodulatory effect of an aqueous extract of *I. obliquus*. Oral administration of the extract increased serum levels of IL-6 and suppressed an NF-α-related pathologic condition. Fan et al. (2012) purified a water-soluble polysaccharide (designated ISP2a) from *I. obliquus*. ISP2a was reported to enhance the proliferation of lymphocytes and increased production of TNF-α. In another study by Li et al. (2010), a polysaccharide isolated from *P. umbellatus* promoted the activation and maturation of murine bone marrow-derived dendritic cells (BMDC) via TLR4. Treatment of BMDC with the sclerotial polysaccharides resulted in enhanced cell-surface expression of CD86, enhanced production of both IL-12 p40 and IL-10, increased T-cell-stimulatory capacity, and decreased phagocytic ability. Further, the sclerotial polysaccharides were found to upregulate the functions of macrophages such as nitric oxide (NO) production and cytokine expression (Li et al., 2011).

The effect of extracts of *L. rhinocerotis* on the immune system have received vast attention. Wong et al. (2009) reported the stimulation of human innate immune cells by sclerotial polysaccharides from *L. rhinocerotis*. The sclerotial polysaccharides were found to upregulate the expression of Dectin-1 in NK-92MI and MD but not $CD56^+$ NK cells, suggesting that Dectin-1 might act as the receptor for binding of sclerotial polysaccharides. In a separate study, Guo et al. (2011) found that the sclerotial hot aqueous extract (25 µg/mL) enhanced the functional activities of RAW 264.7 and murine primary macrophages. The hot aqueous extract also increased the phosphorylation of IKBα, which could trigger the NF-κB signaling pathway for macrophage activation.

7.3.3 ANTIOXIDATIVE ACTIVITY

There are numerous studies directed to analysis of the antioxidant capacity of extracts, fractions, and isolated compounds in both cellular and animal models. Earlier studies have established the correlation between antioxidant capacity and the phenolic contents of various mushrooms, including SFM. Liang et al. (2009) evaluated the antioxidant capacity of a crude ethanol extract of *I. obliquus* and its subfractions (ethyl acetate fraction, *n*-butanol fraction, and aqueous fraction). The results showed that the extent of antioxidant activity is in accordance with the amounts of phenolics and flavonoids. In a study by Lin et al. (2014), the aqueous ethanol extract of *L. tuber-regium* sclerotium showed good antioxidant activity, which might be attributed to the presence of several phenolic compounds, including protocatechuic, chlorogenic, syringic, ferulic, and folic acids. Based on a report by Nakajima et al. (2007), the aqueous methanol extract of *I. obliquus*, which demonstrated strong antioxidative activity, was subjected to further purification that resulted in the isolation of 4-hydroxy-3,5-dimethoxy benzoic acid, 2-hydroxy-1-hydroxymethyl ethyl ester, protocatechic acid, caffeic acid, 3,4-dihybenzaladehyde, 2,5-dihydroxyterephtalic acid, syringic acid, and 3,4-dihydroxybenzalacetone.

The antioxidative compounds from SFM are not limited to phenolics, but there is increasing evidence that polysaccharides might also play a role; for instance, the water-soluble and alkaline-soluble polysaccharides from *P. tuber-regium* were studied for their *in vitro* antioxidant properties (Wu et al., 2014). Results indicated that the alkaline-soluble polysaccharides were stronger than water-soluble polysaccharides in the scavenging of free radicals, as well as in their inhibitory effect on liver lipid peroxidation, liver mitochondria swelling, and red blood cell hemolysis. Tang et al. (2014) prepared polysaccharides (designated PCP-1, PCP-2, and PCP-3) from the degradation of *W. cocos* polysaccharides (PCP) with different concentrations of H_2O_2 solution. These exhibited antioxidant activity, as demonstrated by the scavenging of hydroxyl radicals, ABTS radicals, and ferrous ions. In addition, the polysaccharides demonstrated DNA protective activity. Mu et al. (2012) compared the antioxidant capacity of the water-soluble and alkali-soluble crude polysaccharides (designated IOW and IOA, respectively) isolated from *I. obliquus*, and the carbohydrate-rich fractions IOW-1 and IOA-1 were obtained, respectively, after deproteination and depigmentation. The polysaccharides scavenged DPPH, hydroxyl, and superoxide anion radicals in a dose-dependent manner. These polysaccharides also afforded protection against oxidative stress in PC12 cells.

It is worth pointing out that the contribution of polyphenolic and polysaccharide fractions from SFM is still not conclusive. In a work by Cui et al. (2005), the polyphenolic-rich extracts as well as triterpenoid- and steroid-rich extracts exhibited moderate antioxidant effects by scavenging 1,1-diphenyl-2-picrylhydrazyl (DPPH), superoxide, and peroxyl radicals; however, the polysaccharide extracts were inactive. Nevertheless, only the polyphenolic extract protected the human keratinocyte cell line (HaCaT) from H_2O_2-induced oxidative stress, whereas the polysaccharide, triterpenoid, and steroid extracts were ineffective.

The antioxidant capacity of *Lignosus* spp. has been reported recently. Yap et al. (2013) compared the antioxidant properties of the hot water, cold water, and methanol extracts of the cultivated and wild-type sclerotium of *L. rhinocerotis*. The DPPH, ABTS, and superoxide anion radical scavenging activities of the extracts ranged from 0.52 to 1.12, 0.05 to 0.20, and −0.98 to 11.23 mmol Trolox equivalents/g extract, respectively, whereas FRAP values ranged from 0.006 to 0.016 mmol/min/g extract. Working with *Lignosus tigris*, Yap et al. (2014) reported that FRAP values ranged from 0.002 to 0.041 mmol/min/g extract, while the DPPH, ABTS, and superoxide anion scavenging activities ranged from 0.18 to 2.53, 0.01 to 0.36, and −4.53 to 10.05 mmol Trolox equivalents/g extract, respectively. Taken together, on the basis of antioxidant capacity, preliminary results indicate that both *L. rhinocerotis* and *L. tigris* cultivars showed good prospects of being developed into functional foods, although the extracts of the wild-type samples appeared to be more potent.

In addition, the antioxidant capacity of SFM is associated with potential antiaging effects. A study by Yun et al. (2011) demonstrated that *I. obliquus* exerted protective effects against H_2O_2-induced apoptosis and premature senescence in human fibroblasts. In addition, *I. obliquus* suppressed UV-induced morphologic skin changes, such as skin thickening and wrinkle formation, in hairless mice *in vivo*, and increased collagen synthesis through inhibition of MMP-1 and MMP-9 activities in H_2O_2-treated human fibroblasts. These results seem to provide some indication of potential antiaging effects of *I. obliquus*. In another study, the protective effects of *W. cocos* water extract (designated as PCW) against Aβ1–42-induced cell death were investigated using rat pheochromocytoma (PC12) cells (Park et al., 2009). Pretreatment with PCW (5–125 μg/mL) reduced Aβ1–42-induced cell death, and attenuated cytotoxicity, apoptotic features, and accumulation of intracellular oxidative damage. Moreover, the expression of apoptotic protein Bax and activity of caspase-3 were decreased, but the expression of an antiapoptotic protein Bcl-2 was increased.

7.3.4 ANTI-INFLAMMATORY ACTIVITY

Crude extracts of SFM were found to have anti-inflammatory activity, and bioactive compounds are usually triterpenoids (including steroids). According to Ma et al. (2013), petroleum ether and ethyl acetate fractions of *P. umbellatus* were found to have significant inhibition effects on NO production and NF-κB luciferase activity in RAW 264.7 cells. According to Sun and Yasukawa (2008), compounds isolated from *P. umbellatus*, namely polyporoid A–C and other known ecdysteroids, demonstrated potent anti-inflammatory activity in a test of TPA-induced inflammation (1 μg/ear) in mice. Fuchs et al. (2006) evaluated the anti-inflammatory activity of *W. cocos* extracts on experimentally induced irritant contact dermatitis (ICD) in a repeated sodium lauryl sulfate (SLS) irritation model.

The anti-inflammatory efficacy of the extract on the elicitation phase of the ICD induced by repeated SLS test could be observed, and this might be attributed to its influence on proinflammatory enzymes such as phospholipase A2. In another study by Cuellar et al. (1997), a hydroalcoholic extract from *W. cocos* was examined for oral and topical anti-inflammatory activity. It was shown to be active against carrageenan, arachidonic acid, TPA acute edemas, TPA chronic inflammation, and oxazolone delayed hypersensitivity in mice. Dehydrotumulosic and pachymic acids were identified as active compounds.

There are increasing studies to unravel the molecular mechanism of action; however, it seems that most work was carried out using crude extracts rather than pure compounds; for instance, the ethanol extract of *I. obliquus* inhibited lipopolysaccharide (LPS)-induced inflammation in RAW 264.7 macrophages (Kim et al., 2007).

The inhibition of LPS-induced expression of inducible nitric oxide synthase (iNOS) and cyclooxygenase-2 (COX-2) proteins was mediated by Akt and JNK pathways. Jeong et al. (2014) demonstrated the anti-inflammatory effects of ethanol extract of *W. cocos* in LPS-stimulated RAW 264.7 macrophages. They reported that the extract targeted the inflammatory response of macrophages via inhibition of iNOS, COX-2, IL-1β, and TNF-α through inactivation of the NF-κB signaling pathways.

The anti-inflammatory activity of *L. rhinocerotis* is of interest as the mushroom is traditionally used to relieve cough, asthma, and chronic hepatitis, all of which are presumably related to its anti-inflammatory effect. According to Lee et al. (2014), the cold water extract (CWE) of the sclerotial powder of *L. rhinocerotis* TM02 cultivar possessed potent antiacute inflammatory activity as measured by the carrageenan-induced paw edema test. Nevertheless, CWE at 200 mg/kg did not inhibit transudative and proliferative phases of chronic inflammation, as shown by using the cotton pellet induced granuloma model. The anti-inflammatory activity of CWE was mainly attributed to its HMW fraction, but the precise component has yet to be identified and/or isolated.

7.3.5 ANTIMICROBIAL ACTIVITY

The antimicrobial activities of SFM have received far lesser attention when compared to other medicinal properties. Chemistry-wise, the antimicrobial activities of SFM are mostly attributed to the LMW compounds that are present in the organic solvent extracts rather than the components in aqueous extracts. Current findings seem to indicate that the extracts and fractions of SFM showed inhibitory effect on Gram-positive and negative bacteria but inactive against several pathogenic fungi. In view of its role in fungal survival, it is hypothesized that the sclerotium might produce chemical components (e.g., secondary metabolites) for defense against other antagonistic microorganisms, which might be exploited as antimicrobial compounds.

On the contrary, so far, studies carried out on *L. tuber-regium* seem to suggest that the fruiting body, rather than the sclerotium, is a better source of antimicrobial compounds. Ezeronye et al. (2005) found that only the ethanol extract of the fruiting bodies showed antibacterial activity. The ethanol extract of the sclerotium, as well as the aqueous extracts of the fruiting body and sclerotium, were not active. These findings were confirmed by Jonathan et al. (2008), who reported that the crude methanol and ethyl acetate fractions of the fruiting body of *L. tuber-regium* showed antibacterial activity against *Bacillus cereus*, *Escherichia coli*, *Klebsiella pneumoniae*, and *Proteus vulgaris*. The antimicrobial activity of the sclerotium belonging to *L. rhinocerotis* was demonstrated by Mohanarji et al. (2012). It was proved that the methanol and aqueous extracts showed stronger antibacterial activity than did the petroleum ether and chloroform extracts, suggesting that the active compounds are likely to be polar in nature.

7.3.6 ANTIHYPERTENSIVE ACTIVITY AND RELATED CARDIOVASCULAR COMPLICATIONS

In most studies, *in vitro* inhibition of the angiotensin I-converting enzyme (ACE) is taken as indicative of potential antihypertensive activity *in vivo*. Although there are no reports on the inhibitory effect of mushroom sclerotia against ACE, the antihypertensive effect of sclerotial extracts has been demonstrated. A recent study revealed that the sclerotial aqueous extract of *L. tuber-regium* was reported to lower the systolic, diastolic, pulse, and mean arterial pressures of salt-loaded rats in a dose-dependent manner (Ikewuchi et al., 2014). Chemical analysis indicated that the aqueous extracts contained flavonoids (kaempferol, quercetin, and hesperidin) and sterols; therefore, the hypotensive activity of the aqueous extract might have been due to these compounds. In addition, Ikewuchi et al. (2013) investigated the effect of *L. tuber-regium* aqueous extract on the biochemical and hematological indices in normal and subchronic salt-loaded rats. Their results indicated that, compared to test controls, the treatment lowered the mean cell volume, atherogenic indices (cardiac risk ratio, atherogenic coefficient, and atherogenic index of plasma), plasma alanine and aspartate transaminase activities, mean cell hemoglobin, sodium, bicarbonate, urea, blood urea nitrogen, triglyceride, and total, non-high-density, low-density, and very low-density lipoprotein cholesterol concentrations, as well as neutrophil, monocyte, and platelet counts in the treated animals. However, treatment with the extract increased hemoglobin concentration, mean cell hemoglobin, red cell and lymphocyte counts, plasma high-density lipoprotein cholesterol, calcium, potassium, chloride, creatinine, albumin, and total protein concentrations in the treated animals. Taken together, the results seemed to suggest that the antihypertensive activity of *L. tuber-regium* may be mediated through alteration of plasma levels of sodium and potassium, or increases in muscle tone brought about by changes in plasma calcium levels.

7.3.7 ANTIDIABETIC ACTIVITY

Studies on the antidiabetic activity of SFM are still limited. Earlier, Li et al. (2011) investigated the effect of *W. cocos* on type 2 diabetes. The sclerotial crude extract at 50 mg/kg body weight or more significantly decreased blood glucose levels in db/db mice. Some of the bioactive constituents identified from the chloroform extract and subfractions are dehydrotumulosic acid, dehydrotrametenolic acid, and pachymic acid. Mechanistic study on streptozocin (STZ)-treated mice revealed that the crude extract, dehydrotumulosic acid, dehydrotrametenolic acid, and pachymic acid exhibited different levels of insulin sensitizer activity but did not activate the PPAR pathway.

7.3.8 DIURETIC ACTIVITY

Some of the SFM, namely *P. umbellatus* and *P. cocos*, have been reported to have diuretic activity, while no work has been done on *L. tuber-regium*, *I. obliquus*, and *L. rhinocerotis*. Zhao et al. (2009a,b) evaluated various extracts of *P. umbellatus* sclerotium for diuretic activity, in which oral administration of the *n*-hexane and *n*-butanol extracts led to a significant increase in the urinary volume. Bioassay-guided isolation of the active extracts led to the isolation and identification of ergone, ergosterol, and D-mannitol. All isolated compounds increased the cumulative urinary volume excretion compared to the control. D-mannitol was found to affect the excretion of potassium, sodium, and chloride, whereas ergone did not produce a similar effect. The diuretic activity of ergosterol was suspected to be due to its conversion into ergone by dioxygenase *in vivo*.

Zhang et al. (2010) reported the diuretic activity of the sclerotial aqueous extract of *P. umbellatus*. Oral administration of the aqueous extract led to an increase in the total volume of urine excreted as well as the levels of Na^+, K^+, and Cl^-. In an attempt to gain insight into the possible role of the aquaporin (AQP) family on water reabsorption in different regions of the kidney, their mRNA expression level was investigated. Postoral administration of the sclerotial aqueous extract did not affect the level of AQP1 and AQP3 but downregulated the mRNA expression of AQP2 and V2R relative to the controls in normal rat kidney medulla. Taken together, the sclerotial aqueous extract of *P. umbellatus* exerted a diuretic effect by regulating AQP2, and the downregulation of AQP2 might be due to the downregulation of V2R. The epidermis layer of the sclerotium of *W. cocos* has been claimed to have a diuretic effect.

In a study by Zhao et al. (2012), the sclerotial ethanol and aqueous extracts of *W. cocos* were investigated for potential diuretic activity in saline-loaded rats. It was found that the ethanol extract significantly increased urinary excretion rates compared to the aqueous extract. The ethanol extract also increased the excretion of Na^+ but produced little or no effect on K^+. Due to the remarkable diuretic effect of the ethanol extract, it was fractionated by liquid–liquid partitions to yield fractions of different polarity, i.e., petroleum ether, ethyl acetate, *n*-butanol, and aqueous fractions. The fractions produced different degrees of diuretic activity, but it was noted that the ethyl acetate and *n*-butanol fractions gave the highest activity. Oral administration of the ethyl acetate fraction (400 mg/kg) significantly enhanced the Na^+/K^+ ratio, whereas the *n*-butanol fraction only slightly increased the Na^+/K^+ value. Recently, Wu et al. (2014) found that the scleroderma of *P. cocos* exhibited its diuretic effect by suppression of renal AQP2 expression in rats with chronic heart failure.

Some chemical constituents have been identified as active components from organic solvent extracts, whereas the active components from aqueous extracts have yet to be determined. So far, only steroidal constituents, i.e., ergone and ergosterol, and D-mannitol, a known light diuretic, have been identified as diuretic agents. Chemistry-wise, triterpenoids are abundant in mushroom sclerotia, and tetracyclic triterpenoids that present a structural similarity to aldosterone and its antagonist spironolactone (a potassium saver prototype diuretic). On the assumption that compounds that have similar chemical structures will give the same effect, the triterpenoids were suspected to be the active compounds with diuretic activity in SFM.

7.3.9 NEURITOGENIC ACTIVITY

Studies on the neuritogenic potential of mushroom sclerotia are limited; however, previous work has demonstrated the ability of the extracts of *L. rhinocerotis* sclerotium to stimulate neurite outgrowth using PC-12 Adh (rat pheochromocytoma, adherent variant) and Neuro-2a (mouse neuroblastoma) cell lines as *in vitro* neuronal models (Eik et al., 2012; Phan et al., 2012). Treatment of the cells with sclerotial aqueous and ethanol extracts led to a significant increase in the number of neurite-bearing cells compared to the untreated control. The mechanism of the action has not been fully explored, but preliminary results obtained from work using specific inhibitors seem to indicate the involvement of mitogen-activated protein kinase/extracellular-signal regulated kinase (MEK/ERK1/2) signaling pathways (Seow et al., 2013). The chemical components of *L. rhinocerotis* sclerotium with neuritogenic potential, however, have yet to be identified.

7.4 PERSPECTIVES ON MYCELIAL BIOMASS AS A POTENTIAL SUBSTITUTE FOR SCLEROTIA AND FRUITING BODIES

As the main supply of the sclerotia of the SFM mentioned earlier is from their natural habitat, the difficulty in harvesting sufficient amounts of samples has caused researchers to shift the focus to the mycelia as potential substitutes. While the cultivation of some SFM has been established, large-scale production has not been reached to date. Another aspect to be considered would be the chemical composition of the mycelia and culture broth and the associated medicinal properties.

7.4.1 CULTIVATION

The lack of understanding of the life cycle of SFM has emerged as one of the biggest constraints in improving the cultivation process. The use of solid-substrate fermentation is time consuming, and the fruiting bodies and sclerotia formed are dependent on environmental conditions which might affect the yield and quality of the mushroom products. In most cases, the bottleneck appears to be the way to induce formation of sclerotia in the fastest time possible. Previous work demonstrated that burial of mycelia-colonized substrate bags and continuous watering are some of the commonly used techniques to induce fructification.

The cultivation of *L. rhinocerotis* requires more than a year to obtain the fruiting bodies and sclerotia. Pilot cultivation of *L. rhinocerotis* (Figure 7.1) carried out using an optimized formulation consisting of sawdust, paddy straw, and spent yeast at a ratio of 7.9:1:1 in bags yielded mycelial growth rate of 3.8±0.8 mm/day. Sclerotia formation was induced by burying matured colonized substrate in soil. Sclerotia weighing between 80 and 120 g on a fresh weight basis were formed 3–4 weeks after burial, and this was followed by sporophore formation, 8–12 months after burial (Abdullah et al., 2013). On the contrary, cultivation of some SFM can be relatively easy. Under similar conditions, the fruiting body and sclerotium of *L. tuber-regium* can be produced in a shorter time (Figure 7.2). In fact, *L. tuber-regium* can be cultivated by planting the sclerotium in damp soil, and in some cases a sclerotium can be peeled and both the inner, edible white portion and outer peeling can independently produce mushrooms.

As opposed to solid-substrate fermentation, liquid fermentation enables year-round production in a limited space. Mycelia formed by growing pure cultures in submerged conditions are of constant composition, and submerged culture is considered to be the best technique for obtaining consistent and safe mushroom products. Several authors have reviewed the use of liquid fermentation techniques for the production of mycelial biomass and culture broth (Zhong and Tang, 2004; Gregori et al., 2007; Tang et al.,

FIGURE 7.1

Cultivated *Lignosus rhinocerotis*. (A) Development of fruiting bodies after burial of mycelia-colonized substrate blocks. (B) Mushrooms harvested at different developmental stages.

FIGURE 7.2

Cultivated *Lentinus tuber-regium*. Two fruiting bodies developed from a single sclerotium.

2007; Elisashvili, 2012). The optimization of culture conditions to obtain high mycelial biomass and bioactive compounds, such as endo- and exopolysaccharides has been intensively studied (Wu et al., 2003, 2004). The potential of using liquid fermentation for the production of mycelial biomass is enormous.

7.4.2 CHEMICAL CONSTITUENTS

From a nutritional point of view, the chemical composition of mycelia, in many cases, is comparable to that of the sclerotium. As reported by Lau et al. (2013a), the proximate composition of the mycelium of *L. rhinocerotis* was found to be comparable to that of the sclerotium, in which carbohydrates were dominant with smaller amount of proteins and low levels of fat. However, there are also differences in the bioactive constituents when SFM are cultured by liquid fermentation compared to the fruiting bodies and sclerotia derived from solid-substrate fermentation; for instance, Zheng et al. (2008) reported that phenolics detected in submerged cultures of *I. obliquus* consisted of flavonoids, small amounts of melanins, and trace elements of hispidin analogs, whereas in the sclerotium, hispidin analogs and melanins were dominant.

While the factors that influence the production of intracellular and extracellular polysaccharides have been extensively studied, similar work on the secondary metabolites has received less attention. Studies on the fungal metabolite biosynthesis pathways enable us to understand the production of metabolites under different conditions and to exploit this knowledge to foster the overexpression production of metabolites of interest. The possibility of manipulating culture conditions to alter the fungal biosynthesis pathways for the production of metabolites has been demonstrated by several workers using *I. obliquus* as a model organism. Steroids and phenolics are among the important bioactive compounds in *I. obliquus*, but cultured mycelium has contained lower amount of steroids compared to wild-growing mushrooms (Zheng et al., 2008).

In order to increase the amount of steroids in *I. obliquus* (as a model organism) cultured by submerged fermentation, several strategies have been followed. One plausible method is to mimic its growth condition in the natural habitat by introducing chemical components from host-related species to stimulate mycelial biomass and fungal metabolite production. Wang et al. (2014) reported that aqueous and methanol extracts of birch bark and core resulted in increased steroid content of *I. obliquus* grown in submerged cultures. Production of several steroids, namely botulin, ergosterol, cholesterol, lanosterol, stigmasterol, and sitosterol, was reported to have increased between 45.2% and 166.8% compared to those in the control group, when treated with 0.01 g/L of birch bark aqueous extracts. While the precise molecular mechanism involved in stimulating steroid biosynthesis has not been elucidated, the chemical components of the birch extracts were speculated to play a role.

The production of fungal secondary metabolites is often viewed as a consequence of the environmental stress faced by the fungus in its natural habitat. The production of some metabolites has been associated with a possible role in fungal defense against antagonistic microorganisms. The lack of these environmental stresses in laboratory conditions has been suggested as the likely factor in the lower production of the metabolites. A technique involving coculture of two microorganisms has been studied; the relevance of the presence of another microbe is to induce the biosynthesis of antagonistic metabolites. Coculture of *I. obliquus* and *Phellinus punctatus* was attempted by Zheng et al. (2011b). Under their experimental cultivation, the decrease in mycelial biomass production was accompanied by the accumulation of several metabolites from the classes of phenolics, melanins, and lanostane-type triterpenoids. It was postulated that the gene clusters responsible for production of metabolites with possible roles in defense were upregulated when the mushroom species were cocultured. The ethanol:acetone extracts derived from the mycelia of co-cultures demonstrated higher radical scavenging activities and inhibitory effects against tumor cell proliferation than those from monocultures.

Another method to enhance the production of fungal metabolites is to alter the composition of culture media and conditions. For instance, the effects of culture media on the accumulation of steroids in the mycelium of *I. obliquus* have been studied by Zheng et al. (2008). It was reported that culture media consisting (w/v) of glucose (1.5%), rice powder (0.5%), yeast extract (0.4%), wheat bran (0.1%), KH_2PO_4 (0.01%), and $MgSO_4 \cdot 7H_2O$ (0.05%) with pH adjusted to 6.5 yielded the maximum production of lanosterol and ergosterol. Further, supplementation of metal ions such as Ag^+, Cu^+, and Ca^{2+} had positive effects on the accumulation of the sterols, which was thought to be mediated by enhanced expression of the HMG-CoA reductase (3-hydroxy-3-methyl-glutaryl-CoA reductase) gene. Zhu and Xu (2013) demonstrated that several lignocellulosic materials increased the production of intra- and extracellular phenolic compounds in *I. obliquus*; for instance, the production of phenolics was enhanced by the addition of wheat straw (45.3–151.2%), sugarcane bagasse (26.1–106.9%), and rice straw (38.9–67.6%), and the extracts exhibited higher radical scavenging activity than those of the control media. The supplementation of lignocellulosic materials affected the chemical profiles of *I. obliquus*. Davallialactone and inoscavin B were detected in the extracellular phenolic extracts in the lignocellulosic media but not in the control media. The difference in enhancement among the materials was suggested to be related to the content and degradation rate of cellulose, hemicellulose, and lignin.

The production of phenolics by *I. obliquus* in submerged cultures was also enhanced by imposing oxidative stress. As demonstrated by Zheng et al. (2009a,b,c), the addition of H_2O_2 and simultaneous addition of H_2O_2 and arbutin to the culture medium increased the levels of intracellular phenolics but lowered the production of extracellular phenolics. The production of melanins, on the other hand, was improved by the addition of H_2O_2.

The effect of abiotic and biotic factors has been studied as well. The regulation of light on the biosynthesis of phenolics in *I. obliquus* has been investigated by Zheng et al. (2009a,b,c). It was demonstrated that the production of davallialactone, phelligridins, and other hispidin analogs was inhibited by daylight illumination, whereas their biosynthesis was stimulated under continuous darkness. In addition, both compounds were detected in mycelium grown in blue and red light with levels lower than those found in darkness. Polyphenols produced under daylight conditions were reported to exhibit lower antioxidant capacity than those produced with other light regimes. Zheng et al. (2009a,b,c) investigated the role of NO in the production of phenolics in *I. obliquus*. The use of a fungal elicitor in the form of cell debris from the plant-pathogenic ascomycete *Alternaria alternata* increased the production of NO, and this was accompanied by increases in the levels of hispidin analogs such as inoscavins, phelligridins, davallilactone, and methyldavallialactone. It was deduced that the production of phenolics in *I. obliquus* was mediated by signaling pathways independent of oxylipins or jasmonic acid, a different mechanism from those in some higher plants. Further studies demonstrated that higher NO levels coincided with accumulation of *S*-nitrosothiols (SNO) and higher activity of denitrosylated *S*-nitrosoglutathione reductase (GSNOR) and thioredoxin reductase (TrxR) (Zheng et al., 2011a). The state of *S*-nitrosylation and denitrosylation affects the accumulation of phenolics and the fungal metabolic profile.

7.4.3 COMPARATIVE BIOLOGICAL ACTIVITIES

Several comparative analyses on the biological activities of extracts of the mycelium, culture broth, and/or fruiting bodies of a number of SFM have been reported. Findings have indicated that these extracts exhibited biological activities comparable to the sclerotium; in some cases, these are even more potent than the extracts prepared from the sclerotium; for instance, Zhang et al. (2004) compared the antitumor activities of the hot aqueous extracts of the sclerotium and mycelium of *L. tuber-regium*. The mycelial extracts (65.4%) exhibited higher inhibition ratios than those of the sclerotium (55.3%) when tested against sarcoma 180 solid tumor grown in Balb/c mice at a dosage of 20 mg/kg for 10 days. The mycelial extracts (83–65%) revealed a higher inhibition ratio against HL-60 leukemic cells than sclerotial extract (approximately 30%) at a concentration of 50–200 μg/mL. As for their antioxidative effects, the polysaccharide-enriched extracts of the mycelium and culture broth of *I. obliquus* demonstrated relatively higher antioxidant capacity than that of the natural sclerotia.

The extra- and intracellular polysaccharides showed stronger hydroxyl radical and lipid peroxidation inhibitory effects, but lower DPPH quenching activity, than the sclerotial polysaccharides. Lau et al. (2014) compared the biological activities of the sclerotium with the mycelium and culture broth produced from shaken and static conditions of liquid fermentation. It was demonstrated that the aqueous methanol extracts of the mycelium and culture broth showed either higher or comparable antioxidant capacity compared to the extract of the sclerotium. The extract of culture broth from static liquid fermentation showed the highest radical scavenging, reducing, metal chelating, and inhibitory effects on lipid peroxidation. On the other hand, the effect of the extracts on the cellular viability of mammalian cells was found to be comparable. The extracts were considered to be noncytotoxic (IC_{50} > 200 μg/mL, 72 h) against a panel of tumorigenic and nontumorigenic cell lines. The aqueous extract of *L. rhinocerotis* sclerotium was reported to stimulate neurite outgrowth in PC-12 cells at a low concentration of 20 μg/mL (Eik et al., 2012).

Mycelial extracts of the same species cultured in liquid fermentation also retained their neuritogenic properties, as demonstrated by John et al. (2013). In another study, the polysaccharide extracts and fractions of the sclerotium and mycelium of *I. obliquus* enhanced cell proliferation and stimulated the secretion of cytokines of human PBMC (Xu et al., 2014); again, this seemed to indicate that mycelial extracts might exhibit bioactivities similar to those of the sclerotium. Table 7.1 depicts the biological activities of SFM from different developmental stages, i.e., fruiting body, sclerotium, as well as mycelium and culture broth from liquid fermentation. One important question is whether the observation is due to different active compounds or similar compounds in different concentrations; this remains unclear, and extensive chemical profiling should be carried out to ascertain this.

7.5 FUTURE PERSPECTIVES

Several aspects have been identified for further research. As the medicinal properties of SFM have been scientifically validated, there is a need to explore the active compounds and their mechanisms of action to provide a better understanding. Second, cultivation of SFM should be developed so that higher yields are obtained and shorter amounts of time are needed for fructification and formation of sclerotium. The use of inducers might have an effect on the formation of sclerotium. It was mentioned that mushroom mycelia are "waves of the future" as these represent consistent quality. Third, there seems to be a gap in the current findings on the medicinal properties of SFM from different developmental stages; for instance, most SFM have yet to be screened for biological activities such as neurite-outgrowth stimulatory and antihypertensive activities.

While there is work that describes similar medicinal properties of the mycelia as compared to the sclerotia and fruiting bodies, this cannot be generalized for all species; for example, the diuretic activity

Table 7.1 Summary of the Biological Activities of Selected Sclerotium-Forming Mushrooms at Different Developmental/Morphological Stages

	Sclerotium-Forming Mushrooms														
	Lentinus tuber-regium			*Polyporus umbellatus*			*Inonotus obliquus*			*Wolfiporia cocos*			*Lignosus rhinocerotis*		
Bioactivity	**FB**	**SC**	**MB**	**FB**	**SC**	**MB**	**FB**	**SC**	**MB**	**FB**	**SC**	**MB**	**FB**	**SC**	**MB**
Antitumor	/	/	/	/	/	/	–	/	/	/	/	/	–	/	/
Immunomodulatory	–	/	–	/	/	/	–	/	/	/	/	/	–	/	–
Anti-inflammatory	–	–	/	–	/	–	/	/	–	–	/	/	–	/	–
Antioxidative	–	/	–	–	/	–	–	/	/	–	/	–	–	/	/
Antimicrobial	/	/	–	/	–	/	–	/	–	–	/	–	–	/	–
Antihypertensive	–	/	–	–	–	–	–	–	–	–	–	–	–	–	–
Antidiabetic	–	/	/	–	–	–	–	–	/	–	/	–	–	–	–
Diuretic	–	–	–	–	/	–	–	–	–	–	/	–	–	–	–
Neuritogenic	–	–	–	–	–	–	–	–	–	–	–	–	–	/	/

"/" denotes the presence of scientific reports on the biological activity. "–" denotes the absence of scientific reports on the biological activity. Mushrooms at different developmental stages: FB, fruiting body; SC, sclerotium; MB, mycelium and culture broth from liquid fermentation.

of *P. umbellatus* and *W. cocos* mycelial extracts has yet to be demonstrated. In another study by Zheng et al. (2008), it was reported that the extract of *I. obliquus* sclerotium exhibited two-fold higher capacity in inhibiting CYP-induced reduction of body weight, spleen index, and viability of peripheral lymphocytes than those of the culture filtrate. Since the concentration of some hispidin analogs and melanins was higher in the extract of sclerotium, these were postulated to contribute to the observed effect. Further work is needed to confirm this and to establish the relative potency of extracts and chemical constituents from mycelia as opposed to the sclerotia and fruiting bodies.

7.6 CONCLUSIONS

The vast amount of scientific data provides evidence that SFM are emerging sources of bioactive compounds with various biological activities and that these can be potential lead structures for the development of drugs. The possibility of cultivating SFM on a large scale through solid-substrate fermentation to produce sclerotia and fruiting bodies is high, as proved by the artificial cultivation of *L. rhinocerotis*; however, the length of time required makes mushroom mycelium from liquid fermentation a viable alternative to both native and cultivated sclerotia. The evidence presented above also reveals that mycelium is on par with the sclerotia in terms of chemical composition and bioactivity. This is because mycelium possesses all the outstanding properties of the mushrooms but in even stronger form because it is the substance that creates and feeds the fungal bodies; however, more research is needed to optimize the yield of target compounds found in nature by controlling the time of growth to harvest, physical and chemical requirements of growth, and effects of inducers.

ACKNOWLEDGMENT

The authors would like to acknowledge University of Malaya for research grants (PV097-2011A and RU017-2013), and the Mushroom Research Centre and Glami Lemi Biotechnology Research Centre of University of Malaya for providing the research facilities.

REFERENCES

Abdullah, N., Dzul Haimi, M.Z., Lau, B.F., Annuar, M.S.M., 2013. Domestication of a wild medicinal sclerotial mushroom, *Lignosus rhinocerotis* (Cooke) Ryvarden. Ind. Crops Prod. 47, 256–261.

Afieroho, O.E., Ugoeze, K.C., 2014. Gas chromatography-mass spectroscopic (GC-MS) analysis of *n*-hexane extract of *Lentinus tuber-regium* (Fr) (Polyporaceae) Syn *Pleurotus tuber-regium* Fr sclerotia. Trop. J. Pharm. Res. 13 (11), 1911–1915.

Akindahunsi, A.A., Oyetayo, F.L., 2006. Nutrient and antinutrient distribution of edible mushroom, *Pleurotus tuber-regium* (Fries) Singer. LWT-Food Sci. Technol. 39 (5), 548–553.

Chang, H.H., Yeh, C.H., Sheu, F., 2009. A novel immunomodulatory protein from *Poria cocos* induces Toll-like receptor 4-dependent activation within mouse peritoneal macrophages. J. Agric. Food Chem. 57 (14), 6129–6139.

Chen, X., Xu, X., Zhang, L., Zeng, F., 2009. Chain conformation and anti-tumor activities of phosphorylated (1→3)-β-D-glucan from *Poria cocos*. Carbohydr. Polym. 78 (3), 581–587.

Chen, Y.Y., Chang, H.M., 2004. Antiproliferative and differentiating effects of polysaccharide fraction from fu-ling (*Poria cocos*) on human leukemic U937 and HL-60 cells. Food Chem. Toxicol. 42 (5), 759–769.

Cheung, P.C.K., 2013. Mini-review on edible mushrooms as source of dietary fiber: preparation and health benefits. Food Sci. Hum. Wellness 2 (3-4), 162–166.

Choi, K.D., Lee, K.T., Shim, J.O., Lee, Y.S., Lee, T.S., Lee, S.S., et al., 2003. A new method for cultivation of sclerotium of *Grifola umbellata*. Mycobiology 31 (2), 105–112.

Choong, Y.K., Xu, C.H., Lan, J., Chen, X.D., Jamal, J.A., 2014. Identification of geographical origin of lignosus samples using Fourier transform infrared and two-dimensional infrared correlation spectroscopy. J. Mol. Struct. 1069, 188–195.

Coley-Smith, J.R., Cooke, R.C., 1971. Survival and germination of fungal sclerotia. Annu. Rev. Phytopathol. 9, 65–92.

Cuellar, M.J., Giner, R.M., Recio, M.C., Just, M.J., Manez, S., Rios, J.L., 1997. Effect of the Basidiomycete *Poria cocos* on experimental dermatitis and other inflammatory conditions. Chem. Pharm. Bull. 45 (3), 492–494.

Cui, Y., Kim, D.S., Park, K.C., 2005. Antioxidant effect of *Inonotus obliquus*. J. Ethopharmacol. 96, 79–85.

Deng, C., Yang, X., Gu, X., Wang, Y., Zhou, J., Xu, H., 2000. A β-D-glucan from the sclerotia of *Pleurotus tuber-regium* (Fr.) Sing. Carbohydr. Res. 328 (4), 629–633.

Eik, L.F., Naidu, M., David, P., Wong, K.H., Tan, Y.S., Sabaratnam, V., 2012. *Lignosus rhinocerus* (Cooke) Ryvarden: a medicinal mushroom that stimulates neurite outgrowth in PC-12 cells. Evid. Based Complement. Altern. Med. Article ID 320308, http://dx.doi.org/10.1155/2012/320308.

Elisashvili, V., 2012. Submerged cultivation of medicinal mushrooms: bioprocesses and products (review). Int. J. Med. Mushrooms 14 (3), 211–239.

Ezeronye, O.U., Okwujiako, D.A.S., Onumajuru, I.A.I.C., 2005. Antibacterial effect of crude polysaccharide extracts from sclerotium and fruitbody (sporophore) of *Pleurotus tuber-regium* (Fried) Singer on some clinical isolates. Int. J. Mol. Med. Adv. Sci. 1 (3), 202–205.

Fan, L., Ding, S., Ai, L., Deng, K., 2012. Antitumor and immunomodulatory activity of water-soluble polysaccharide from *Inonotus obliquus*. Carbohydr. Polym. 90 (2), 870–874.

Fasidi, I.O., Ekuere, U.U., 1993. Studies on *Pleurotus tuber-regium* (Fries) Singer: cultivation, proximate composition and mineral contents of sclerotia. Food Chem. 48 (3), 255–258.

Fasidi, I.O., Olorunmaiye, K.S., 1994. Studies on the requirements for vegetative growth of *Pleurotus tuber-regium* (Fr.) Singer, a Nigerian mushroom. Food Chem. 50 (4), 397–401.

Feng, Y.L., Lei, P., Tian, T., Yin, L., Chen, D.Q., Chen, H., et al., 2013. Diuretic activity of some fractions of the epidermis of *Poria cocos*. J. Ethnopharmacol. 150 (3), 1114–1118.

Fuchs, S.M., Heinemann, C., Schliemann-Willers, S., Hartl, H., Fluhr, J.W., Elsner, P., 2006. Assessment of anti-inflammatory activity of *Poria cocos* in sodium lauryl sulphate-induced irritant contact dermatitis. Skin Res. Technol. 12 (4), 223–227.

Giavasis, I., 2014. Bioactive fungal polysaccharides as potential functional ingredients in foods and nutraceuticals. Curr. Opin. Biotechnol. 26, 162–173.

Gregori, A., Svagelj, M., Pohleven, J., 2007. Cultivation techniques and medicinal properties of *Pleurotus* spp. Food Technol. Biotechnol. 45 (3), 238–249.

Guo, C., Wong, K.H., Cheung, P.C.K., 2011. Hot water extract of the sclerotium of *Polyporus rhinocerus* Cooke enhances the immune functions of murine macrophages. Int. J. Med. Mushrooms 13 (3), 237–244.

Handa, N., Yamada, T., Tanaka, R., 2010. An unusual lanostane-type triterpenoid, spiroinonotsuoxodiol, and other triterpenoids from *Inonotus obliquus*. Phytochemistry 71, 1774–1779.

Ikewuchi, J.C., Ikewuchi, C.C., Ifeanacho, M.O., Igboh, N.M., Ijeh, I.I., 2013. Moderation of hematological and plasma biochemical indices of sub-chronic salt-loaded rats by aqueous extract of the sclerotia of *Pleurotus tuberregium* (Fr) Sing's: implications for the reduction of cardiovascular risk. J. Ethnopharmacol. 150 (2), 466–476.

Ikewuchi, J.C., Ikewuchi, C.C., Ifeanacho, M.O., 2014. Blood pressure lowering activity of a flavonoid and phytosterol rich extract of the sclerotia of *Pleurotus tuberregium* (Fr) Sing in salt-loaded rats. Biomed. Prev. Nutr. 4 (2), 257–263.

Isikhuemhen, O.S., Okhuoya, J.A., 1995. A low-cost technique for the cultivation of *Pleurotus tuberregium* (Fr.) Singer in developing tropical countries. Mushroom Growers Newslett. 4, 2–4.

Isikhuemhen, O.S., Nerud, F., Vigalys, R., 2000. Cultivation studies on wild and hybrid strains of *Pleurotus tuber-regium* (Fr.) Sing. on wheat straw substrate. World J. Microbiol. Biotechnol. 16 (5), 431–435.

Jeong, J.W., Lee, H.H., Han, M.H., Kim, G.Y., Hong, S.H., Park, C., et al., 2014. Ethanol extract of *Poria cocos* reduces the production of inflammatory mediators by suppressing the NF-kappaB signaling pathway in lipopolysaccharide-stimulated RAW 264.7 macrophages. BMC Complement. Altern. Med. http://dx.doi.org/10.1186/1472-6882-14-101.

John, P.A., Wong, K.H., Naidu, M., Sabaratnam, V., David, P., 2013. Combination effects of curcumin and aqueous extract of *Lignosus rhinocerotis* mycelium on neurite outgrowth stimulation activity in PC-12 cells. Nat. Prod. Commun. 8 (6), 711–714.

Jonathan, S.G., Kigigha, L.T., Ohimain, E., 2008. Evaluation of the inhibitory potentials of eight higher Nigerian fungi against pathogenic microorganisms. Afr. J. Biomed. Res. 11, 197–202.

Kikuchi, T., Uchiyama, E., Ukiya, M., Tabata, K., Kimura, Y., Suzuki, T., et al., 2011. Cytotoxic and apoptosis-inducing activities of triterpene acids from *Poria cocos*. J. Nat. Prod. 74 (2), 137–144.

Kim, H.G., Yoon, D.H., Kim, C.H., Shrestha, B., Chang, W.C., Lim, S.Y., et al., 2007. Ethanol extract of *Inonotus obliquus* inhibits lipopolysaccharide-induced inflammation in RAW 264.7 macrophage cells. J. Med. Food 10 (1), 80–89.

Kim, Y.R., 2005. Immunomodulatory activity of the water extract from medicinal *mushroom Inonotus obliquus*. Mycobiology 33 (3), 158–162.

Kubo, T., Terabayashi, S., Takeda, S., Sasaki, H., Aburada, M., Miyamoto, K.I., 2006. Indoor cultivation and cultural characteristics of *Wolfiporia cocos* sclerotia using mushroom culture bottles. Biol. Pharm. Bull. 29 (6), 1191–1196.

Kuforiji, O.O., Fasidi, I.O., 2009. Biodegradation of agro-industrial wastes by an edible mushroom *Pleurotus tuber-regium* (Fr.). J. Environ. Biol. 30 (3), 355–358.

Lai, C.K.M., Wong, K.H., Cheung, P.C.K., 2008. Antiproliferative effects of sclerotial polysaccharides from *Polyporus rhinocerus* Cooke (Aphyllophoromycetideae) on different kinds of leukemic cells. Int. J. Med. Mushrooms 10 (3), 255–264.

Lau, B.F., Abdullah, N., Aminudin, N., 2013a. Chemical composition of the tiger's milk mushroom, *Lignosus rhinocerotis* (Cooke) Ryvarden, from different developmental stages. J. Agric. Food Chem. 61 (20), 4890–4897.

Lau, B.F., Abdullah, N., Aminudin, N., Lee, H.B., 2013b. Chemical composition and cellular toxicity of ethnobotanical-based hot and cold aqueous preparations of the tiger's milk mushroom (*Lignosus rhinocerotis*). J. Ethnopharmacol. 150 (1), 252–262.

Lau, B.F., Abdullah, N., Aminudin, N., Lee, H.B., Yap, K.C., 2014. The potential of mycelium and culture broth of *Lignosus rhinocerotis* as substitutes for the naturally occurring sclerotium with regard to antioxidant capacity, cytotoxic effect, and low-molecular-weight chemical constituents. PLoS One 9 (7), e102509. http://dx.doi.org/10.1371/journal.pone.0102509.

Lee, K.Y., Jeon, Y.J., 2003. Polysaccharide isolated from *Poria cocos* sclerotium induces NF-κB/Rel activation and iNOS expression in murine macrophages. Int. Immunopharmacol. 3 (10–11), 1353–1362.

Lee, M.L., Tan, N.H., Fung, S.Y., Tan, C.S., Ng, S.T., 2012. The antiproliferative activity of sclerotia of *Lignosus rhinocerus* (tiger milk mushroom). Evid. Based Complement. Altern. Med. Article ID 697603, http://dx.doi.org/10.1155/2012/697603.

Lee, S.S., Chang, Y.S., Noraswati, M.N.R., 2009. Utilization of macrofungi by some indigenous communities for food and medicine in Peninsular Malaysia. Forest Ecol. Manag. 257 (10), 2062–2065.

Lee, S.S., Tan, N.H., Fung, S.Y., Sim, S.M., Tan, C.S., Ng, S.T., 2014. Anti-inflammatory effect of the sclerotium of *Lignosus rhinocerotis* (Cooke) Ryvarden, the tiger milk mushroom. BMC Complement. Altern. Med. 14, 359. http://dx.doi.org/10.1186/1472-6882-14-359.

Li, G., Xu, M.L., Lee, C.S., Woo, M.H., Chang, H.W., Son, J.K., 2004. Cytotoxicity and DNA topoisomerases inhibitory activity of constituents from the sclerotium of *Poria cocos*. Arch. Pharm. Res. 27 (8), 829–833.

Li, T.H., Hou, C.C., Chang, C.L.T., Yang, W.C., 2011. Anti-hyperglycemic properties of crude extract and triterpenes from *Poria cocos*. Evid. Based Complement. Altern. Med. Article ID 128402, http://dx.doi.org/10.1155/2011/128402.

Li, X., Xu, W., Chen, J., 2010. Polysaccharide purified from *Polyporus umbellatus* (Per) Fr induces the activation and maturation of murine bone-derived dendritic cells via toll-like receptor 4. Cell. Immunol. 265 (1), 50–56.

Liang, L., Zhang, Z., Wang, H., 2009. Antioxidant activities of extracts and subfractions from *Inonotus obliquus*. Int. J. Food Sci. Nutr. 60 (Suppl. 2), 175–184.

Lin, S., Lai, T.C., Chen, L., Kwok, H.F., Lau, C.B., Cheung, P.C., 2014. Antioxidant and antiangiogenic properties of phenolic extract from *Pleurotus tuber-regium*. J. Agric. Food Chem. 62 (39), 9488–9498.

Ma, C.Y., Chang, W.C., Chang, H.M., Wu, J.S.B., 2010. Immunomodulatory effect of the polysaccharide-rich fraction from sclerotium of medicinal mushroom *Poria cocos* F.A. Wold (Aphyllophoromycetideae) on Balb/c mice. Int. J. Med. Mushrooms 12 (2), 111–121.

Ma, L., Chen, H., Dong, P., Lu, X., 2013. Anti-inflammatory and anticancer activities of extracts and compounds from the mushroom *Inonotus obliquus*. Food Chem. 139, 503–508.

Mizushina, Y., Akihisa, T., Ukiya, M., Murakami, C., Kuriyama, I., Xu, X., et al., 2004. A novel DNA topoisomerase inhibitor: dehydroebriconic acid, one of the lanostane-type triterpene acids from *Poria cocos*. Cancer Sci. 95 (4), 354–360.

Mohanarji, S., Dharmalingam, S., Kalusalingam, A., 2012. Screening of *Lignosus rhinocerus* extracts as antimicrobial agents against selected human pathogens. J. Pharm. Biomed. Sci. 18 (11), 1–4.

Moradali, M.F., Mostafavi, H., Ghods, S., Hedjaroude, G.A., 2007. Immunomodulating and anticancer agents in the realm of macromycetes fungi (macrofungi). Int. Immunopharmacol. 7 (6), 701–724.

Mu, H., Zhang, A., Zhang, W., Cui, G., Wang, S., Duan, J., 2012. Antioxidative properties of crude polysaccharides from *Inonotus obliquus*. Int. J. Mol. Sci. 13 (7), 9194–9206.

Nakajima, Y., Satom, Y., Konishi, T., 2007. Antioxidant small phenolic ingredients *in Inonotus obliquus* (person) Pilat (Chaga). Chem. Pharm. Bull. 55 (8), 1222–1226.

Nakata, T., Yamada, T., Taji, S., Ohishi, H., Wada, S., Tokuda, H., et al., 2007. Structure determination of inonotsuoxides A and B and *in vivo* anti-tumor promoting activity of inotodiol from the sclerotia of *Inonotus obliquus*. Bioorgan. Med. Chem. 15 (1), 257–264.

Ning, X., Luo, Q., Li, C., Ding, Z., Pang, J., Zhao, C., 2014. Inhibitory effects of a polysaccharide extract from the chaga medicinal mushroom, *Inonotus obliquus* (higher Basidiomycetes) on the proliferation of human neurogliocytoma cells. Int. J. Med. Mushrooms 16 (1), 29–36.

Nwokolo, E., 1987. Composition of nutrients in the sclerotium of the mushroom Pleurotus tuber regium. Plant Foods Hum. Nutr. 37 (2), 133–139.

Okhuoya, J.A., Etugo, J.E., 1993. Studies of the cultivation of *Pleurotus tuberregium* (FR) sing. An edible mushroom. Bioresour. Technol. 44 (1), 1–3.

Oso, B.A., 1977. *Pleurotus tuber-regium* from Nigeria. Mycologia 67, 271–279.

Park, Y.H., Son, I.H., Kim, B., Lyu, Y.S., Moon, H.I., Kang, H.W., 2009. *Poria cocos* water extract protects PC12 neuronal cells from beta-amyloid-induced cell death through antioxidant and antiapoptotic functions. Pharmazie 64 (11), 760–764.

Patel, S., 2015. Chaga (*Inonotus obliquus*) mushroom: nutraceuticals assessment based on latest findings. In: Patel, S. (Ed.), Emerging Bioresources with Nutraceutical and Pharmaceutical Prospects Springer International Publishing, Switzerland, pp. 115–126.

Patel, S., Goyal, A., 2012. Recent developments in mushrooms as anti-cancer therapeutics: a review. 3 Biotech 2 (1), 1–15.

Phan, C.W., Wong, W.L., David, P., Naidu, M., Sabaratnam, V., 2012. *Pleurotus giganteus* (Berk.) Karunarathna & K.D. Hyde: Nutritional value and *in vitro* neurite outgrowth activity in rat pheochromocytoma cells. BMC Complement. Altern. Med. 12, 102. http://dx.doi.org/10.1186/1472-6882-12-102.

Ren, L., Perera, C., Hemar, Y., 2012. Antitumor activity of mushroom polysaccharides: a review. Food Funct. 3, 1118–1130.

Rios, J.L., 2011. Chemical constituents and pharmacological properties of *Poria cocos*. Planta Med. 77 (7), 681–691.

Ruthes, A.C., Smiderle, F.R., Iacomini, M., 2015. D-Glucans from edible mushrooms: a review on the extraction, purification and chemical characterization approaches. Carbohydr. Polym. 117, 753–761.

Seow, S.L.S., Naidu, M., David, P., Wong, K.H., Sabaratnam, V., 2013. Tiger's milk mushrooms – Nature's hidden treasure that promotes neuro health. International Functional Food Conference, 18–20th August 2013. Cyberjaya Resort and Spa, Cyberjaya, Malaysia, p. 38.

Shashkina, M.Y., Shashkin, P.N., Sergeev, A.V., 2006. Chemical and medicobiological properties of Chaga (review). Pharm. Chem. J. 40 (10), 560–568.

Smith, M.E., Henkel, T.W., Rollins, J.A., 2015. How many fungi make sclerotia? Fungal Ecol. 13, 211–220.

Song, F.Q., Liu, Y., Kong, X.S., Chang, W., Song, G., 2013. Progress on understanding the anticancer mechanisms of medicinal mushroom: *Inonotus obliquus*. Asian Pac. J. Cancer Prev. 14 (3), 1571–1578.

Sun, Y., Yasukawa, K., 2008. New anti-inflammatory ergostane-type ecdysteroids from the sclerotium of *Polyporus umbellatus*. Bioorgan. Med. Chem. Lett. 18 (11), 3417–3420.

Sun, Y., Yin, T., Chen, X.H., Zhang, G., Curtis, R.B., Lu, Z.H., et al., 2011. *In vitro* antitumor activity and structure characterization of ethanol extracts from wild and cultivated Chaga medicinal mushroom, *Inonotus obliquus* (Pers.:Fr.) Pilát (Aphyllophoromycetideae). Int. J. Med. Mushrooms 13 (2), 121–130.

Tang, J., Nie, J., Li, D., Zhu, W., Zhang, S., Ma, F., et al., 2014. Characterization and antioxidant activities of degraded polysaccharides from *Poria cocos* sclerotium. Carbohydr. Polym. 105, 121–126.

Tang, Y.J., Zhu, L.W., Li, H.M., Li, D.S., 2007. Submerged culture of mushrooms in bioreactors – Challenges, current state-of-the-art, and future prospects. Food Technol. Biotechnol. 45 (3), 221–229.

Tao, Y., Zhang, L., Cheung, P.C.K., 2006. Physicochemical properties and antitumor activities of water-soluble native and sulfated hyperbranched mushroom polysaccharides. Carbohydr. Res. 341 (13), 2261–2269.

Tao, Y., Zhang, Y., Zhang, L., 2009. Chemical modification and antitumor activities of two polysaccharide-protein complexes from *Pleurotus tuber-regium*. Int. J. Biol. Macromol. 45 (2), 109–115.

Townsend, B.B., Willetts, H.J., 1954. The development of sclerotia of certain fungi. Trans. Br. Mycol. Soc. 37 (3), 213–221.

Wang, L.X., Lu, Z.M., Geng, Y., Zhang, X.M., Xu, G.H., Shi, J.S., et al., 2014. Stimulated production of steroids in *Inonotus obliquus* by host factors from birch. J. Biosci. Bioeng. 118 (6), 728–731.

Wang, Y., Zhang, L., Li, Y., Hou, X., Zeng, F., 2004. Correlation of structure to antitumor activities of five derivatives of a β-glucan from *Poria cocos* sclerotium. Carbohydr. Res. 339 (15), 2567–2574.

Wang, Y.Z., Zhang, J., Zhao, Y.L., Li, T., Shen, T., Li, J.Q., et al., 2013. Mycology, cultivation, traditional uses, phytochemistry and pharmacology of *Wolfiporia cocos* (Schwein.) Ryvarden et Gilb.: a review. J. Ethnopharmacol. 147 (2), 265–276.

Wasser, S., 2002. Medicinal mushrooms as a source of antitumor and immunomodulating polysaccharides. Appl. Microbiol. Biotechnol. 60 (3), 258–274.

Wasser, S.P., Weis, A.L., 1999. Medicinal properties of substances occurring in higher Basidiomycetes mushrooms: current perspectives (review). Int. J. Med. Mushrooms 1 (1), 31–62.

Willets, H.J., Bullock, S., 1992. Developmental biology of sclerotia. Mycol. Res. 96, 801–816.

Wong, K.H., Cheung, P.C.K., 2005a. Dietary fibers from mushroom sclerotia: 1. Preparation and physicochemical and functional properties. J. Agric. Food Chem. 53 (24), 9395–9400.

Wong, K.H., Cheung, P.C.K., 2005b. Dietary fibers from mushroom sclerotia: 2. *In vitro* mineral binding capacity under sequential simulated physiological conditions of the human gastrointestinal tract. J. Agric. Food Chem. 53 (24), 9401–9406.

Wong, K.H., Cheung, P.C.K., 2008a. Sclerotium of culinary-medicinal king tuber oyster mushroom, *Pleurotus tuberregium* (Fr.) Singer (Agaricomycetideae): its cultivation, biochemical composition, and biopharmacological effects (review). Int. J. Med. Mushrooms 10 (4), 303–313.

Wong, K.H., Cheung, P.C.K., 2008b. Sclerotia: emerging functional food derived from mushrooms. In: Cheung, P.C.K. (Ed.), Mushrooms as Functional Foods John Wiley & Sons, Inc, Hoboken, New Jersey, pp. 111–146.

Wong, K.H., Cheung, P.C.K., Wu, J.Z., 2003. Biochemical and microstructural characteristics of insoluble and soluble dietary fiber prepared from mushroom sclerotia *of Pleurotus tuber-regium*, *Polyporus rhinocerus*, and *Wolfiporia cocos*. J. Agric. Food Chem. 51 (24), 7197–7202.

Wong, K.H., Wong, K.Y., Kwan, H.S., Cheung, P.C.K., 2005. Dietary fibers from mushroom sclerotia: 3. *In vitro* fermentability using human fecal microflora. J. Agric. Food Chem. 53 (24), 9407–9412.

Wong, K.H., Katsumata, S., Masuyama, R., Uehara, M., Suzuki, K., Cheung, P.C.K., 2006. Dietary fibers from mushroom sclerotia. 4. *In vivo* mineral absorption using ovariectomized rat model. J. Agric. Food Chem. 54 (5), 1921–1927.

Wong, K.H., Lai, C.K.M., Cheung, P.C.K., 2009. Stimulation of human innate immune cells by medicinal mushroom sclerotial polysaccharides. Int. J. Med. Mushrooms 11 (3), 215–223.

Wong, K.H., Lai, C.K.M., Cheung, P.C.K., 2011. Immunomodulatory activities of mushroom sclerotial polysaccharides. Food Hydrocoll. 25 (2), 150–158.

Wong, S.M., Wong, K.K., Chiu, L.C.M., Cheung, P.C.K., 2007. Non-starch polysaccharides from different developmental stages of *Pleurotus tuber-regium* inhibited the growth of human acute promyelocytic leukemia HL-60 cells by cell-cycle arrest and/or apoptotic induction. Carbohydr. Polym. 68 (2), 206–217.

Wu, G.H., Hu, T., Li, Z.Y., Huang, Z.L., Jiang, J.G., 2014. *In vitro* antioxidant activities of the polysaccharides from *Pleurotus tuber-regium* (Fr.) Sing. Food Chem. 148, 351–356.

Wu, J.Z., Cheung, P.C.K., Wong, K.H., Huang, N.L., 2003. Studies on submerged fermentation of *Pleurotus tuber-regium* (Fr.) Singer—part 1: physical and chemical factors affecting the rate of mycelial growth and bioconversion efficiency. Food Chem. 81 (3), 389–393.

Wu, J.Z., Cheung, P.C.K., Wong, K.H., Huang, N.L., 2004. Studies on submerged fermentation of *Pleurotus tuber-regium* (Fr.) Singer. Part 2: effect of carbon-to-nitrogen ratio of the culture medium on the content and composition of the mycelial dietary fibre. Food Chem. 85 (1), 101–105.

Wu, Z.L., Ren, H., Lai, W.Y., Lin, S., Jiang, R.Y., Ye, T.C., et al., 2014. Sclederma of *Poria cocos* exerts its diuretic effect via suppression of renal aquaporin-2 expression in rats with chronic heart failure. J. Ethnopharmacol. 155 (1), 563–571.

Xing, Y.M., Zhang, L.C., Liang, H.Q., Lv, J., Song, C., 2013. Sclerotial formation of *Polyporus umbellatus* by low temperature treatment under artificial conditions. PLoS One 8 (2), e56190. http://dx.doi.org/10.1371/journal.pone.0056190.

Xu, L., Li, J., Hu, Y., 2014. Polysaccharides from *Inonotus obliquus* sclerotia and cultured mycelia stimulate cytokine production of human peripheral blood mononuclear cells *in vitro* and their chemical characterization. Int. Immunopharmacol. 21 (2), 269–278.

Yap, H.Y., Aziz, A.A., Fung, S.Y., Ng, S.T., Tan, C.S., Tan, N.H., 2014. Energy and nutritional composition of Tiger milk mushroom (*Lignosus tigris* Chon S. Tan) sclerotia and the antioxidant activity of its extracts. Int. J. Med. Sci. 11 (6), 602–607.

Yap, Y.H., Tan, N., Fung, S., Aziz, A.A., Tan, C., Ng, S., 2013. Nutrient composition, antioxidant properties, and anti-proliferative activity of *Lignosus rhinocerus* Cooke sclerotium. J. Sci. Food Agric. 93 (12), 2945–2952.

Youn, M.J., Kim, J.K., Park, S.Y., Kim, Y., Kim, S.J., Lee, J.S., et al., 2008. Chaga mushroom (*Inonotus obliquus*) induces G0/G1 arrest and apoptosis in human hepatoma Hep G2 cells. World J. Gastroenterol. 14 (4), 511–517.

Youn, M.J., Kim, J.K., Park, S.Y., Kim, Y., Park, C., Kim, E.S., et al., 2009. Potential anticancer properties of the water extract of *Inonotus obliquus* by induction of apoptosis in melanoma B16-F10 cells. J. Ethnopharmacol. 121 (2), 221–228.

Yun, J.S., Pahk, J.W., Lee, J.S., Shin, W.C., Lee, S.Y., Hong, E.K., 2011. *Inonotus obliquus* protects against oxidative stress-induced apoptosis and premature senescence. Mol. Cells 31 (5), 423–429.

Zhang, G., Zeng, X., Li, C., Li, J., Huang, Y., Han, L., et al., 2011. Inhibition of urinary bladder carcinogenesis by aqueous extract of sclerotia of *Polyporus umbellatus* fries and polyporus polysaccharide. Am. J. Chinese Med. 39 (1), 135–144.

Zhang, M., Cheung, P.C.K., Zhang, L., 2001. Evaluation of mushroom dietary fiber (nonstarch polysaccharides from sclerotia of *Pleurotus tuber-regium* (Fries) Singer as a potential antitumor agent. J. Agric. Food Chem. 49 (10), 5059–5062.

Zhang, M., Zhang, L., Cheung, P.C.K., Ooi, V.E.C., 2004. Molecular weight and anti-tumor activity of the water-soluble polysaccharides isolated by hot water and ultrasonic treatment from the sclerotia and mycelia of *Pleurotus tuber-regium*. Carbohydr. Polym. 56 (2), 123–128.

Zhang, M., Cheung, P.C.K., Chiu, L.C.M., Wong, E.Y.L., Ooi, V.E.C., 2006. Cell-cycle arrest and apoptosis induction in human breast carcinoma MCF-7 cells by carboxymethylated β-glucan from the mushroom sclerotia of *Pleurotus tuber-regium*. Carbohydr. Polym. 66 (4), 455–462.

Zhang, M., Cui, S.W., Cheung, P.C.K., Wang, Q., 2007. Antitumor polysaccharides from mushrooms: a review on their isolation process, structural characteristics and antitumor activity. Trends Food Sci. Technol. 18 (1), 4–19.

Zhang, G., Zeng, X., Han, L., Wei, J.A., Huang, H., 2010. Diuretic activity and kidney medulla AQP1, AQP2, AQP3, V2R expression of the aqueous extract of sclerotia of Polyporus umbellatus FRIES in normal rats. J. Ethnopharmacol. 128 (2), 433–437.

Zhao, F., Mai, Q., Ma, J., Xu, M., Wang, X., Cui, T., et al., 2015. Triterpenoids from *Inonotus obliquus* and their antitumor activities. Fitoterapia 101, 34–40.

Zhao, Y.Y., 2013. Traditional uses, phytochemistry, pharmacology, pharmacokinetics, and quality control of *Polyporus umbellatus* (Pers.) Fries: a review. J. Ethnopharmacol. 149 (1), 35–48.

Zhao, Y.Y., Xie, R.M., Chao, X., Zhang, Y., Lin, R.C., Sun, W.J., 2009a. Bioactivity-directed isolation, identification of diuretic compounds from *Polyporus umbellatus*. J. Ethnopharmacol. 126 (1), 184–187.

Zhao, Y.Y., Yang, L., Wang, M., Wang, L., Cheng, X., Zhang, Y., et al., 2009b. 1B-hydroxylfriedelin, a new natural pentacyclic triterpene from the sclerotia of *Polyporus umbellatus*. J. Chem. Res. 11, 699–701.

Zhao, Y.Y., Chao, X., Zhang, Y., Lin, R.C., Sun, W.J., 2010. Cytotoxic steroids from *Polyporus umbellatus*. Planta Med. 76 (15), 1755–1758.

Zhao, Y.Y., Shen, X., Chao, X., Ho, C.C., Cheng, X.L., Zhang, Y., et al., 2011. Ergosta-4,6,8(14),22-tetraen-3-one induces G2/M cell cycle arrest and apoptosis in human hepatocellular carcinoma Hep G2 cells. Biochim. Biophys. Acta 1810 (4), 384–390.

Zhao, Y.Y., Feng, Y.L., Du, X., Xi, Z.H., Cheng, X.L., Wei, F., 2012. Diuretic activity of the ethanol and aqueous extracts of the surface layer of *Poria cocos* in rat. J. Ethnopharmacol. 144 (3), 775–778.

Zheng, W., Zhang, M., Zhao, Y., Wang, Y., Miao, K., Wei, Z., 2009a. Accumulation of antioxidant phenolic constituents in submerged cultures of *Inonotus obliquus*. Bioresour. Technol. 100 (3), 1327–1335.

Zheng, W., Zhang, M., Zhao, Y., Miao, K., Jiang, H., 2009b. NMR-based metabonomic analysis on effect of light on production of antioxidant phenolic compounds in submerged cultures of *Inonotus obliquus*. Bioresour. Technol. 100 (19), 4481–4487.

Zheng, W., Miao, K., Zhang, Y., Pan, S., Zhang, M., Jiang, H., 2009c. Nitric oxide mediates the fungal-elicitor-enhanced biosynthesis of antioxidant polyphenols in submerged cultures of *Inonotus obliquus*. Microbiology 155 (10), 3440–3448.

Zheng, W., Miao, K., Zhao, Y., Zhang, M., Pan, S., Dai, Y., 2010. Chemical diversity of biologically active metabolites in the sclerotia of *Inonotus obliquus* and submerged culture strategies for up-regulating their production. Appl. Microbiol. Biotechnol. 87 (4), 1237–1254.

Zheng, W., Liu, Y., Pan, S., Yuan, W., Dai, Y., Wei, J., 2011a. Involvements of S-nitrosylation and denitrosylation in the production of polyphenols by *Inonotus obliquus*. Appl. Microbiol. Biotechnol. 90 (5), 1763–1772.

Zheng, W., Zhao, Y., Zheng, X., Liu, Y., Pan, S., Dai, Y., et al., 2011b. Production of antioxidant and antitumor metabolites by submerged cultures of *Inonotus obliquus* cocultured with *Phellinus punctatus*. Appl. Microbiol. Biotechnol. 89 (1), 157–167.

Zheng, W.F., Zhao, Y.X., Zhang, M.M., Yin, Z.J., Chen, C.F., Wei, Z.W., 2008. Phenolic compounds from *Inonotus obliquus* and their immune-stimulating effects. Mycosystema 27 (4), 574–581.

Zheng, Y., Yang, X.W., 2008. Two new lanostane triterpenoids from *Poria cocos*. J. Asian Nat. Prod. Res. 10, 323–328.

Zhong, J.J., Tang, Y.J., 2004. Submerged cultivation of medicinal mushrooms for production of valuable bioactive metabolites. Adv. Biochem. Eng. Biotechnol. 87, 25–59.

Zhong, X.H., Ren, K., Lu, S.J., Yang, S.Y., Sun, D.Z., 2009. Progress of research on *Inonotus obliquus*. Chin. J. Integr. Med. 15 (2), 156–160.

Zhou, W.W., Lin, W.H., Guo, S.X., 2007. Two new polyporusterones isolated from the sclerotia of *Polyporus umbellatus*. Chem. Pharm. Bull. 55 (8), 1148–1150.

Zhu, L., Xu, X., 2013. Stimulatory effect of different lignocellulosic materials for phenolic compound production and antioxidant activity from *Inonotus obliquus* in submerged fermentation. Appl. Biochem. Biotechnol. 169 (7), 2138–2152.

CHAPTER

MEDICINAL MUSHROOMS WITH ANTI-PHYTOPATHOGENIC AND INSECTICIDAL PROPERTIES

8

Gayane S. Barseghyan, Avner Barazani and Solomon P. Wasser
Institute of Evolution, Haifa University, Mt. Carmel Haifa, Israel

8.1 INTRODUCTION

In the early 1950s, the agrochemical industry provided agriculture with a vast array of chemicals for crop protection, including virucides, fungicides, bactericides, insecticides, and nematocides. Pathogenic organisms are mainly controlled chemically; however, the use of synthetic compounds is limited due to several undesirable aspects, such as carcinogenicity, teratogenicity, acute toxicity, and the requirement of an extended degradation period with consequent development of environmental pollution problems (Soković et al., 2013). The new awareness of modern consumers about these problems has created a "green" consumer profile that demands the removal of synthetic chemicals from food production and preservation, together with extended shelf life for the majority of food products. Altogether, this demand forces the scientific community and agroindustrial and pharmaceutical companies to search for natural compounds that will satisfy the consumer (Harvey, 2008). Furthermore, there is growing concern about chemical protection because of their undesirable side effects in humans and other target organisms, and their behavior and fate in the environment (Jespers, 1994). Recently, interest has been growing in natural products derived from mushrooms due to their availability, fewer side effects, and lower toxicity as well as better biodegradability, which is important in the agricultural industry.

The number of mushroom species on Earth is currently estimated at 140,000, yet perhaps only 10% (approximately 14,000 named species) are known to science; 2000 of these are safe for human consumption, and about 660 possess medicinal properties. In the second half of the twentieth century, mushroom-producing technologies have grown enormously, and the value of world mushroom production in 2011 was estimated to be worth about US$60 billion. Many pharmaceutical substances with potent and unique properties have recently been extracted from mushrooms. In particular, and most importantly for modern medicine, medicinal mushrooms contain an unlimited source of polysaccharides and polysaccharide–protein complexes with anticancer and immunostimulating properties. Many, if not all, higher Basidiomycetes mushrooms contain biologically active polysaccharides in their fruit bodies, cultured mycelia, and cultured broth. The data on mushroom polysaccharides today have been summarized for 660 species and intraspecific taxa from 182 higher Hetero- and Homobasidiomycetes.

Mushroom Biotechnology. DOI: http://dx.doi.org/10.1016/B978-0-12-802794-3.00008-4

Numerous excellent scientific investigations and review articles have been published on the subject of biologically active secondary metabolites from higher Basidiomycetes (Anke, 1989; Lorenzen and Anke, 1998; Wasser and Weis, 1999a,b; Wasser, 2002; Brandt and Piraino, 2000; Reshetnikov et al., 2001; Abraham, 2001; Zjawiony, 2004; Rai et al., 2005; Robles-Hernández et al., 2008, Barros et al., 2007; Fagade and Oyelade, 2009, Hearst et al., 2009; Beattie et al., 2010; Fu et al., 2010; Saddiqe et al., 2010; Yu et al., 2011; Li et al., 2012; Wang et al., 2012; Chang and Wasser, 2012; Younis et al., 2015). These investigations have focused exclusively on higher Basidiomycetes, which are considered by many authors to be a major source of pharmacologically active natural products. According to Soković et al. (2013), from higher Basidiomycetes—exemplified by species of the genera *Ganoderma*, *Lactarius*, or *Agaricus*—about 2000 active compounds have been derived.

Medicinal mushrooms offer an advantage in that their active components are safe for humans. Many compounds such as β-D-glucans, heteropolysaccharides, glycoproteins, lectins, and terpenoids inhibit tumor cells and have not shown negative effects on treated patients. The antimicrobial properties of certain Basidiomycetes provide human and plant pathogen control that is generally safe and effective. Several species of Basidiomycetes mushrooms have demonstrated antibacterial activity against human pathogens; others have shown antifungal activity against both human and plant pathogens, while others have inhibited phytopathogenic nematodes.

In this context, systematic screening of secondary metabolites of higher Basidiomycetes may result in the discovery of novel and unique sets of compounds with the potential to address agricultural and medicinal challenges.

This paper gives an overview on the activity of mushroom compounds as well as their chemical composition and potential uses.

8.2 ANTIBACTERIAL METABOLITES

There are numerous publications describing the antibacterial properties of secondary metabolites isolated from various higher Basidiomycetes (Barros et al., 2007; Fagade and Oyelade, 2009, Hearst et al., 2009; Beattie et al., 2010; Fu et al., 2010; Saddiqe et al., 2010; Yu et al., 2011; Li et al., 2012; Wang et al., 2012). The antibacterial activity of some Basidiomycetes mushrooms provides efficient and low-cost methods for human and plant disease control. The highest antibacterial activity occurred among members of the Ganodermatales, Poriales, Agaricales, and Stereales, and these may constitute a good source for developing new antibiotics. But the effect of Basidiomycetes secondary metabolites has been investigated mainly on human and animal disease pathogens. Unfortunately, the publications describing antibacterial properties of isolated secondary metabolites of mushrooms on plant bacteria models are very limited. Despite this, several interesting publications have shown the potential of Basidiomycetes as valuable producers of substances that can be used successfully in the agricultural sector.

Two Basidiomycete mushrooms, *Ganoderma lucidum* and *Laetiporus sulphureus*, showed strong antibacterial activity against *Agrobacterium rhizogenes*, *Agrobacterium tumefaciens*, *Erwinia carotovora* subsp. *carotovora*, *Pseudomonas syringae* pv. *syringae*, and *Xanthomonas campestris* pv. *campestris* (Robles-Hernández, 2004).

Extracts from *Clytocybe geotropa* have shown the broadest range of inhibition against *Ralstonia solanacearum*, *E. carotovora* subsp. *carotovora*, *P. syringae* pv. *syringae*, *X. campestris* pv. *vesicatoria*, and *Clavibacter michiganensis* subsp. *sepedonicus*. Purified protein, **clitocypin** from *C. geotropa*, showed inhibition against *C. michiganensis* subsp. *sepedonicus* when tested on agar plates (Dreo et al., 2007).

Coprinol, isolated from culture fluids of *Coprinus* spp., shows antibacterial activity against most of the plant pathogenic Gram-positive bacteria *in vitro* (Johansson et al., 2001).

The fungicide **strobilurin** F 500 enhances resistance of tobacco to the wild fire pathogen *Pseudomonas syringae* pv. *tabaci*. The mechanism of action of **strobilurin** F 500 is by inducing cellular responses to the pathogen attack. It induces the production of endogenous salicylic acid and pathogenesis-related proteins that usually are used as molecular markers for disease resistance (Herms et al., 2002).

8.3 ANTIFUNGAL AND HERBICIDAL METABOLITES

Naturally occurring substances found in medicinal mushrooms have revealed important sources of molecules with antifungal properties. For example, the secondary metabolites **strobilurin** (Table 8.1, N1) and **oudemansin** (Table 8.1, N6), isolated from mycelia of the Basidiomycete fungi *Strobilurus tenacellus* (Anke and Oberwinkler, 1977) and *Oudemansiella mucida* (Musilek et al., 1969), showed high antifungal activity against phytopathogenic fungi at very low concentrations (Lorenzen and Anke, 1998). They have a unique mode of action, selectively inhibiting the respiration of fungi by interfering with the ubiquinol oxidation center of the mitochondrial bc1 complex (Lorenzen and Anke, 1998). **Oudemansin X** (Table 8.1, N8) was isolated from *Oudemansiella radicata*, and showed high fungicidal and bactericidal activities against many plant pathogens (Anke et al., 1990). The **chlorinated strobilurins B** (Table 8.1, N7) are also isolated from *Mycena alkalina*, *Mycena avenacea*, *Mycena crocata*, *Xerula longipes*, and *Xerula melanotricha*. Basidiocarps of the genera *Agaricus*, *Favolaschia*, and *Filoboletus* produced **strobilurins A** (Table 8.1, N1), **E** (Table 8.1, N3), and **F1** (Table 8.1, N4), **9-methoxystrobilurins A** (Table 8.1, N5), and **oudemansin A** (Table 8.1, N6) (Zjawiony, 2004). According to Clough (1993), the **strobirulin E** and **oudemansins** also were isolated from the Basidiomycetes fungus *Crepidotus fulvotomentosus*. These compounds have served as natural product prototypes for the design of synthetic analogs **Azoxystrobin** (ICI5504) and **kresoxim-methyl** (BAS490F), which were sold for the first time in 1996 (Knight et al., 1997). Their lack of mammalian toxicity has made them good lead compounds for the development of commercial agricultural fungicides.

Very similar aromatic compounds, such as **anisaldehyde** (Table 8.1, N9) and **(4-methoxyphenyl)-1,2–propanediol** (Table 8.1, N10), showing weak antifungal activity, were isolated from *Pleurotus pulmonarius* and *Bjerkandera adusta* (De Jong et al., 1994).

Another fungal metabolite, **favolon** (Table 8.1, N11), produced from a culture of an Ethiopian *Favolaschia* species, is an unusual ergosterone with a B/C-*cis* ring junction. This compound displayed strong antifungal activity against numerous fungal pathogens, with the strongest inhibitions in the agar diffusion assay for *Mucor miehei*, *Paecilomyces varioti*, and *Penicillium islandicum* (Anke et al., 1995). According to Aqueveque et al. (2005) the new biologically active triterpenoid **favolon B** (Table 8.1, N12), isolated from fermentation broth of *Mycena* spp., showed antifungal activity toward *Botrytis cinerea*, *M. miehei*, *P. variotii*, and *Penicillium notatum*.

A potent fungicidal N-hydroxylated maleimide derivative, **himanimide C** (Table 8.1, N14), was isolated from Basidiomycete *Serpula himantoides* (Aqueveque et al., 2002). **Himanimide C** exhibited

Table 8.1 Naturally Occurring Metabolites of Higher Basidiomycetes Fungi with Antiphytoviral, Antibacterial, Antifungal, Herbicidal, Insecticidal, and Nematocidal Activities

N	Formula	Name	Source	Properties	References
1.	MeOOC, OMe	Strobilurin A, Mucidin, Mucidermin	*Strobilurus tenacellus*, *S. conigenoides*, *S. esculentus*, *S. stephanocystis*, *Oudemansiella mucida*, *Ou. radicata*, *Agaricus* spp., *Favolaschia* spp., *Filoboletus* spp., *Cyphellopsis anomala*, *Hydropus scabripes*, *Mycena aetites*, *M. atromarginata*, *M. capillaripes*, *M. fagetorum*, *M. galopus*, *M. galopus* var. *alba*, *M. oregonensis*, *M. purpureofusca*, *M. rosella*, *M. zephira*, *Pterula* spp., *Xerula melanotricha*	Fungicidal	Anke and Oberwinkler (1977), Lorenzen and Anke (1998), and Zjawiony (2004)
2.	MeO, Cl, MeOOC, OMe	Chlorinated strobilurin B	*Strobilurus tenacellus*, *Mycena alkalina*, *M. avenacea*, *M. crocata*, *Xerula longipes*, *Xerula melanotricha*, *M. crocata*, *M. vitilis*, *Xerula melanotricha*	Fungicidal, antibiotic, respiration inhibitor	Anke and Oberwinkler (1977)
3.	O, O, O, O, MeOOC, OMe	Strobilurin E	*Crepidotus fulvotomentosus*, *Agaricus* spp., *Favolaschia* spp., *Filoboletus* spp.	Fungicidal	Lorenzen and Anke (1998), Weber et al. (1990), Clough (1993), and Zjawiony (2004)
4.	HO, O, MeOOC, OMe	Strobilurin F1	*Agaricus* spp., *Favolaschia* spp., *Filoboletus* spp., *Cyphellopsis anomala*	Fungicidal	Lorenzen and Anke (1998) and Zjawiony (2004)
5.	OMe, MeOOC, OMe	9-Methoxystrobilurin A	*Favolaschia* spp.	Fungicidal, cytostatic	Zapf et al. (1993) and Zjawiony (2004)
6.	OMe, CH_3, MeOOC, OMe	Oudemansin A	*Oudemansiella mucida*, *Favolaschia* spp., *M. polygramma*, *Pterula* spp.	Antibiotic, fungicidal	Anke and Oberwinkler (1977)

7.	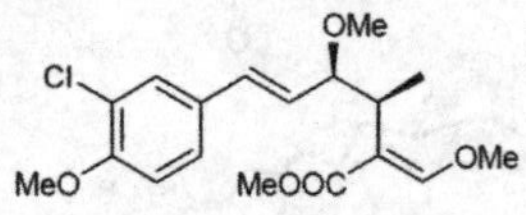	Chlorinated oudemansine B	*Xerula longipes, X. melanotricha*	Fungicidal, inhibitor of eukariotic respiration	Anke et al. (1983)
8.	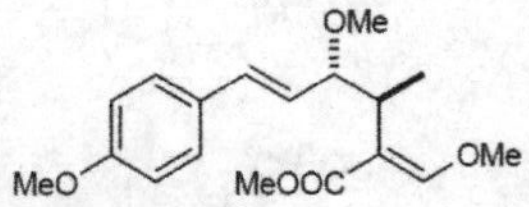	Oudemansin X	*Oudemansiela radicata*	Fungicidal, antibacterial	Anke et al. (1990)
9.	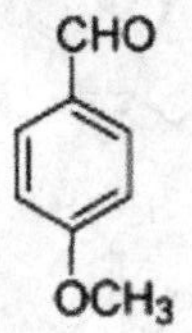	Anisaldehyde	*Pleurotus pulmonarius*, *Bjerkandera adusta*	Fungicidal	De Jong et al. (1994)
10.	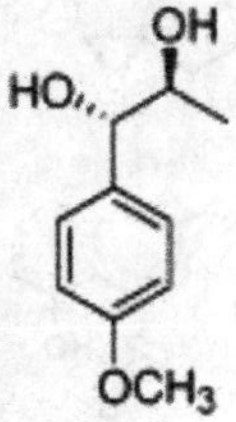	(4-methoxyphenyl)-1,2–propanediol	*Pleurotus pulmonarius*, *Bjerkandera adusta*	Fungicidal	De Jong et al. (1994)
11.	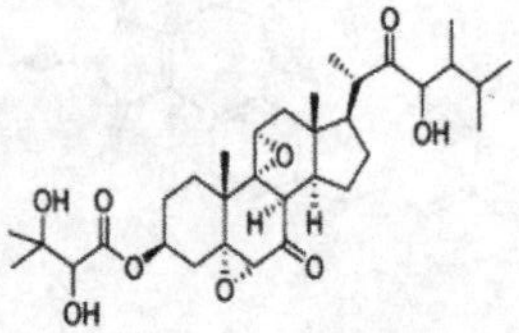	Favolon	*Favolaschia* spp.	Fungicidal	Anke et al. (1995)
12.	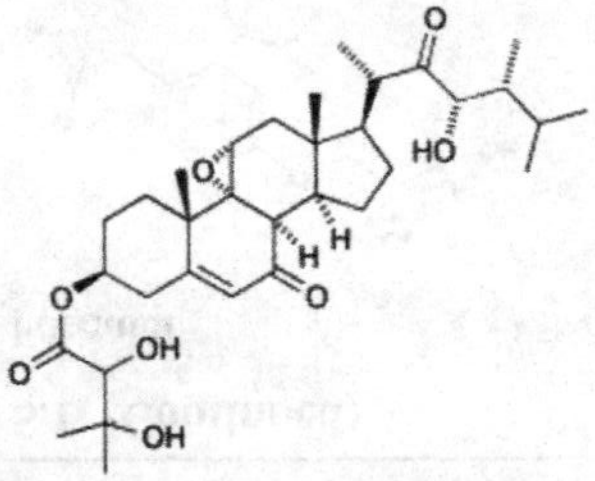	Favolon B	*Mycena* spp.	Fungicidal, cytostatic	Aqueveque et al. (2005)

(*Continued*)

Table 8.1 (Continued)

N	Formula	Name	Source	Properties	References
13.	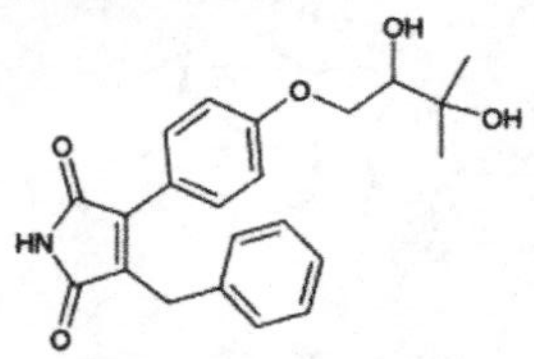	Himanimide B	*Serpula himantoides*	Herbicidal	Aqueveque et al. (2002)
14.	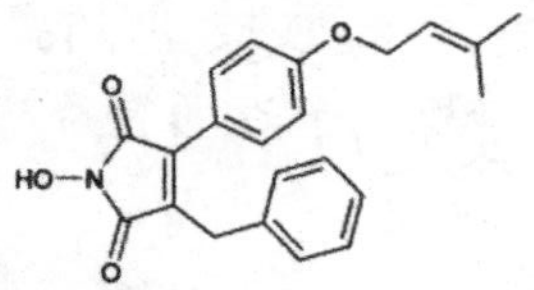	Himanimimide C	*Serpula himantoides*	Fungicidal	Aqueveque et al. (2002)
15.	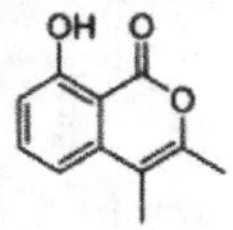	Oospolactone	*Gleophyllum sepiarium*	Fungicidal	Nakajima et al. (1976)
16.	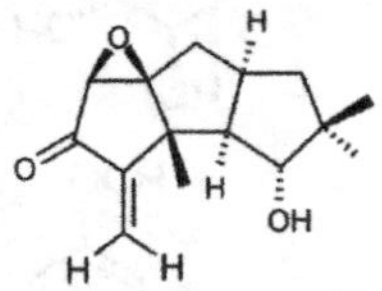	Hypnophilin	*Pleurotellus hypnophillus*	Herbicidal	Kupka et al. (1981)
17.	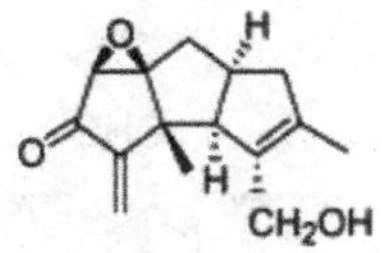	Pleurotellol	*Pleurotellus hypnophillus*	Herbicidal	Kupka et al. (1981)
18.	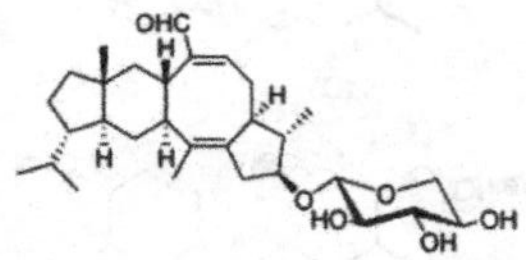	Aleurodiscal	*Aleurodiscus mirabilis*	Fungicidal	Lauer et al. (1989)
19.	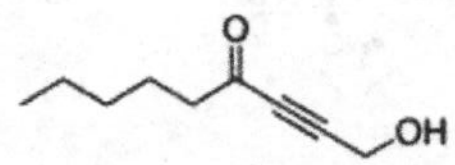	1-hydroxy-2-nonyn-3-one	*Ischnoderma benzoinum*	Fungicidal	Lorenzen and Anke (1998)

20.	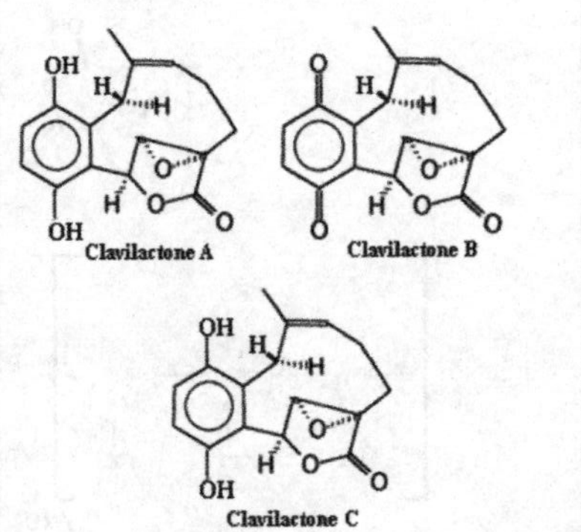 Clavilactone A; Clavilactone B; Clavilactone C	Clavilactone A, B, C	*Clytocybe claviceps*	Fungicidal	Arnone et al. (1994)
21.	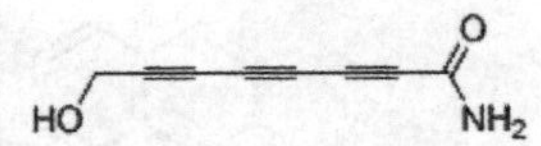	Agrocybin	*Agrocybe dura*, *Marasmius oreades*	Herbicidal	Ayer and Craw (1989)
22.	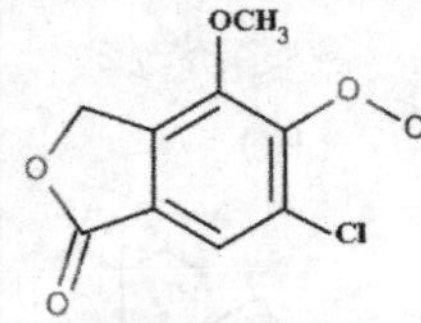	7-chloro-4,6-dimethoxy-1(3H)-isobenzofuranon	*Leucoagaricus carnefolia*	Fungicidal	Lorenzen and Anke (1998)
23.	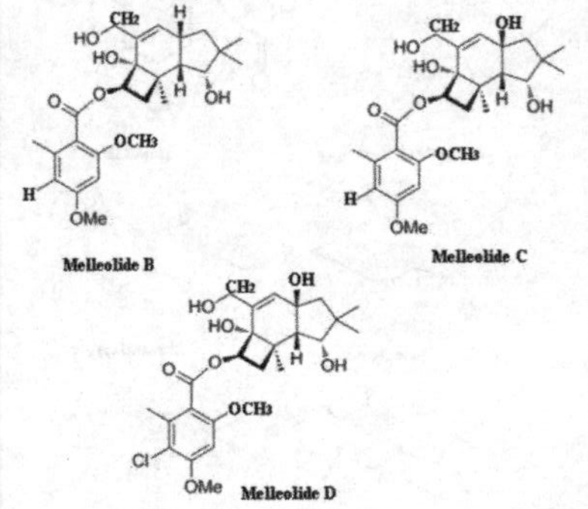 Melleolide B; Melleolide C; Melleolide D	Melleolides (B, C, D)	*Armillaria mellea*	Fungicidal	Lorenzen and Anke (1998)
24.	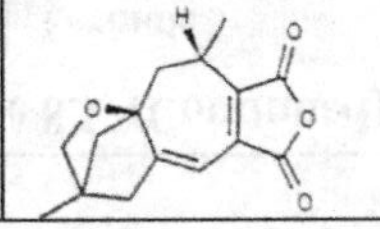	Omphalodin	*Lentinellus* spp.	Herbicidal, weak fungicidal	Stärk et al. (1991)

(*Continued*)

Table 8.1 (Continued)

N	Formula	Name	Source	Properties	References
25.	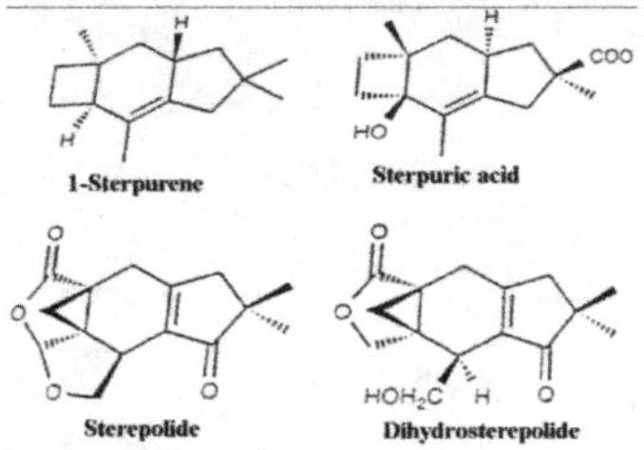 1-Sterpurene; Sterpuric acid; Sterepolide; Dihydrosterepolide	1-sterpurene, sterpuric acid, sterepolide, dihidrosterepolide	*Stereum purpureum*	Phytotoxic	Ayer and Saeedi-Ghomi (1981) and Ayer et al. (1984)
26.	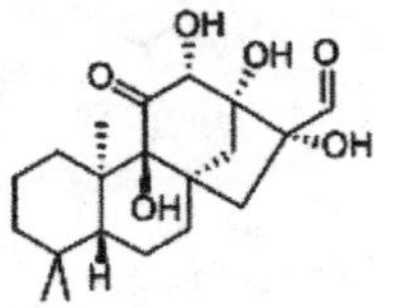	Phlebiakauranol aldehyde	*Phlebia strigosozonata*	Antimicrobial, cytotoxic	Lorenzen and Anke (1998)
27.	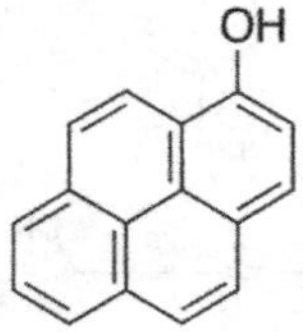	1-hydroxypyrene	*Crinipellis stipitaria*	Fungicidal, antimicrobial, nematicidal	Lambert et al. (1995)
28.	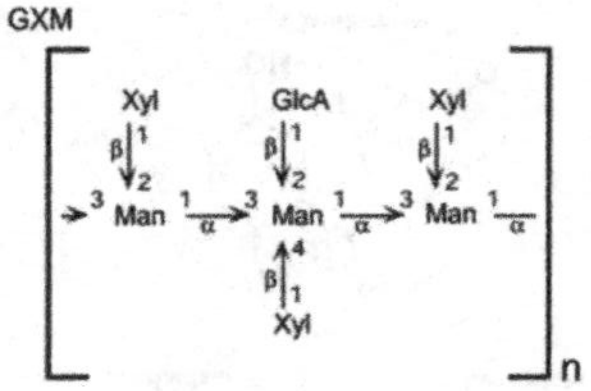 GXM	Glucuronoxylomannan (GXM)	*Tremella mesenterica*	Antiphytoviral	Vinogradov et al. (2004)
29.	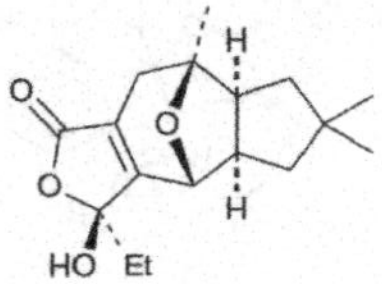	Sescuiterpene lactones	*Lactarius fuliginosus, L. fumosus*	Insecticidal	Dowd and Miller (1990)

	Structure	Compound	Source	Activity	Reference
30.		Clitocine	*Clitocybe inversa*	Insecticidal	Kubo et al. (1986)
31.		Anisaldehyde, p-anisaldehyde	*Pleurotus pulmonarius*, *Bjerkandera adusta*, *Hypholoma fasciculare*, *Pholiota squarrosa*	Nematocidal, weak fungicidal	Stadler et al. (1994) and De Jong et al. (1994)
32.		Linoleic acid	*Pleurotus pulmonarius*	Nematocidal	Lorenzen and Anke (1998)
33.		1,2 dihydroxymintlactone	*Cheimonophyllum candidissimum*	Nematocidal, herbicidal	Stadler et al. (1995) and Lorenzen and Anke (1998)
34.	5-Pentyl-2-furaldehyde 5-(4-Pentenyl)-2-furaldehyde	5-pentyl-2-furaldehyde, 5(4-penteny)-2-furaldehyde	*Irpex lacteus*	Nematocidal	Hayashi et al. (1981)
35.		Omphalotin	*Omphalotus olearius*	Nematocidal	Mayer et al. (1997)

fungicidal effects, especially against *Alternaria porri*, *Aspergillus ochraceus*, and *Pythium irregular* from a concentration of 25 μg/mL. Fungistatic effects were observed against *Absidia glauca*, *Cladosporium cladosporiodes*, *Curvularia lunata*, *Zygorhynchus moelleri*, *Nadsonia fulvescens*, and *Saccharomyces cerevisiae*. **Himanimide B** (Table 8.1, N13) exhibited a weak inhibition of root formation of *Lepidium sativum* at the concentration of 100 μg/disc.

The antifungal isocumarin **oospolactone** (Table 8.1, N15) was identified as a secondary metabolite of *Gleophyllum sepiarium* (Nakajima et al., 1976). This compound was most active against strains of the asexual ascomycete *Alternaria*, showing MIC values of 12.5–25 μg/mL.

Hypnophilin (Table 8.1, N14), together with **pleurotellol** (Table 8.1, N17), was isolated from fermentation of the agaricoid Basidiomycete *Pleurotellus hypnophillus* (Kupka et al., 1981). **Hypnophilin** and **pleurotellol** both act as plant growth inhibitors.

The antifungal sesterterpene β-D-xyloside **aleurodiscal** (Table 8.1, N18) was isolated from the wood-rotting polypore *Aleurodiscus mirabilis* (Lauer et al., 1989). **Aleurodiscal** is selectively active against Zygomycetes, especially against *M. miehei*.

An interesting acetylenic compound exhibiting antifungal activity is **1-hydroxy-2-nonyn-3-one** (Table 8.1, N19), isolated from the biomass after fermentation of the polypore *Ischnoderma benzoinum* (Lorenzen and Anke, 1998).

The cultivation of *Clitocybe claviceps* led to the isolation of three metabolites with pronounced antifungal activities, the **clavilactones A**, **B**, and **C** (Table 8.1, N20). In addition to its antifungal activities, **clavilactone B** exhibits antibacterial and herbicidal effects (Arnone et al., 1994).

Agrocybin (Table 8.1, N21) was isolated from *Marasmius oreades*. This metabolite, previously isolated from *Agrocybe dura*, is suggested to be the principal herbicide responsible for the killing of grass caused by the fairy ring mushroom *M. oreades* (Ayer and Craw, 1989).

A new chlorinated phthalide**, 7-chloro-4,6-dimethoxy-1(3H)-isobenzofuranon** (Table 8.1, N22), was isolated from *Leucoagaricus carneifolia* (Lorenzen and Anke, 1998), which exhibited selective activity against *Botrytis cinerea*.

The **melleolides** (Table 8.1, N23) **B**, **C**, and **D** are the first terpenoid orsellinates isolated from *Armillaria mellea*. Elsewhere in nature, many esters of orsellinic acid have been isolated from lichens or their mycobionts. The melleolides exhibited activity against the fungus *Cladosporium cucumerinum* (Lorenzen and Anke, 1998).

Omphalodin (Table 8.1, N24) containing a succinic acid anhydride element was isolated from *Lentinellus* spp. This herbicidal compound delayed the germination and reduced the growth of *L. sativum* by 80% at a concentration of 100 μg/mL. Weak antifungal activity against *Naematospora coryli* was observed (Stärk et al., 1991).

1-Sterpurene, together with **sterpuric acid**, **sterepolide**, and **dihydrosterepolide** (Table 8.1, N25), was isolated from *Stereum purpureum* (Ayer and Saeedi-Ghomi, 1981). **Sterpuric acid** was reported to possess phytotoxic properties and is thought to be the causative agent of silver leaf disease in many fruit and ornamental trees (Ayer et al., 1984).

Phlebiakauranol, an ent-kaurane, was first described as a metabolite from *Phlebia strigosozonata*. **Phlebiakauranol aldehyde** (Table 8.1, N26) was derived by heating **phlebiakauranol**. **Phlebiakauranol aldehyde** exhibited strong antimicrobial and cytotoxic activities. In greenhouse trials with *Phytophtora infestans* on tomatoes and *Plasmopara viticola* of grapevines, positive protective effects were obtained (Lorenzen and Anke, 1998).

The methanol extract of *Laetiporus sulphureus* indicated complete inhibition of *Aspergillus flavus* growth in tomato paste for 15 days. An inhibition rate of 99.83% was achieved with 0.15 µg/mL of extract. Complete fungicide activity (100%) and no spore survival in the tomato product was recorded using 0.25 µg/mL of *L. sulphureus* extract in tomato medium. Since *L. sulphureus* is widely consumed as an edible mushroom, its use as a natural preservative in tomato paste can be considered safe (Stojković et al., 2011).

The initial transformation product of pyrene metabolization by *Crinipellis stipitaria*, **1-hydroxypyrene** (Table 8.1, N27), showed significant antifungal activity. Growth of most fungi was inhibited at 5 to 25 µg/mL (Lambert et al., 1995).

8.4 ANTIVIRAL METABOLITES

Brandt and Piraino divided the antiviral compounds from fungi into two major classes: (i) those that act indirectly as biological response modifiers (usually from polysaccharide fractions) and (ii) those that act directly as viral inhibitors (Brandt and Piraino, 2000). In higher Basidiomycetes, especially polypores, several polysaccharide fractions display direct inhibitory effects on various viruses. Most commonly, the effect of Basidiomycete polysaccharides has been investigated using only human and animal virus models (Gao et al., 1996, 2004). The influence of Basidiomycete metabolites on plant pathogenic viruses compared with human and animal pathogens has been poorly studied. Only a few articles on anthiphytoviral activities of higher Basidiomycetes have been published.

A pink quinone, tentatively identified as **β-L-glutaminyl-3,4-benzoquinone**, present in sporophores of *Agaricus bisporus* is a potent systemic inhibitor of plant virus infections (Tavantzis and Smith, 1982). A high level of resistance (98.5% or 81.5% reduction in members of lesions) has been observed when cowpea or pinto beans were mechanically inoculated with tobacco ringspot virus (TRSV) or tobacco mosaic virus (TMV).

Aqueous extracts from *Agaricus brasiliensis* and *Lentinula edodes* fruiting bodies showed antiviral activity against infection of passionflower with cowpea aphid-borne mosaic virus (Di Piero et al., 2010).

The filtrate from cultured biomass of the polypore *Fomes fomentarius*, "tinder conk," is highly active against the mechanical transmission of tobacco mosaic virus (TMV), with an IC_{50} value of 10 µg/mL, and it has similar effects against TMV infection on bell pepper and tomato plants (Lorenzen and Anke, 1998).

The screening of *G. lucidum* and *Ganoderma applanatum* strains for antiviral properties of their metabolites has shown that all of them were able to inhibit TMV development. The activity of preparations increased with the increase of concentration. *Ganoderma lucidum* and *G. applanatum* at a concentration of 1000 µg/mL inhibited viral infection to 65–70% (Kovalenko et al., 2008).

A new lectin, named **AAL**, has been purified from the fruiting bodies of the edible mushroom *Agrocybe aegerita* (Sun et al., 2003). It showed inhibition activity to infection of TMV on *Nicotiana glutinosa*. The result of IEF suggested that **AAL** attached to TMV particles.

According to their chemical properties, two types of polysaccharides, neutral and acid polysaccharides, have been isolated from the culture liquid and fruit bodies of higher Basidiomycetes. In particular, the acid polysaccharide of **glucuronoxylomannan (GXM)** (Table 8.1, N28), produced by *Tremella mesenterica*, consists of a linear backbone of β-(1→2)(1→4)-linked oligosaccharides of xylose and glucuronic acid, which produces the polyanion properties (Kakuta et al., 1979; Kovalenko, 1993; Vinogradov et al., 2004). According to Kovalenko et al. (2009), the neutral and acid polysaccharides have different characteristics of antiphytoviral activity. Neutral polysaccharides inhibited the development of local lesions induced by TMV on *Datura* plants by 80% and 99.4% (in concentrations of 100–1000 μg/mL). **GXM** was considerably less active, and, in this case, the total preparation occupied an intermediate position, revealing evidence that the total preparation activity relative to the infectivity of TMV is induced to a greater extent by neutral polysaccharides (Kovalenko et al., 2009).

8.5 INSECTICIDAL AND NEMATOCIDAL METABOLITES

Fruiting bodies and mycelium of Basidiomycete species, such as those belonging to the genera *Lepista*, *Clitocybe*, and *Cantharellus*, are never inhabited by insects. Extracts from numerous mushroom fruiting bodies have been demonstrated to possess insecticidal properties, and several of these fungi are edible, which makes them valuable sources of new candidate insecticides. The insecticidal properties of these mushrooms were attributed to proteins such as lectins or hemolysins (Meir et al., 1996; Wang et al., 2002). The first protein isolated from a Basidiomycete mushroom showing insecticidal activity was a lectin from the red cracking bolete (*Xerocomus chrysenteron*), named lectin XCL, which is the third known member of the new saline-soluble lectin family present in fungi (Triguéros et al., 2003). This protein, purified from mushrooms, was found to be toxic to some insects, such as the dipteran *Drosophila melanogaster* and the hemipteran *Acyrthosiphon pisum*.

Several species of *Lactarius* (Russulaceae) contain **sescuiterpene lactones** (Table 8.1, N29) that deter insects from feeding (Nawrot et al., 1986). The European *Lactarius fuliginosus* contains a variety of **chromenes** (Conca et al., 1981; Allievi et al., 1983). **Chromenes** are toxic to various *Lepidoptera* species or cause anti-hormone effects that induce a precocious metamorphosis (Bowers, 1976). Extracts of *L. fuliginosus* and *L. fumosus* var. *fumosus* showed the strongest toxic effects against the large milkweed bug *Oncopeltus fasciatus*, and in some cases caused precocious development (Dowd and Miller, 1990).

Glucose-, galactose-, sucrose-, lactose-, and sepharose-binding lectines have been isolated from fruiting bodies of *Clytocybe nebularis* (Pohleven et al., 2011). Sucrose-binding lectin showed the strongest activity against *D. melanogaster*, followed by lactose- and galactose-binding lectins. A feeding bioassay with the Colorado potato beetle revealed that *C. nebularis* extract exhibited high antinutritional activity against the insect; of those tested, only lactose-binding lectin, named **CNL**, showed the effect. *C. nebularis* lectines could thus have potential use as natural insecticides.

The nucleoside antibiotic **clitocine** (Table 8.1, N30) was isolated from *C. inversa* as an insecticidal compound (Kubo et al., 1986).

According to Bücker et al. (2013), crude extracts from the Basidiomycete *Pycnoporus sanguineus* have high larvicidal activity against the mosquitoes *Aedes aegypti* and *Anopheles nuneztovari*, and have potential for the production of bioactive substances against larvae of these two tropical disease vectors, with *An. nuneztovari* being more susceptible to the extracts.

According to Chelela et al. (2014), the *Lactarius gymnocarpoides* crude ethanol extract exhibited the highest larvicidal activity against the mosquito *A. aegypti*, with an LC_{50} of 10.75 μg/mL after 72 h of exposure. *L. densifolius* chloroform extract was effective against *Anopheles gambiae* (LC_{50} = 91.33 μg/mL) and moderately effective against *Culex quinquefasciatus* (LC_{50} = 181.16 μg/mL), respectively.

Anisaldehyde (Table 8.1, N31), **3-chloro-anisaldehyde**, and **(4-methoxyphenyl)-1,2-propandiol** were isolated from fungal fermentation products and from natural substances of several common wood and forest-litter degrading fungi, e.g., *P. pulmonarius*, *B. adusta*, *Hypholoma fasciculare*, and *Pholiota squarrosa* (Stadler et al., 1994; De Jong et al., 1994). For **p-anisaldehyde** and **(4-methoxyphenyl)-1,2-propandiol**, weak antifungal and nematocidal properties have been described. Fatty acids, e.g., **S-coriolic acid** or **linoleic acid** (Table 8.1, N32), isolated from *P. pulmonarius* exhibit nematocidal effects against the saprophytic nematode *Caenorhabditis elegans*, with LD_{50} values of 10 and 5 μg/mL, respectively (Stadler et al., 1994). These effects depend on the degree of unsaturation, the position of the double bonds, and the length of the fatty acid (Lorenzen and Anke, 1998). Also, **1,2-dihydroxy-mintlactone** (Table 8.1, N33), a nematocidal monoterpene, was isolated from the wood-inhabiting Basidiomycete *Cheimonophyllum candidissimum* (Stadler et al., 1995). The LD_{50} toward the saprophytic nematode *C. elegans* was 25 μg/mL, and herbicidal effects against *Setaria italic* and *L. sativum* were detected at concentrations starting from 50 μg/mL (Lorenzen and Anke, 1998). Six new bisabolane type sesquiterpenes, the **cheimonophyllons** and **cheimonophyllal**, were isolated from the wood-inhabiting mushroom *Ch. candissimum*. These sesquiterpenes exhibit nematocidal activities against nematode *C. elegans*, with LD_{50} values of 10–25 μg/mL (Lorenzen and Anke, 1998).

The furaldehydes **5-pentyl-2-furaldehyde** and **5(4-penteny)-2-furaldehyde** (Table 8.1, N34) were isolated from *Irpex lacteus*. They exhibited nematocidal activity against *Aphelencoides besseyi*, with IC_{50} values of 25–50 μg/mL (Hayashi et al., 1981).

A nematocidal cyclic peptide, **omphalotin** (Table 8.1, N35), was isolated from biomass after fermentation of *Omphalotus olearius* (Mayer et al., 1997). The LD_{50} against the plant-pathogenic nematode *Meloidogyne incognita* was determined at 0.75 μg/mL, while only weak effects against the saprophytic nematode *C. elegans* were detectable (LD_{50} 25 μg/mL). **Omphalotin** is a promising candidate for the development of an agricultural nematocide.

1-Hydroxypyrene (Table 8.1, N27), derived from *C. stipitaria*, showed very strong nematocidal activity against the saprotrophic soil-inhabiting nematode *C. elegans*. The effects of 1-hydroxypyrene were visible after 1 h, with immobilization of nematodes by 1 mg/mL (Lambert et al., 1995).

The cultural filtrates from *Amauroderma macer*, *Laccaria tortilis*, *Peziza* spp., *O. mucida*, *Pleurotus pulmatus*, and *Tylopilus striatulus* showed high nematocidal activity against the pine wood nematode *Bursaphelenchus xylophilus*, with over 80% pathogenicity within 72 h of exposure being observed (Dong et al., 2006).

8.6 CONCLUSIONS

Deadly chemical pesticides have been the go-to fix for the conventional agricultural industry, but new research suggests that biopesticides made from fungi could be a safer, more planet-friendly alternative.

The present review focuses on isolated biologically active compounds of mushrooms from all over the world and their potential as sources of biopesticides. This information certainly will be useful for future scientific studies. Natural product-based virucides, fungicides, bactericides, insecticides, and nematicides are generally considered safer than synthetic products because of their relatively short environmental half-life and lack of harmful effects. Many if not all bioactive metabolites presented may lead to the discovery of novel target sites or to new classes of chemicals that can be developed for pathogen management.

The advantage of biopesticides compared with synthetic ones is not only their nontoxic characteristics but also the low cost for their production. Growing culinary-medicinal mushrooms is well established and, in most cases, economically justified. Identification and isolation of active compounds from mushrooms has been clearly elaborated. Mushroom-derived biopesticides usually operate at very small concentrations, and for further application only small amounts are needed. Most importantly, mushroom-derived biopesticides are relatively cheap and are available as biocontrol agents.

The data available from the literature clearly demonstrate that culinary-medicinal mushrooms and their metabolites present great potential for developing biopesticides. Future studies of bioactive compounds and synergistic combinations from mushrooms may further the development of virucides, fungicides, bactericides, insecticides, and nematicides, as well as applications for food and crop management.

REFERENCES

Abraham, W.R., 2001. Bioactive sesquiterpenes produced by fungi: are they useful for humans as well? Curr. Med. Chem. 8, 583–606.

Allievi, C., DeBernardi, M., Demarchi, F., Mellerio, G., 1983. Chromatographic analysis of 2,2-dimethylchromene derivatives. J. Chromatogr. 261 331–314.

Anke, T., 1989. Basidiomycetes: a source for new bioactive secondary metabolites. Prog. Ind. Microbiol. 27, 51–66.

Anke, T., Oberwinkler, F., 1977. The strobilurins new antifungal antibiotics from the Basidiomycete *Strobilurus tenacellus*. J. Antibiot. 30, 806–810.

Anke, T., Besl, H., Mocek, U., Steglich, W., 1983. Antibiotics from Basidiomycetes. XVIII. Strobilurin C and oudemansin B, two new antifungal metabolites from *Xerula* species (Agaricales). J. Antibiot. 36, 661–666.

Anke, T., Werle, A., Bross, M., Steglich, W., 1990. Antibiotics from Basidiomycetes. XXXII. Oudemansin X, a new antifungal E-β-methoxyacrylate from *Oudemansiella radicata* (Relhan ex Fr.) Sing. J. Antibiot. 43, 1010–1011.

Anke, T., Werle, A., Zapf, S., Velten, R., Steglich, W., 1995. Favolon, a new antifungal triterpenoid from *Favolaschia* species. J. Antibiot. 48 (7), 725–726.

Aqueveque, P., Anke, T., Sterner, O., 2002. The himanimides, new bioactive compounds from *Serpula himantoides*. Z. Naturforsch. 57, 257–262.

Aqueveque, P., Anke, T., Anke, H., Sterner, O., Becerra, J., Silva, M., 2005. Favolon B, a new triterpenoid isolated from the Chilean *Mycena* sp. strain 96180. J. Antibiot. 58 (1), 61–64.

Arnone, A., Cardillo, R., Meille, S.V., Nasini, G., Tolazzi, M., 1994. Secondary mold metabolites. 47. Isolation and structure elucidation of clavilactones AC, new metabolites from the fungus *Clitocybe clavipes*. J. Chem. Soc. –Perkin Trans. 15, 2165–2168.

Ayer, W.A., Saeedi-Ghomi, M.H., 1981. The sterepolides: new isolactaranes from *Stereum purpureum*. Tetrahedron Lett. 22, 2071–2074.

Ayer, A.W., Craw, P., 1989. Metabolites of the fairy ring fungus, *Marasmius oreades*. Part 2. Norsesquiterpenes, further sesquiterpenes, and agrocybin. Can. J. Chem. 67 (9), 1371–1380.

Ayer, W.A., Nakashima, T.T., Hossein Saeedi-Ghomi, M., 1984. Studies on the biosynthesis of the sterpurenes. Can. J. Chem. 62 (3), 531–533.

Barros, L., Baptista, P., Estevinho, L.M., Ferreira, I.C.F.R., 2007. Effect of fruiting body maturity stage on chemical composition and antimicrobial activity of *Lactarius* sp. mushrooms. J. Agric. Food Chem. 55, 8766–8771.

Beattie, K.D., Rouf, R., Gander, L., May, T.W., Ratkowsky, D., Donner, C.D., et al., 2010. Antibacterial metabolites from Australian macrofungi from the genus *Cortinarius*. Phytochemistry 71, 948–955.

Bowers, W.S., 1976. Discovery of insect antiallatotropins. In: Gilbert, L.I. (Ed.), The Juvenile Hormones Plenum Press, New York, NY, pp. 394–408.

Brandt, C.R., Piraino, F., 2000. Mushroom antivirals. Recent Res. Dev. Antimicrob. Agents Chemother. 4, 11–26.

Bücker, A., Bücker, N.C.F., Lima de Souza, A.Q., Matos da Gama, A., Rodrigues-Filho, E., Medeiros da Costa, F., et al., 2013. Larvicidal effects of endophytic and Basidiomycete fungus extracts on *Aedes* and *Anopheles* larvae (Diptera, Culicidae). Rev. Soc. Bras. Med. Trop. 46 (4), 411–419.

Chang, S.-T., Wasser, S.P., 2012. The role of culinary-medicinal mushrooms on human welfare with a pyramid model for human health. Int. J. Med. Mushrooms 14 (2), 95–134.

Chelela, B.L., Chacha, M., Matemu, A., 2014. Larvicidal potential of wild mushroom extracts against *Culex quinquefasciatus* Say, *Aedes aegypti* and *Anopheles gambiae* Giles S.S. Am. J. Res. Commun. 2 (8), 105–114.

Clough, J.M., 1993. The strobirulins, oudemansins, and myxothiazols, fungicidal derivatives of beta methoxyacrilic acid. Nat. Prod. Rep. 10 (6), 565–574.

Conca, E., DeBernardi, M., Fronza, G., Girometta, M.A., Mellerio, G., Vidari, G., et al., 1981. Fungal metabolites 10. New chromenes from *Lactarius fuliginosus* and *Lactarius picinus* Fries. Tetrahedron Lett. 22, 4327–4330.

De Jong, E., Cazemier, A.E., Field, J.A., de Bont, J.A.M., 1994. Physiological role of chlorinated aryl alcohols-biosynthesized de novo by the white rot fungus *Bjerkandera* sp strain BOS55. Appl. Environ. Microbiol. 60, 271–277.

Di Piero, R.M., de Novaes, Q.S., Pascholati, S.F., 2010. Effect of *Agaricus brasiliensis* and *Lentinula edodes* mushrooms on the infection of passionflower with cowpea aphid-born mosaic virus. Braz. Arch. Biol. Technol. 53 (2), 269–278.

Dong, J.Y., Li, X.P., Li, L., Li, G.H., Liu, Y.J., Zhang, K.Q., 2006. Preliminary results on nematocidal activity from culture filtrates of Basidiomycetes against the pine wood nematode, *Bursaphelenchus xylophilus* (Aphelenchoididae). Ann. Microbiol. 56, 163–166.

Dowd, P.F., Miller, O.K., 1990. Insecticidal properties of *Lactarius fuliginosus* and *Lactarius fumosus*. Entomologia Experimentalis et Applicata 57, 23–28.

Dreo, T., Želko, M., Scubic, J., Brzin, J., Ravnikar, M., 2007. Antibacterial activity of proteinaceous extracts of higher Basidiomycetes mushrooms against plant pathogenic bacteria. Int. J. Med. Mushrooms 9 (3&4), 226–227.

Fagade, O.E., Oyelade, A.A., 2009. A comparative study of the antibacterial activities of some wood-decay fungi to synthetic antibiotic discs. Electron. J. Environ. Agric. Food Chem. 8 (3), 184–188.

Fu, L.Q., Guo, X.S., Liu, X., He, H.L., Wang, Y.L., Yang, Y.S., 2010. Synthesis and antibacterial activity of C-2(S)-substituted pleuromutilin derivatives. Chinese Chem. Lett. 21, 507–510.

Gao, Q.P., Seljelid, R., Chen, H., Jiang, R., 1996. Characterization of acidic heteroglycans from *Tremella fuciformis* Berk. with cytokine stimulating. Carbohydr. Res. 288, 135–142.

Gao, Y., Zhou, Sh., Huang, M., 2004. Antibacterial and antiviral value of *Ganoderma* P. Karst. species (Aphyllophoromycetideae): a review. Int. J. Med. Mushrooms 6 (1), 96–106.

Harvey, A.L., 2008. Natural products in drug discovery. Drug Disc. Today 13 (19-20), 894–901.

Hayashi, M., Wada, K., Munakata, K., 1981. New nematicidal metabolites from a fungus, *Irpex lacteus*. Agric. Biol. Chem. 45, 1527–1529.

Hearst, R., Nelson, D., McCollum, G., Millar, B.C., Maeda, Y., Goldsmith, C.E., et al., 2009. An examination of antibacterial and antifungal properties of constituents of shiitake (*Lentinula edodes*) and oyster (*Pleurotus ostreatus*) mushrooms. Complement. Ther. Clin. Pract. 15, 5–7.

Herms, S., Seehaus, K., Koehle, H., Conrath, U., 2002. Strobilurin fungicide enhances the resistance of tobacco against tobacco mosaic virus and *Pseudomonas syringaes* pv. *tabaci*. Plant Physiol. 130, 120–127.

Jespers, A., 1994. Mode of action of the phenylpyrole fungicide fenpiclonil in *Fusarium sulphureum*. Ph.D. Thesis, Wageningen, Holland.

Johansson, M., Sterner, O., Labischinski, H., Anke, T., 2001. Coprinol, a new antibiotic cuparane from *Coprinus* species. Z. Naturforsch. 56, 31–34.

Kakuta, M., Sone, Y., Umeda, T., Misaki, A., 1979. Comparative structural studies on acidic heteropolysaccharides isolated from "Shirokikurage", fruit body of *Tremella fuciformis* Berk., and the growing culture of its yeast-like cells. Agric. Biol. Chem. 43, 1659–1668.

Knight, S.C., Anthony, V.M., Brady, A.M., Greenland, A.J., Heaney, S.P., Murray, D.C., et al., 1997. Rational and perspectives on the development of fungicides. Ann. Rev. Phytopathol. 35, 349–372.

Kovalenko, A.G., 1993. The protein-carbohydrate interaction in realization of plant resistance to viruses. J. Microbiol. 55, 74–91.

Kovalenko, O.G., Polishchuk, O.M., Krupodorova, T.A., Bisko, N.A., Buchalo, A.S., 2008. Screening of metabolites produced by strains of *Ganoderma lucidum* (Curt.:Fr.) P. Karst and *Ganoderma applanatum* (Pirs.:Waller) Pat. for their activity against tobacco mosaic virus. Visnik Nacionalnogo Universiteta Imeni Tarasa Shevchenko, pp. 32–34.

Kovalenko, O.G., Polishchuk, O.N., Wasser, S.P., 2009. Virus resistance induced by glucuronoxylomannan isolated from submerged cultivated yeast-like cell biomass of medicinal yellow brain mushroom *Tremella mesenterica* Ritz.:Fr. (Heterobasidiomycetes) in hypersensitive host plants. Int. J. Med. Mushrooms 11 (2), 199–205.

Kubo, I., Kim, M., Wood, W.F., Naoki, H., 1986. Clitocine, new insecticidal nucleoside from the mushroom *Clitocybe inversa*. Tetrahedron Lett. 27, 4277–4280.

Kupka, J., Anke, T., Gianetti, B.M., Steglich, W., 1981. Antibiotics from Basidiomycetes. XIV. Isolation and biological characterisation of hypnophilin, pleurotellol, and pleurotellic acid from *Pleurotellus hypnophilus* (Berk.) Sacc. Arch. Microbiol. 130, 223–227.

Lambert, M., Kremer, S., Anke, H., 1995. Antimicrobial, phytotoxic, nematicidal, cytotoxic, and mutagenic activities of 1-Hydroxypyrene, the initial metabolite in Pyrene metabolism by the Basidiomycete *Crinipellis stipitaria*. Bull. Environ. Contamination Toxicol. 55, 521–257.

Lauer, U., Anke, T., Shledrick, W.S., Scherer, A., Steglich, W., 1989. Antibiotics from Basidiomycetes. XXXI. Aleurodiscal: an antifungal sesquiterpenoid from *Aleurodiscus mirabilis* (Berk. & Curt.) HÖhn. J. Antibiot. 42, 875–882.

Li, W.-J., Nie, S.-P., Liu, X.-Z., Zhang, H., Yang, Y., Yu, Q., et al., 2012. Antimicrobial properties, antioxidant activity and cytotoxicity of ethanol-soluble acidic components from *Ganoderma atrum*. Food Chem. Toxicol. 50, 689–694.

Lorenzen, K., Anke, T., 1998. Basidiomycetes as source for new bioactive natural products. Curr. Org. Chem. 2, 329–364.

Mayer, A., Sterner, O., Anke, H., 1997. Omphalotin, a new cyclic peptide with potent nematicidal activity from *Omphalotus olearius* 1. Fermentation and biological activity. Nat. Prod. Lett. 10, 25–33.

Meir, N., Klaebc, A., Chavant, L., Fournier, D., 1996. Insecticidal properties of mushroom and toadstool carpophores. Phytochemistry 41 (5), 1293–1299.

Musilek, V., Cerna, J., Sasek, V., Semerzieva, M., Vondracek, M., 1969. Antifungal antibiotic from the Basidiomycete *Oudemansiella mucida*. I. Isolation and cultivation of a producing strain. Folia Microbiol. 14, 377–388.

Nakajima, S., Kawai, K., Yamada, S., 1976. Studies on fungal products. –IV. Isolation of oospolactone as the antifungal principle of *Gloephyllum sepiarium*. Agric. Biol. Chem. 40, 811–812.

Nawrot, J., Bloszyk, E., Harmatha, J., Ladislav, N., Dohdan, B., 1986. Action of antifeedants of plant origin on beetles infecting stored products. Acta Entomol. Bohemoslovaca 83, 327–335.

Pohleven, J., Brzin, J., Vrabec, L., Leonardi, A., Čokl, A., Štrukelj, B., et al., 2011. Basidiomycete *Clytocybe nebularis* is rich in lectins with insecticidal activities. Appl. Microbiol. Biotechnol. 91, 1141–1148.

Rai, M., Tidke, G., Wasser, S.P., 2005. Therapeutic potential of mushrooms. Nat. Prod. Radiance 4 (4), 246–257.

Reshetnikov, S.V., Wasser, S.P., Tan, K.K., 2001. Higher Basidiomycota as a source of antitumor and immunostimulating polysaccharides. Int. J. Med. Mushrooms 3, 361–394.

Robles-Hernández, L., 2004. Novel antimicrobial activities of *Ganoderma lucidum* and *Laetiporus sulphureus* for agriculture. Dissertation, University of Idaho, 113 p.

Robles-Hernández, L., Cecilia-Gonzáles-Franco, A., Soto-Parra, J.M., Montes-Domínguez, F., 2008. Review of agricultural and medicinal applications of Basidiomycete mushrooms. Technociencia 2 (2), 95–107.

Saddiqe, Z., Naeem, I., Maimoona, A., 2010. A review of the antibacterial activity of *Hypericum perforatum* L. J. Ethnopharmacol. 131, 511–512.

Soković, M.D., Glamočlija, J.M., Ćirić, A.D., 2013. Natural products from plants and fungi as fungicides. Licensee InTech, 185–232.

Stadler, M., Mayer, A., Anke, H., 1994. Fatty acid and other compounds with nematicidal activity from cultures of Basidiomycetes. Planta Med. 3, 509–510.

Stadler, M., Fouron, J.-Y., Sterner, O., Anke, H., 1995. 1,2-Dihydroxymintlactone, a new nematicidal monoterpene isolated from the Basidiomycete *Cheimonophyllum candidissimum* (Berk. & Curt.) Sing. Z. Naturforsch. 50, 473–475.

Stärk, A., Anke, T., Mocek, U., Steglich, W., 1991. Antibiotics from Basidiomycetes. 42. omphalone, an antibiotically active benzoquinone derivative from fermentation of *Lentinellus omphalodes*. Z. Naturforsch. C-A J. Biosci. 46, 989–992.

Stojković, D., Šiljegović, J., Nicolić, M., Ćirić, A., Glamočlija, J., Soković, M., et al., 2011. Sulphur Polypore, *Laethiporus sulphureus* extract as natural preservative for the in vivo control of *Aspergillus flavus* in tomato paste. The 6th International Medicinal Mushroom Conference, Book of Abstract, Zagreb, Croatia, pp. 63–64.

Sun, H., Zhao, C.G., Tong, X., Qi, Y.P., 2003. A lectin with mycelia differentiation and antiphytovirus activities from the edible mushroom *Agrocybe aegerita*. J. Biochem. Mol. Biol. 36 (2), 214–222.

Tavantzis, S.M., Smith, S.H., 1982. Isolation and evaluation of a plant-virus-inhibiting quinone from sporophores of *Agaricus bisporus*. Physiol. Biochem. 72 (6), 619–621.

Triguéros, V., Lougarre, A., Ali-Ahmed, D., Rahbe, Y., Guillot, J., Chavant, L., et al., 2003. *Xerocomus chrysenteron* lectin: identification of a new pesticidal protein. Biochim. Biophys. Acta 1621 (3), 292–298.

Vinogradov, E., Petersen, B.O., Duus, J., Wasser, S.P., 2004. The structure of the glucuronoxylomannan produced by culinary-medicinal yellow brain mushroom (*Tremella mesenterica* Ritz.:Fr., Heterobasidiomycetes) grown as one cell biomass in submerged culture. Carbohydr. Res. 339, 1483–1489.

Wang, M., Trigueros, V., Paquereau, L., Chavant, L., Fournier, D., 2002. Proteins as active compounds involved in insecticidal activity of mushroom fruitbodies. J. Econ. Entomol. 95 (3), 603–607.

Wang, Y., Bao, L., Li, L., Li, S., Gao, H., Yao, X.-S., et al., 2012. Bioactive sesquiterpenoids from the solid culture of the edible mushroom *Flammulina velutipes* growing on cooked rice. Food Chem. 132, 1346–1353.

Wasser, S.P., 2002. Medicinal mushrooms as a source of antitumor and immunomodulating polysaccharides. Appl. Microbiol. Biotechnol. 60, 258–274.

Wasser, S.P., Weis, A.L., 1999a. Medicinal properties of substances occurring in higher Basidiomycetes mushrooms: current perspective. Int. J. Med. Mushrooms 1, 31–62.

Wasser, S.P., Weis, A.L., 1999b. Therapeutic effects of substances occurring in higher Basidiomycetes mushrooms: a modern perspective. Crit. Rev. Immunol. 19, 65–96.

Weber, W., Anke, T., Steffan, B., Steglich, W., 1990. Antibiotics from Basidiomycetes. XXXII. Strobilurin E: a new cytostatic and antifungal (E)-beta-methoxyacrylate antibiotic from *Crepidotus fulvotomentosus* Peck. J. Antibiot. 43 (2), 207–212.

Younis, A.M., Wu, F.-Sh., El Shikh, H.H., 2015. Antimicrobial activity of extracts of oyster culinary-medicinal mushroom, *Pleurotus ostreatus* (Higher Basidiomycetes) and identifi cation of a new antimicrobial compound. Int. J. Med. Mushrooms 17 (6), 579–590.

Yu, J.-Q., Lei, J.-Ch., Zhang, X.-Q., Yu, H.-D., Tian, D.-Z., Liao, Z.-X., et al., 2011. Anticancer, antioxidant and antimicrobial activities of the essential oil of *Lycopus lucidus* Turcz. var. *hirtus* Regel. Food Chem. 126, 1593–1598.

Zapf, S., Anke, T., Dasenbrock, H., 1993. Antifungal metabolites from *Agaricus* sp. 89139. Bioengineering 1, 92.

Zjawiony, J.K., 2004. Active compounds from Aphyllophorales (Polypore) fungi. J. Nat. Prod. 67, 300–310.

CHAPTER

CULTIVATION OF MEDICINAL FUNGI IN BIOREACTORS

9

Marin Berovic[1] and Bojana Boh Podgornik[2]

[1]Faculty of Chemistry and Chemical Technology, University of Ljubljana, Ljubljana, Slovenia [2]Faculty of Natural Sciences and Engineering, University of Ljubljana, Ljubljana, Slovenia

9.1 INTRODUCTION

According to Wasser (2002), the number of mushrooms on Earth is estimated at 140,000, yet perhaps only 10% of species are known and named. Basidiomycete mushrooms comprise a vast but largely untapped source of new pharmaceutical products in the fruit bodies, cultured mycelium, and culture broth. They are of varying chemical composition, such as polysaccharides, glycopeptide–protein complexes, proteoglycans, proteins, and triterpenoids, with most scientific attention focused on the group of noncellulosic β-glucans with β-(1–3) linkages in the main chain of the glucan, and additional β-(1–6) branch points, which are characteristic for antitumor and immunostimulating action.

Mushroom polysaccharides do not attack cancer cells directly, but produce their antitumor effects by activating various immune responses in the host. Structurally different β-glucans have different affinities toward receptors and thus generate different host responses (Chen and Seviour, 2007).

Immunomodulating and antitumor activities of these metabolites are related to immune cells such as hematopoietic stem cells, lymphocytes, macrophages, T cells, dendritic cells, and natural killer cells, which are involved in the innate and adaptive immunity, resulting in the production of biologic response modifiers.

The clinical evidence for antitumor and other medicinal activities comes primarily from some commercialized purified polysaccharides, such as lentinan from *Lentinula edodes*, krestin from *Coriolus versicolor*, grifolan from *Grifola frondosa*, and schizophyllan from *Schizophyllum commune* (Wasser, 2002; Chen and Seviour, 2007), but polysaccharide preparations of some other medicinal mushrooms in bioreactors represent an interesting direction in comprehensive biotechnology for the near future.

9.2 CULTIVATION TECHNOLOGIES

9.2.1 OVERVIEW OF CULTIVATION TECHNOLOGIES

Since medicinal mushrooms are scarce in nature, cultivation of these fungal species on artificial media has been introduced. The traditional cultivation of mushrooms on wood logs has been known for centuries. Over time, cultivation methods have been diversified, modified, and developed. In addition to

Mushroom Biotechnology. DOI: http://dx.doi.org/10.1016/B978-0-12-802794-3.00009-6

wood logs, fruit bodies are being produced on sawdust substrates in trays or beds, and in sterilized plastic bags or in bottles. In addition, production of fungal mycelia has been developed in bioreactors, utilizing submerged cultivation technologies in liquid media or solid-state substrates (Wasser, 2002).

9.2.2 PRODUCTION OF BIOMASS IN BIOREACTORS

The traditional cultivation of mushrooms to get their fruit bodies takes several months, and it is difficult to control the quality of products. In order to decrease cultivation time and to improve the quality, controlled cultivation in bioreactors has been developed, using both submerged and solid-state processes. Cultivation of pharmaceutically active biomass of higher fungi in bioreactors with optimized substrate compositions and controlled process parameters enables a shorter cultivation time and large-scale production under full process control. Two main aerobic technologies are used—submerged cultivation, in a liquid medium, and solid-state substrate bioprocessing. Compared the traditional farming production to get fungal fruit bodies, which requires months for cultivation, submerged and solid-state bioprocessing (SSB) produce greater amounts of fungal mycelia and pharmaceutically active products in a much shorter time.

The submerged cultivation of fungal biomass takes place in aerobic conditions in liquid media. Although far from the natural environment of higher fungi, this technology enables fast and large-scale production of fungal mycelia and its extracellular (EPS) and intracellular (IPS) polysaccharides, protein–glucans, sterols, steroids, and other compounds in 2–3 weeks. Large amounts of fungal biomass and metabolites can be produced, suitable for human consumption. Solid-state cultivation in bioreactors enables the production of medicinal fungi on various secondary raw material sources, obtained from wood, agriculture, and food industry wastes. The final product is a delignified solid substrate, overgrown with fungal mycelia, enriched with protein-rich biomass, with immobilized pharmaceutically active compounds. This type of cultivation is suitable primarily for veterinary uses. Drying, pulverization, and pelletization of pulverized material enable the inexpensive production of animal feed for ruminants.

9.2.3 SUBMERGED BIOPROCESSING

Submerged bioprocessing represents a method of liquid-state cultivation with the introduction of inlet air as a source of oxygen for aerobic processes. The bioreactors for submerged cultivation are based on various mixing principles, with various kinds of radial or axial agitators, pneumatic mixing systems, or circulation pumps. Usually, the production size of bioreactors ranges from 50 to 400 m^3, and they are fully equipped with comprehensive instrumental control. The cultivation of liquid substrate inside the bioreactor is either batch or continuous. In submerged cultivation the consistency of the substrate and secretion of fungal polysaccharides influence the formation of pseudoplastic bioprocess in broth rheology, and consequently limit transport processes, namely oxygen and heat transfer.

In recent decades, submerged cultivation has become a promising alternative for efficient production of valuable mushroom metabolites, such as fungal polysaccharides. The influence of agitation rate and the shearing action of impellers on the morphology and productivity of filamentous fungi has received considerable scientific attention. There seems to be an inverse relationship between agitation intensity and pellet size. Vigorous agitation appears to prevent pellet formation, while elevated agitation results in small and compact pellets. The outstanding problems in submerged production of medicinal mushroom biomass in bioreactors are: (i) intensive growth on fungal biomass on the bioreactor head space, sensors, and bioreactor wall; (ii) secretion of fungal polysaccharides changes the cultivation liquid rheology; (iii) changes of mass and heat transfers (Berovic et al., 2003).

9.2.4 SOLID-STATE BIOPROCESSING

SSB involves the growth of microorganisms or fungi in beds of moist solid materials, in which the interparticle spaces contain a continuous gas phase and little or no free water. The main sources of water, carbohydrates, phosphorus, nitrogen, and sulfur are intrapartically bounded; therefore, the culture applied must possess the ability to access water and essential element sources in the solid matrix. Heat removal is a major design consideration for SSB bioreactors, both static and mixed. Internal and external heat exchangers have been incorporated in many bioreactor designs, including packed beds and bioreactors that provide agitation during bioprocessing. Since the late 1970s, there has been interest in using SSB to produce various bioproducts, such as enzymes, organic acids, and antibiotics.

A major difference between submerged cultivation and solid-state cultivation is thus the amount of free liquid in the substrate. In submerged cultivation, the amount of dissolved solids rarely exceeds 5–10%, whereas in solid-state cultivation the solids typically represent between 20% and 70% of the total substrate mass. The water activity (a_w) of substrates is an important aspect of fungal cultivation, because the growth of fungi is controlled by the level of a_w. This means that the water activity of the substrate determines the lower limit of available water for fungal growth (Mitchell and Berovic, 2014).

9.3 CULTIVATION OF MEDICINAL MUSHROOMS IN BIOREACTORS

Almost unknown in Western scientific research, some of the wood-degrading Basidiomycetes have been intensely and systematically studied in bioreactors due to their promising pharmacological effects (Wasser, 2002). Medicinal mushrooms of greatest interest, cultivated in various kinds of bioreactors, have been *Ganoderma lucidum*, *G. frondosa*, *Trametes versicolor*, *Hericium erinaceus*, and *Cordyceps militaris*.

9.3.1 SUBMERGED CULTIVATION OF *G. LUCIDUM*

Recently, submerged cultivation technology for mycelial cultures of *G. lucidum* has been developed, and has attracted much attention as a promising cultivation alternative (Berovic et al., 2003). In commercial applications, most market products emphasize both the amount of fungal mycelium and dose of polysaccharides. Therefore, an efficient cultivation process combines the production of cellular biomaterials and released EPS. Optimization of substrate composition and cultivation conditions is vital for enhancing production efficiency in a submerged culture.

Cultivation parameters, including inoculation density (Fang et al., 2002), pH (Fang et al., 2002a), temperature, mixing intensity (Yang and Liau, 1998), and medium composition (Yang et al., 2000; Fang and Zhong, 2002b), have been reported to affect the efficiency of cultivation of *G. lucidum*. However, few investigations have taken into consideration the optimization of cultivation conditions while simultaneously studying interactions among various operational factors, especially when both mycelium and polysaccharides were produced. Chang et al. (2006) applied the Taguchi experimental design to optimize the medium composition for submerged culture of *G. lucidum*.

9.3.1.1 Inoculum preparation

There are two main strategies for the preparation of mycelial inocula in submerged cultivation: either preparation of the seed culture on agar, or preparation by submerged cultivation. Zore et al. (1998) used the inocula of 120h shaken vegetative cultures grown on a substrate consisting of potato dextrose, 20g/L of glucose, and 2% (v/v) of olive oil at pH 5.8. A stirred tank bioreactor (STB) was inoculated at 30°C with 17% (v/v) of the vegetative inoculum. Cultivation lasted 160h, the concentration of dissolved oxygen was maintained by aeration at 10L/min, and the mixing speed was 400rpm. Minimal oxygen partial pressure was 30% (v/v) at pH 4.2.

Simonic et al. (2008) inoculated the bioreactor with 5.0% (v/v) of the homogenized seed culture, and cultivated it at 30°C and initial pH of 4. The optimum inoculum density for biomass and polysaccharide production in the case of *G. lucidum* HAI 447 was 40%. These results were different from the results of Berovic et al. (2003), using strain *G. lucidum* MZKI, which showed that 17% of inoculum was the most suitable. Using 14% inoculum was not sufficient to support growth intensively, due to the low production of EPS activators and growth factors. An inoculum of 20% concentration was too high, inducing competition in substrate consumption. Fang et al. (2002b) and Fang et al. (2002) obtained at an inoculation density of 330mg dry weight/L a maximal cell concentration of 15.7g dry cell weight/L. Within an inoculation density range of 70–670mg dry weight/L, a larger inoculation density led to small pellet size and high production of EPS and IPS, while a relatively large pellet size and high accumulation of ganoderic acids were observed at a low inoculation density. The conclusion was that small pellet size resulted in high polysaccharide production, while large pellet size led to high production of ganoderic acids. Similar results were reported by Hsieh et al. (2006b).

9.3.1.2 The effect of medium initial pH

A medium initial pH can greatly affect cell membrane function, cell growth, morphology and structure, salt solubility, the ionic state of substrates, uptake of various nutrients, and product biosynthesis. Yang and Liau (1998) found that pH values from 4.0 to 4.5 in a glucose-NH_4Cl medium in a range were optimal for *G. lucidum* production of EPS (1.6mg/mL); with decreasing pH, the amount of EPS declined. Lee et al. (1999) reported that pH control affected *G. lucidum* mycelial growth and exopolysaccharide production. Bistage pH control retained the desirable morphologies of the mycelia during cultivation, and resulted in low viscosity of the culture broth.

Fang and Zhong (2002a) investigated the effects of initial pH of 3.5–7.0 on production of *G. lucidum* polysaccharides and ganoderic acids. The initial pH value had a significant effect on the cell growth and product biosynthesis. At a pH of 6.5, a biomass of 17.3 ± 0.12g/L was obtained, and the production of ganoderic acid was 1.20 ± 0.03mg. Gradually lowering the initial pH from 6.5 to 3.5 led to higher production of EPS and IPS. In *G. lucidum* CCRC36021 cultivation in a medium composed of thin stillage and wastewater resulting from the rice wine production, the lowest level of biomass and exopolysaccharide production was at pH 4.0, and the highest at pH 5.0 (7.8 and 7.5g/L); with further pH increase, the yields decreased (Hsieh et al., 2005). Hsieh et al. (2006a,b) reported that exopolysaccharide and specific IPS production increased with a decrease in the initial medium pH. This can be explained by the fact that cells continuously metabolized the carbon source and secreted various ganoderic and lucidenic acids in the medium during the first few days of cultivation (Berovic et al., 2003; Hsieh et al., 2005). Using a liquid substrate based on whey permeate, optimum mycelial growth was observed at pH 4.4 and cultivation temperature of 30°C (Song et al., 2007).

9.3.1.3 The influence of aeration and agitation

The oxygen in a liquid substrate affects cell growth, cellular morphology, nutrient uptake, and metabolite biosynthesis, while agitation and aeration have a great influence on biomass and polysaccharide production, particularly EPS (Yang and Liau, 1998; Tang and Zhong, 2003). Stationary conditions seem to be more suitable for IPS production (Simonic et al., 2008). According to Yang and Liau (1998), while higher agitation speeds enhanced mixing efficiency and polysaccharide release, the higher shear stress had a detrimental effect on mycelial growth and polysaccharide formation. Tang and Zhong (2003) reported on the effects of oxygen supply on submerged cultivation of *G. lucidum* in a 3.5-L bioreactor with two six-bladed turbine impellers for agitation. An initial volumetric oxygen transfer coefficient (k_La) value within the range of 16.4–96.0/h had a significant effect on cell growth, cellular morphology, and metabolite biosynthesis. An increase in the initial coefficient led to a larger size of mycelia aggregates and a higher production of ganoderic acids. Mycelial growth was significantly inhibited when dissolved oxygen tension was about 10% of air saturation. Hsieh et al. (2006b) observed optimum polysaccharide production under conditions of sufficient oxygen during the first few days of cultivation. Zhang and Zhong (2010) studied the effects of oxygen concentration of 21–100% in gaseous phase on the morphology and ganoderic acid production by *G. lucidum* in liquid stationary culture. A higher oxygen concentration increased individual ganoderic acid production, and more spores and higher total ganoderic acid content were obtained at an oxygen level of 80%.

9.3.1.4 The influence of substrate composition

The substrate compositions for submerged cultivation of *G. lucidum* mycelia differ from author to author. For instance, an optimized medium for a shake-flask submerged culture of *G. lucidum*, as reported by Chang et al. (2006), consisted of 1.88 g/L $CaCO_3$, 71.4 g/L brown sugar, 12.1 g/L malt extract, 2.28 g/L yeast extract, 18.4 g/L skim milk, 3.44 g/L safflower seed oil, and 3.96 g/L olive oil, at pH 6.5. Compared to an unoptimized substrate, mycelium formation was markedly improved, from 1.70 to 18.70 g/L, and polysaccharide production increased from 0.140 to 0.420 g/L. Yang and Liau (1998) studied the influence of cultivation parameters on polysaccharide formation by *G. lucidum* in submerged cultures. The substrate consisted of 50 g/L glucose, 0.5 g/L K_2HPO_4, 0.5 g/L KH_2PO_4, 0.5 g/L $MgSO_4{\cdot}7H_2O$, 1 g/L yeast extract, and 4 g/L ammonium chloride. Optimal temperature was 30–35°C and the pH was 4–4.5. Tang and Zhong (2003) used a medium consisting of 35 g/L lactose, 5 g/L peptone, 2.5 g/L yeast extract, 1 g/L $KH_2PO_4{\cdot}H_2O$, 0.5 g/L $MgSO_4{\cdot}7H_2O$, and 0.05 g/L vitamin B_1. Cultivation was conducted at 30°C in the dark. Numerous studies have shown that biomass and polysaccharide production obtained using various *G. lucidum* strains were dependent on the type and concentration of carbon source, such as glucose, maltose, or saccharose. Glucose and maltose were confirmed as suitable carbon sources for mycelial biomass production, but not saccharose (Song et al., 2007). Using complex substrates, Fang and Zhong (2002a) noted the highest biomass (16.7 g/L) and IPS (1.19 g/L) production with a glucose-peptone-yeast extract medium, at initial glucose concentration of 50 g/L; at higher concentrations the yields were lower due to high osmotic pressure. However, EPS production of 1.08 g/L increased with an increase of glucose concentration, and the maximum value was observed at glucose concentration of 65 g/L. Berovic et al. (2003) observed significant EPS and IPS production by *G. lucidum* in a liquid potato dextrose-olive oil medium (9.6 and 6.3 g/L, respectively). Hsieh et al. (2005) reported that biomass production increased significantly with an increase of molasses concentration, but the contrary was observed for EPS and IPS production. According to Xu et al. (2008), a composition of 16 g/L glucose, 2.93 g/L peptone, 20.93 g/L corn flour, and 6.44 g/L soybean powder resulted in 496 mg/L ganoderic acid production. Zhu et al. (2008) studied the stimulating effects of fungal elicitors on the production of ganoderic acids and polysaccharides in a submerged culture. Lee et al. (2003a,b) used deproteinated cheese whey for cultivating mycelia of *G. lucidum* in a bioreactor by submerged cultivation. However, proper mixing with other solid media was necessary to adjust the C/N ratio.

9.3.1.5 Influences of carbon and nitrogen sources, and C/N ratio

Tang and Zhong (2002) studied the effects of carbon sources and initial sugar concentration on the production of ganoderic acids and polysaccharides in a *G. lucidum* cultivation process. Sucrose was suitable as a carbon source for EPS polysaccharide production, although the cells did not grow well. Lactose was beneficial for cell growth and production of ganoderic acid and IPS. When the initial lactose concentration exceeded 35 g/L, ganoderic acid accumulation decreased. However, ganoderic acid production improved remarkably with pulse feeding of lactose, when the residual concentration was between 10 and 5 g/L. Fang and Zhong (2002a,b) studied the effects of nitrogen source and initial glucose concentration in submerged cultivation of *G. lucidum* for simultaneous production of bioactive ganoderic acids and polysaccharides. The cells could not grow well when either yeast extract or peptone was used as the sole nitrogen source. However, addition of 5 g/L of yeast extract and 5 g/L of peptone was optimal for cell growth and metabolite production. Initial glucose concentration within 20–65 g/L greatly affected cell growth and product biosynthesis. Different responses of polysaccharide production were observed under different limitations of nutrients. Of the carbon sources examined by Chang et al. (2006), brown sugar and malt extract most effectively promoted mycelial growth and the production of polysaccharides, while yeast extract and skim milk were the most effective nitrogen sources. By adding safflower seed oil in the medium, formation of mycelia increased significantly.

9.3.1.6 The effects of nitrogen sources and concentrations

Several studies have shown that organic nitrogen sources were more suitable than inorganic ones for biomass and polysaccharide production, due to the fact that essential amino acids could not be synthesized from inorganic nitrogen sources. This observation was reconfirmed in a cultivation of *G. lucidum* HAI 447 (Simonic et al., 2008), where peptone was the optimum nitrogen source for biomass and for EPS production. According to Fang and Zhong (2002a), biomass and IPS production was maximal (15.8 and 0.9 g/L) with a combined usage of yeast extract and peptone in a ratio of 1:2, and EPS production in a ratio of 1:1 (0.8 g/L). However, in cases when either yeast extract or peptone was used as the sole nitrogen source, production was not optimal, and glucose consumption was incomplete. Fang and Zhong (2002a) explained this observation by the fact that yeast extract and peptone had complementary effects in promoting the cell growth. Hsieh et al. (2006a,b) noted the best biomass production (5.6 g/L) and the lowest EPS production (1.4 g/L) in a medium with a corn-steep powder as the nitrogen source, at a concentration of 5.0 g/L. Biomass production increased with an increase of nitrogen source concentration, while polysaccharide production was higher at a lower nitrogen concentration.

9.3.1.7 The influence of macro- and microelements

Macroelements, especially P and Mg, have significant effects on biomass and polysaccharide production, as they are cofactors for some enzymes, catalyzing polysaccharide synthesis (Hsieh et al., 2006a,b). However, Chang et al. (2006) reported that Ca addition caused a decrease in biomass and an increase in polysaccharide production. The results of Simonic et al. (2008) on Fe influence on biomass and polysaccharide production were in accordance with the previous results of Park et al. (2001). The best results were obtained in the medium containing maltose, peptone, K, Na, and Mg, inoculated with 20 mL of inoculum at pH 5.5 and under constant aeration. In a study by Song et al. (2007), when the cultivation medium was enriched with K, Na, and Mg at concentrations of 6.0, 3.0, and 2.0 mM, respectively, a mycelial biomass 29.2 g/L and EPS production of 2.9 mg/mL were obtained. IPS production was increased with an increase of K and Na concentration, while in the case of Mg it was 53.3 mg/g.

9.3.1.8 The effects of plant oils and fatty acids

Several plant oils, which can be used as antifoam agents, may be beneficial to fungal growth. Lee et al. (1999) used a commercial brand of an antifoaming agent at a volume fraction of 0.5%, while Habijanic and Berovic (2002) used olive oil at a volume fraction of 2%. The advantages of plant oils are their low cost and the possibility of higher growth yields due to their stimulatory properties. The effects of soy, peanut, safflower, corn, sunflower, and olive oils were investigated by Yang et al. (2000) at a volume fraction of 1%. EPS production was highest with safflower oil. In studying the effects of fatty acids, no general relationship could be identified between either the length of the carbon chain or the extent of its unsaturation and the level of growth stimulation.

9.3.1.9 The effect of polymer additives

Through the addition of a polymer into the medium, very uniform and tiny pellets could be formed during inoculum preparation. The addition of polymers, such as Carbopol 934 (carboxypolymethylene), Junlon (polyacrylic acid), and Hostacerin (sodium polyacrylate), reduced agglomeration, and hence decreased pellet formation in many filamentous fungi (Jones et al., 1988). Junlon was found to be effective in preventing mycelia aggregation. Polymer additives and a novel agitation scheme were employed by Yang et al. (2009) to vary mycelia morphology, which in turn led to higher mycelium production rates. The effect of various polymer additives, including sodium alginate, polyacrylmide, and agar, on mycelia growth was investigated.

9.3.1.10 **Ganoderma lucidum** *cultivation in an STB*

The optimum agitation rate represents a balance between oxygen transfer into the medium and shear forces, both of which increase with the agitation rate. Yang and Liau (1998) suggested that higher shaking speeds favored EPS production, because they decreased the adsorption of the secreted EPS on the cell wall, providing stimulus for further EPS synthesis. However, a mechanism by which this stimulus would occur was not proposed. Berovic et al. (2003) cultivated *G. lucidum* in an STB using a liquid substrate based on potato dextrose and olive oil. A cultivation temperature of 30°C, mixing at 300 rpm, aeration of 1 L/min, pH 5.8–4.2, oxygen partial pressure of 70–80%, and redox potential Eh 300–400 mV were applied.

9.3.1.11 **Ganoderma lucidum** *cultivation in airlift bioreactor*

Lee et al. (1999) cultivated *G. lucidum* in an airlift bioreactor at 25°C with an aeration rate of 2.5 vvm under batch conditions. With the bistage pH control technique, in which pH was shifted from 3 to 6 at the initial phase of exponential growth, 20.1 g/L of exopolysaccharides were produced; desirable mycelia morphologies were retained during the cultivation, and a low viscosity and yield stress in the culture broth were maintained.

9.3.2 SOLID-STATE CULTIVATION OF *G. LUCIDUM*

Solid-state cultivation of *G. lucidum* has several advantages, such as low investment cost, easier operation, and minimal waste production, as well as some disadvantages, including longer cultivation time and the difficulty of separating/isolating the biomass and pharmaceutically active compounds from the solid substrate. Data on *G. lucidum* solid-state cultivation in bioreactors are scarce. Habijanic and Berovic (2002) patented a process of growing *G. lucidum* on a solid cultivation substrate using solid-state cultivation in a horizontal stirred bioreactor. Beech sawdust was used as the cultivation substrate. Large quantities of biomass were produced, including pharmaceutically applicable products.

9.3.2.1 The influence of substrate composition

For the production of *G. lucidum* mycelia in a bioreactor by Habijanic and Berovic (2000, 2002), the substrate consisted of beech sawdust, olive oil, $(NH_4)_2SO_4$, KH_2PO_4, $CaCl_2{\cdot}2H_2O$, $MgSO_4{\cdot}7H_2O$, $FeSO_4{\cdot}7H_2O$, and distilled water. During cultivation, the content of EPS in the solids increased rapidly over the first 7 days, remained relatively constant until 21 days, and then decreased, suggesting that the polysaccharide was actually degraded in the latter stages of the process. The period during which the polysaccharide content decreased corresponded with the time in which the water mass fraction was falling rapidly from the values of 70–80% that were maintained during the first 21 days to 20% at 35 days (Habijanic and Berovic, 2002). Song et al. (2007) used a whey permeate as an alternative additive to a growth medium for the cultivation *G. lucidum* mycelia in a solid-state cultivation. Optimal conditions were found at pH 4.4 and 29°C. The results showed that the cultivation of *G. lucidum* mycelia could be a potential cost-effective solution for treatment of cheese whey permeates. However, proper mixing with other solid media was necessary to adjust the C/N ratio. In a rare report, cultivation was carried out in a horizontal STB with a total working volume of 30L (Habijanic and Berovic, 2002). Conditions were controlled as follows: temperature 30°C, airflow 2L/min, agitation rate 80rpm for 2min every second day during the first 7 days, and every day during the latter stages of the cultivation. The effect of initial moisture content was evaluated. An a_w of at least 0.85 was necessary to get satisfactory rates of cell growth and exopolysaccharide production.

9.3.3 SUBMERGED CULTIVATION OF *G. FRONDOSA*

9.3.3.1 Inoculum

In a study carried out by Bae et al. (2005), the stock culture was maintained on potato dextrose agar slants, incubated at 27°C for 5 days, and then stored at 4°C. The seed culture was grown at 27°C on a rotary shaker incubator at 120rpm for 5 days in a 250-mL flask containing 50mL of the following medium: 30g/L glucose, 6g/L yeast extract, 2g/L polypeptone, 0.5g/L $MgSO_4{\cdot}7H_2O$, 0.5g/L K_2HPO_4, and 0.2g/L $MnSO_4{\cdot}5H_2O$. The cultivation media were inoculated with 3% (v/v) of the seed culture and then cultivated at 27°C in a 5-L STB.

9.3.3.2 The effect of initial pH

Lee et al. (2004) reported that pH 5.5 was the optimal initial level for both mycelial biomass and EPS production with *G. frondosa* in a submerged culture. These results were reconfirmed by Shih et al. (2008), and were consistent with the fact that many fungi have acidic pH optima during submerged cultivation (Yang and Liau, 1998; Fang and Zhong, 2002a; Hsieh et al., 2006a,b).

9.3.3.3 The effects of carbon and nitrogen sources

The influence of carbon sources for mycelial biomass and EPS production was studied by Shih et al. (2008). In media containing various carbon sources, i.e., sucrose, maltose, glucose, fructose, and molasses, each carbon source was added to the basal medium at 20 or 40 g/L. Maltose and glucose were preferred carbon sources for high mycelial production. The concentration of EPS was doubled (0.76–1.49 g/L) after 13 days of cultivation when added glucose increased from 2 to 4 g/L. Yeast extract in combination with corn-steep powder (0.4:0.6 or 0.8:1.2 g/L) were preferred nitrogen sources for high production of mycelia and EPS. The conclusion was that an increase of carbon source concentration led to an increase in EPS production for all carbon sources tested; in contrast, mycelium dry weight showed no substantial increase when the concentration of carbon source increased (Zhu, 2006).

9.3.3.4 The effects of plant oils and surfactants

Under high oxygen concentration supply in an STB, the production of mycelial biomass and EPS was greatly enhanced with 1% olive oil addition in 21% O_2 and 40% O_2 (Hsieh et al., 2006a). In a further study by Hsieh et al. (2008), the effects of adding various plant oils and surfactants were examined in terms of cell growth and production of EPS and IPS in a submerged culture of *G. frondosa* with 2% and 4% glucose medium. Olive, safflower seed, soy, and sunflower oils were favorable plant oil sources for mycelial growth. The highest cell growth, 12.6 g/L, was obtained after 13 days of cultivation in a medium containing 1% of plant oils. EPS polysaccharide production was slightly enhanced by olive oil, but significantly inhibited by safflower seed oil and sunflower oil after 13 days of cultivation. Among four plant oil sources examined, cell growth yielded relatively high mycelial biomass of 11.22 g/L in 4% glucose medium with 0.5% soybean oil. High EPS polysaccharide production of 2.24 g/L and slightly lower cell growth were found with 4% glucose media with olive oil addition. Tween 80 and Span 80 addition showed increased cell growth. A maximal cell concentration of 9.10 g/L was obtained with 1% Span 80 addition. However, both EPS and IPS production decreased with all the tested concentrations of Tween 80 and Span 80 addition. The worst results were obtained with Tween 20 and Span 20 addition—these were shown to have serious inhibiting effects on cell growth of *G. frondosa* and also on polysaccharide production (Yang et al., 2000).

9.3.3.5 The effects of oxygen concentration

Hsieh et al. (2006a) examined the effects of oxygen concentrations of 21%, 30%, and 40% on the production of mycelial biomass and polysaccharides with *G. frondosa* in a 5-L STB. The 40% concentration of oxygen was found to inhibit both cell growth and polysaccharide production, with a higher rate of glucose consumption. The highest cell concentration of 9.29 g/L was found at day 9 with 21% O_2. Oxygen concentrations also affected intercellular polysaccharide production. With 21% O_2, intercellular polysaccharides reached a maximum concentration 19.5 mg/g at 11 days with 30% O_2, and 21.9 mg/g at 8 days and 28.2 mg/g with 40% O_2 at 7 days. On the other hand, EPS production increased from 0.7–0.9 to 2.24–3.00 g/L at 13 days with 21% O_2 and 40% O_2 aeration, respectively.

9.3.3.6 **Grifola frondosa** *cultivation in an STB*

To find the optimal operational parameters in a 5-L STB, the self-directing optimization technique was used by Lee et al. (2004), with culture pH 5.06, aeration rate of 1.16 vvm, and agitation rate of 166 rpm. Mycelial morphology was significantly altered by culture pH, aeration rate, and hydrodynamic characteristics, which subsequently affected the yield of EPS production. While compact pellets were formed under low aeration conditions, freely suspended mycelial growth was observed at high aeration conditions. Under optimal culture conditions, maximum biomass concentration and EPS production in an STB were 16.8 and 5.3 g/L, respectively, which were significantly greater than results prior to optimization (13 and 4 g/L, respectively).

Bae et al. (2005) used a 5-L STB with a working volume of 3 L, a 3% (v/v) inoculum of 5-day-old seed culture, and cultivation parameters of 27°C, aeration rate 1.0 vvm, agitation speed 150 rpm, and pH 5.5. Biomass concentration at 4 days was 24.8 g/L, and EPSs were 7.2 g/L. Xing et al. (2006) investigated the effects of nutritional parameters in batch and fed-batch cultivation in a 5-L STB. Cui et al. (2006) applied a 15-L STB equipped with three six-bladed Rusthon disc impellers. Batch cultivation was carried out under inoculation volume 10% (v/v), temperature 25°C, initial pH 5.5, aeration rate 8.0 vvm, and agitation speed 80 rpm. A maximum mycelial biomass yield of 22.50 g/L was achieved. Habijanic et al. (2009) cultivated *G. frondosa* in a 5-L airlift reactor on substrate that consisted of 45 g/L glucose, 6.5 g/L peptone, 3 g/L KH_2PO_4, and 7 g/L olive oil. Cultivation was carried out at 28°C and aeration of 50 L/h; 4.5 g/L dry biomass, 183 mg/L of EPS, and 480 mg/L IPS were obtained.

9.3.3.7 Grifola frondosa *cultivation in airlift bioreactor*

In a comparative study by Suzuki et al. (1989), *G. frondosa* was cultured in a 5 L airlift bioreactor. Mycelial biomass yields of 10 g/L and EPS yields of 4.53 g/L were lower than those produced in an STB. Lin and Liu (2006), in a 2-L airlift reactor, used 150 mL of seed culture for inoculating a medium consisting of 40 g of glucose, 10 g of malt extract, and 5 g of yeast extract per 1 L of distilled water, at pH 4.5. The cultivation was carried out at an aeration rate of 0.5 vvm and 25°C for 14 days. A maximum 12.8 g/L of biomass was produced.

9.3.4 SOLID-STATE CULTIVATION OF *G. FRONDOSA*

9.3.4.1 Substrates

Cheap substrates composed of secondary raw materials, such as agricultural, food, and wood industry residues, with little pretreatment or enrichment with mineral salts ($MgSO_4 \cdot 7H_2O$, K_2HPO_4, and $MnSO_4 \cdot 5H_2O$), can be used for *G. frondosa* cultivation (Berovic et al., 2008). Xing et al. (2006) used a substrate consisting of (dry weight basis) 75% of oak sawdust (25% humidity), 23% of corn bran (15% humidity), 1% sucrose (2% humidity), and 1% of calcium carbonate. The moisture content was then adjusted to 62%. Svagelj et al. (2008) cultivated *G. frondosa* in a 15-L horizontal STB using various substrates based on beech, Norwegian spruce, European larch, or common grape vine (*Vitis vinifera*) sawdust with 75% humidity. The best results were obtained with milled whole corn plant straw and the addition of (20% dry weight) wheat bran, 2% $CaCO_3$, 2% olive oil, and olive press cake supplemented with mineral additives at 28°C. Periodic mixing at 80 rpm was applied for 2 min/day and aeration of 5 L/h throughout the cultivation. Maintenance of the moisture content in the solid substrate was found to be of crucial importance. Moistures higher than 70% promoted the growth of *G. frondosa* mycelium and polysaccharide production. Four fractions of pure EPS β-D-glucans with a total mass 127.2 mg and four fractions of IPS with total mass 47.2 mg were isolated (Svagelj et al., 2008, Gregori et al., 2009).

9.3.5 CULTIVATION OF *T. VERSICOLOR*

9.3.5.1 Submerged cultivation of T. versicolor

Trametes versicolor cultivation was conducted on milk whey and beer worth (Klecak et al., 2009), tomato pomace (Freixo et al., 2008), barley husk, oak and cornbeam sawdust (Tisma et al., 2012), and bread crumbs in a submerged culture (Ivanova et al., 2014). The influence of various nutrient sources on EPS production and biomass yield by submerged culture of *T. versicolor* was investigated. The best EPS production by *T. versicolor* was achieved after stimulation with sucrose ranging within 1.00–1.80 and 1.80–2.00 mg/L. The lowest EPS production was recorded using xylose and glucose. Xylose and sorbitol supported the highest biomass yield (1.9 g dry weight/L and 2.0 g dry weight/L, respectively) by *T. versicolor*, and yeast extract induced the highest EPS production (7.83 mg/L). The EPS ranged within 0.70–0.78 mg/L. Yeast extract and $NaNO_3$ induced the highest biomass yield by *T. versicolor*; ranging within 1.0–1.6 g dry weight/L and 2.0–1.9 g dry weight/L. The least stimulatory amino acids were alanine and glutamate, while the best amino acids for biomass production by the isolates were aspartic acid and asparagines, in ranges within 1.0–15.0 g dry weight/L and 1.0–16.0 g dry weight/L, respectively.

9.3.5.2 The effects of carbon sources on biomass and EPS

Bolla et al. (2010) used a basal medium containing $MgSO_4 \cdot 7H_2O$ (0.2 g), K_2HPO_4 (1.0 g), NH_4SO_4 (5.0 g), D-glucose (9.75 g), yeast extract (3.0 g), peptone (1.0 g), and 1.0 L of distilled water. Also, 97.5 g of carbon of each carbon source (lactose, glucose, sucrose, maltose, mannose, galactose, fructose, xylose, sorbitol, and starch) were supplements to the basal medium. The pH of the basal medium was adjusted to 6. The basal medium without any carbon source served as control. Further, 100 mL of each basal medium was dispensed into a corresponding 500-mL flask and autoclaved at 121°C for 15 min. In order to suppress bacterial growth, 1 mL of streptomycin was added. Then the flasks were inoculated with 10% of 5-day-old actively growing cultures of *T. versicolor*. Finally, the flasks were incubated at 25°C for 2, 4, 7, and 14 days. The cultivation medium was then analyzed for biomass and EPS production.

9.3.5.3 The effects of nitrogen and amino acid sources on biomass and EPS

Bolla et al. (2010) used a basal medium made up of 10.0 g glucose, 0.1 g NaCl, 0.1 g $CaCl_2$, 0.5 g KH_2PO_4, 0.5 g $MgSO_4 \cdot 7H_2O$, 0.5 mg thiamine hydrochloride, and 1.00 L distilled water. The medium was supplemented separately with amino acids and inorganic nitrogen sources at a rate of 1.0 g per liter. Complex nitrogen sources (casein, urea, yeast extract, and peptone) were supplemented at a concentration of 2.0 g/L. The liquid without any nitrogen source served as control. Further, 100 mL of each liquid medium was dispensed into a conical flask and treated as described in the carbon medium experiment.

9.3.6 SOLID-STATE CULTIVATION OF *T. VERSICOLOR*

Knežević et al. (2013) used wheat straw and oak sawdust as carbon sources, and 10 mL of modified synthetic medium without glucose, with NH_4NO_3 and pH 6.5. Stoilova et al. (2010) used wheat bran and oat straw as carbon sources. Solid-state cultivation was carried out using a medium consisting of 4.0 g wheat bran, 2.5 g oat straw, and 2.5 g beetroot press in 300-mL flasks. The moisture of the substrate was adjusted to 60% by mineral salt solution containing 0.14% $(NH_4)_2SO_4$, 0.2% KH_2PO_4, 0.03% $MgSO_4{\cdot}7H_2O$, 0.03% $CaCl_2$, $FeSO_4{\cdot}7H_2O$, $ZnSO_4{\cdot}7H_2O$, $MnSO_4{\cdot}7H_2O$, and 0.002% $CoCl_2$ at pH 4.5. Rakus et al. (2015) cultivated *T. versicolor* mycelia in solid-state cultivation on corn straw in a 15-L horizontal STB. The moisture content of the substrate before inoculation was 2.33 (w/w dry biomass). The inoculum consisted of twenty-five 1 cm^2 cuts mixed in a sterile grinder together with 700 ml of sterilized distilled water. The cultivation of *T. versicolor* biomass was carried out at 24°C and the airflow was 5 L/min. Periodic mixing at 80 rpm per 2 min (in the first 7 days every second day, and 2 minutes every day in the last part of cultivation) was used. In the end, 5.95 g/L IPS were isolated.

9.3.7 SUBMERGED CULTIVATION OF *H. ERINACEUS*

Krzyczkowski et al. (2009) optimized the production of IPS and EPS of *H. erinaceus* in a submerged culture. Various factors, including carbon and nitrogen sources, vitamins, mineral elements, and initial pH, had significant effects when tested using a 2^4 central composite rotatable design (CCRD). Under optimal culture conditions, the maximum yield of biomass reached 14.24 g/L, 1.85-fold higher than in the basal medium. The kinetics of EPS biosynthesis in a bioreactor showed that the highest yield of 2.75 g/L of EPS could be obtained on the eighth day of cultivation. The process of biosynthesizing high-molecular-weight polysaccharides proceeded through 14 days of cultivation after the depletion of the carbon source in the medium. The results are very helpful in the large-scale production of bioactive polysaccharides from *H. erinaceus* (Krzyczkowski et al., 2009).

In the second study by Krzyczkowski et al. (2010), the most favorable combination of nutrient medium constituents, ensuring the highest erinacine production, was found to be 69.87 g/L glucose, 11.17 g/L casein peptone, 1.45 g/L NaCl, 55.24 mg/L $ZnSO_4$, 1.0 g/L KH_2PO_4, and pH 4.5. The kinetics of metabolite biosynthesis were examined during cultivation in a 10 L bioreactor. The biomass was 13.52 g/L, while the production of erinacine A was 192.42 mg/L. The best results were obtained after 8 days of cultivation. A new diterpene from the fungal mycelia of *H. erinaceus* was found by Zhang et al. (2015). The effects of medium composition and cultivation parameters on the simultaneous production of mycelial biomass and EPS by *H. erinaceus* CZ-2 were investigated in submerged cultures by a one-factor-at-a-time method using an orthogonal array design. The most suitable carbon, nitrogen, and mineral sources, and cofactors for the mycelial biomass and exopolymer production were corn flour combined with 1% glucose, yeast extract, KH_2PO_4, and corn steep liquor. The effects of nutritional requirements on the mycelial growth of *H. erinaceus* CZ-2 were in the regular sequence of corn flour combined with 1% glucose > yeast extract > corn steep liquor > KH_2PO_4, and those on exopolymer production were in the order of corn flour combined with glucose > KH_2PO_4 >yeast extract > corn steep liquor. A maximum of 16.07 g/L biomass was obtained when the composition of the culture medium was 30 g/L corn flour, 10 g/L glucose, 3 g/L yeast extract, 1 g/L KH_2PO_4, 0.5 g/L $CaCO_3$, and 15 mL/L corn steep liquor; 1.31 g/L exopolymer yield was achieved when the composition of medium was 30 g/L corn flour, 10 g/L glucose, 5 g/L yeast extract, 3 g/L KH_2PO_4, 0.5 g/L $CaCO_3$, and 15 g/L mL/L corn steep liquor. In the 15-L STB cultivation, a maximum of 20.50 g/L mycelial biomass was achieved using the optimized medium (Huang et al., 2007).

9.3.8 SOLID-STATE CULTIVATION OF *H. ERINACEUS*

The potential for using several agricultural by-products as supplements of sawdust substrate was evaluated by Ko et al. (2005). Gerbec et al. (2015) cultivated *H. erinaceus* in a 15-L horizontal STB (HSTB) using the design and construction of Berovic et al. (2012). The cultivation was carried out at 24°C and 5 L/min airflow. Periodic mixing at 80 rpm for 2 min (in the first 7 days every second day, and every day in the last part of cultivation) was used (Gerbec et al., 2015). Han (2003) studied *H. erinaceus* solid-state cultivation on corn meal degrading starch, and found an upgrading of its nutritional value.

9.3.9 SUBMERGED CULTIVATION OF *C. MILITARIS*

A study by Mao and Zhong (2006), using a chemically defined medium, indicated that NH_4^+ played an important role in cordycepin biosynthesis. In a complex medium, peptone was identified as the best nitrogen source for cordycepin biosynthesis (Song et al., 1998). Park et al. (2001) reported that exopolysaccharide production and mycelial growth in the submerged culture of *C. militaris* were almost tripled by supplementation with vegetable oils in the medium (Park et al., 2001). Mao and Zhong (2004) investigated submerged cultivation of *C. militaris* and the effect of oxygen supply on cordycepin production in a 5-L STB. Initial volumetric oxygen transfer coefficient (k_La) within the range of 11.5–113.8 h^{-1} had a significant influence on cordycepin production. A cordycepin concentration of 167.5 mg/L was obtained at an initial k_La value of 54.5 h^{-1}, where a moderate dissolved oxygen (DO) pattern was observed throughout cultivation. A higher cordycepin production of 201.1 mg/L was achieved by a proposed DO control strategy. Similar results were obtained by Park et al. (2002). Mao and Zhong (2006) also found in a fed-batch culture in a 3.5-L STB that NH_4^+ feeding with peptone enhanced cordycepin production. By optimizing the feeding time and feeding amount of NH_4^+, a maximal cordycepin concentration of 420.55 mg/L was obtained, which was 70% higher than in batch cultivation (Mao and Zhong, 2006).

9.3.10 SOLID-STATE CULTIVATION OF *C. MILITARIS*

Substrates of rye seeds and spent brewing grains were mixed in different ratios (9:1, 8:2, 7:3, 6:4, 5:5, 4:6). Water was added to the mixture to achieve 65% moisture content and 100 g of substrate. To date, concentrations of 10.42 mg/g of cordycepin have been reported (Gregori, 2014).

9.3.11 CULTIVATION OF OTHER MEDICINAL MUSHROOM SPECIES IN BIOREACTORS

Besides the cultures presented in this chapter, there are some other medicinal mushrooms, such as *Agaricus brasiliensis*, *Agrocybe argerita*, *Antrodia camphorata*, *Auricularia polytricha*, *Inonolus obliquus*, *Collybia marculate*, *Phelinius gilvus*, and *Sarcodon asparatus*, suitable to be cultivated in laboratory-scale STBs. *Schizophyllum commune* was cultivated in airlift and bubble column reactors. Although several medicinal mushroom species are suitable for large-scale production, only *A. camphorata* has reportedly been cultivated in a 500-L pilot-scale stirred reactor having 350 L of working volume (Tang et al., 2007).

9.4 CONCLUSIONS

The reports on pharmacological activity of extracts, partly purified preparations, and isolated compounds from medicinal mushroom biomass of *G. lucidum*, *G. frondosa*, *T. versicolor*, *H. erinaceus*, and *C. militaris* are based on experimental research. Their biotechnological cultivation in bioreactors has mostly been studied in laboratory- and pilot-scale installations. There is abundant scientific evidence that triterpenoids, polysaccharides, and proteoglycans have been effective in *in vitro* and *in vivo* testing. The synergistic effects of mixtures of active components have been known, but their biological activities need further assessment before they can be fully accepted, not only by traditional Asian medicine, but also by Western science and medicine. In this respect, modern biotechnological cultivation methods in bioreactors enable fast, efficient, and economic production of medicinal fungi in sufficient quantities for future potential production of pharmaceuticals at a large industrial level.

REFERENCES

Bae, J.T., Sim, G.S., Lee, D.H., Lee, B.C., Pyo, H.B., Choe, T.B., et al., 2005. Production of exopolysaccharide from mycelial culture of *G. frondosa* and its inhibitory effect on matrix metalloproteinase-1 expression in UV-irradiated human dermal fibroblasts. FEMS Microbiol. Lett. 251, 347–354.

Berovic, M., 2007. Bioreactor types. In: Encyclopedia of Life Support Systems. Eolss Publishers, Oxford, UK, 31 pp. [online] http://www.eolss.net/outlinecomponents/Biotechnology.aspx.

Berovic, M., Habijanic, J., Zore, I., Wraber, B., Hodzar, D., Boh, B., et al., 2003. Submerged cultivation of *G. lucidum* biomass and immunostimulatory effects of fungal polysaccharides. J. Biotechnol. 103, 77–86.

Berovic, M., Boh, B., Menard, A., Simcic, S., Wraber, B., 2008. Solid-state cultivation of *G. frondosa* (Dicks: Fr.) S.F. Gray biomass and immunostimulatory effects of fungal intra- and EPS β-polysaccharides. New Biotechnol. 25, 150–156.

Berovic, M., Habijanic, J., Boh, B., Wraber, B., Petravic-Tominac, V., 2012. Production of lingzhi or reishi medicinal mushroom, *G. lucidum*, biomass and polysaccharides by solid state cultivation. Int. J. Med. Mushrooms 14, 513–520.

Bolla, K., Gopinath, B.V., Shaheen, S.Z., Singara Charya, M.A., 2010. Optimization of carbon and nitrogen sources in *T. versicolor* submerged culture. Int. J. Biotechnol. Mol. Biol. Res. 1, 15–21.

Chang, M.Y., Tsai, G.J., Houng, J.Y., 2006. Optimization of the medium composition for the submerged culture of *Ganoderma lucidum* by Taguchi array design and steepest ascent method. Enzyme Microb. Technol. 38, 407–414.

Chen, J.Z., Seviour, R., 2007. Medicinal importance of fungal β-(1-3), (1-6)-glucans. Mycol. Res. 111, 635–652.

Cui, F.J., Li, Y., Xu, Z.H., Xu, H.Y., Sun, K., Tao, W.Y., 2006. Optimization of the medium composition for production of mycelial biomass and exo-polymer by *G. frondosa* GF9801 using response surface methodology. Bioresour. Technol. 10, 1209–1216.

Fang, Q.H., Zhong, J.J., 2002a. Effect of initial pH on production of ganoderic acid and polysaccharide by submerged cultivation of *G. lucidum*. Process Biochem. 37, 769–774.

Fang, Q.H., Zhong, J.J., 2002b. Submerged cultivation of higher fungus *Ganoderma lucidum* for production of valuable bioactive metabolites—ganoderic acid and polysaccharide. Biochem. Eng. J. 10, 61–65.

Fang, Q.H., Tang, Y.J., Zhong, J.J., 2002. Significance of inoculation density control in production of polysaccharide and ganoderic acid by submerged culture of Ganoderma lucidum. Process Biochem. 37, 1375–1379.

Freixo, M.R., Karmali, A., Arteiro, J.M., 2008. Production of polygalacturonase from *Coriolus versicolor* grown on tomato pomace its chromatographic behavior on immobilized metal chelates. J. Ind. Microbiol. Biotechnol. 35, 475–484.

Gerbec, B., Tavčar Benković, E., Gregori, A., Kreft, S., Berovič, M., 2015. Solid state cultivation of Hericium erinaceus biomass and erinacine: a production. J. Bioprocess. Biotechniques 5 (3), 1–5. http://www.omicsonline.org/open-access/solid-state-cultivation-of-hericium-erinaceus-biomass-and-erinacine-a-production-2155-9821.1000210.php?aid=40841, http://dx.doi.org/10.4172/2155-9821.1000210.

Gregori, A., 2014. Cordycepin production by *Cordyceps militaris* cultivation on spent brewery grains. Acta Biol. Slov. 57, 45–52.

Gregori, A., Svagelj, M., Berovic, M., Liu, Y., Zhang, J., Pohleven, F., et al., 2009. Cultivation and bioactivity assessment of *Grifola frondosa* fruiting bodies on olive oil press cakes substrates. New Biotechnol. 26, 260–262.

Habijanic, J., Berovic, M., 2000. The relevance of solid-state substrate moisturing on *Ganoderma lucidum* biomass cultivation. Food Technol. Biotechnol. 38, 225–228.

Habijanic, J., Berovic, M., 2002. Process of Cultivation of Fungus *Ganoderma lucidum* on a Solid Cultivation Substrate. Patent SI 20923, December 31, 2002.

Habijanic, J., Švagelj, M., Berovic, M., Boh, B., Wraber, B., 2009. Submerged and solid-state cultivation of bioactive extra- and intracellular polysaccharides of medicinal mushrooms *Ganoderma lucidum* (W. Curt.: Fr.) P. Karst. and *Grifola frondosa* (Dicks.: Fr.) S. F. Gray (Aphyllophoromycetideae). Int. J. Med. Mushrooms 11, 1–10.

Han, J., 2003. Solid-state cultivation of cornmeal with the basidiomycete *Hericium erinaceus* for degrading starch and upgrading nutritional value. Int. J. Food Microbiol. 80, 61–66.

Hsieh, C., Hsu, T.H., Yang, F.C., 2005. Production of polysaccharides of *Ganoderma lucidum* (CCRC36021) by reusing thin stillage. Process Biochem. 40, 909–916.

Hsieh, C., Liu, C.J., Tseng, M.H., Lo, C.T., Yang, Y.C., 2006a. Effect of olive oil on the production of mycelial biomass and polysaccharides of *Grifola frondosa* under high oxygen concentration aeration. Enzyme Microb. Technol. 39, 434–439.

Hsieh, C., Tseng, M.H., Liu, C.J., 2006b. Production of polysaccharides from *Ganoderma lucidum* (CCRC 36041) under limitations of nutrients. Enzyme Microb. Technol. 38, 109–117.

Hsieh, C., Wang, H.L., Chen, C.C., Hsu, T.H., Tseng, M.H., 2008. Effect of plant oil and surfactant on the production of mycelial biomass and polysaccharides in submerged culture of Grifola frondosa. Biochem. Eng. J. 38, 198–205.

Huang, D., Cui, F.-J., Li, Y., Zhang, Z., Zhao, J., Han, X.-M., et al., 2007. Nutritional requirements for the mycelial biomass and exopolymer production by *Hericium erinaceus* CZ-2. Food Technol. Biotechnol. 45, 389–395.

Ivanova, T.S., Bisko, N.A., Krupodorova, T.A., Barshteyn, V.Y., 2014. Breadcrumb as a substrate for *T. versicolor* and *S. commune* polysaccharides and protein in submerged cultivation. Korean J. Microbiol. Biotechnol. 42, 1–6.

Jones, P., Shabab, B.A., Trinci, A.P.J., Moore, D., 1988. Effect of polymeric additives, especially Junlon and Hostacerin, on growth of some basidiomycetes in submerged culture. Trans. Br. Mycol. Soc. 90, 577–583.

Klechak, I.R., Mitropolskaya, N.J., Antonenko, L.O., Nyshporska, O.I., 2009. The specificity of *Coryolus versicolor* growth in a deep culture. Res. Bull. NTUU "KPI" 20, 128–133.

Knežević, A., Milovanović, I., Stajić, M., Vukojević, J., 2013. *Trametes suaveolens* as ligninolytic enzyme producer. J. Nat. Sci. (Novi Sad) 124, 437–444.

Ko, H.G., Gu, H., Park, S.H., Choi, C.W., Kim, S.H., Park, W.M., 2005. Comparative study of mycelial growth and basidiomata formation in seven different species of the edible mushroom genus *Hericium*. Bioresour. Technol. 96, 439–1444.

Krzyczkowski, W., Malinowska, E., Suchocki, P., Kleps, J., Olejnik, M., Merold, F., 2009. Isolation and quantitative determination of ergosterol peroxide in varios eidble mushroom species. Food Chem. 113, 351–355.

Krzyczkowski, W., Malinowska, E., Herold, F., 2010. Erinacine A biosynthesis in submerged cultivation of *Hericium erinaceus*: quantification and improved cultivation. Eng. Life Sci. 10, 446–457.

Lee, B.C., Bae, J.T., Pyo, H.B., Choe, T., Kim, S.W., Hwang, H.J., et al., 2004. Submerged culture conditions for the production of mycelial biomass and EPS by the edible Basidiomycete *Grifola frondosa*. Enzyme Microb. Technol. 35, 369–376.

Lee, H., Song, M., Yu, Y., Hwang, S., 2003a. Optimizing bioconversion of deproteinated cheese whey to mycelia of *Ganoderma lucidum*. Process Biochem. 38, 1685–1693.

Lee, H., Song, M., Yu, Y., Hwang, S., 2003b. Production of *Ganoderma lucidum* mycelium using cheese whey as an alternative substrate: response surface analysis and biokinetics. Biochem. Eng. J. 15, 93–99.

Lee, K.M., Lee, S.Y., Lee, H.Y., 1999. Bistage control of pH for improving exopolysaccharide production from mycelia of *Ganoderma lucidum* in an air lift fermentor. J. Biosci. Bioeng. 88, 646–650.

Lin, J.T., Liu, W.H., 2006. O-orsellinaldehyde from the submerged culture of the edible mushroom Grifola frondosa exhibits selective cytotoxic effect against Hep 3B cells through apoptosis. J. Agric. Food Chem. 54, 7564–7569.

Mao, X.B., Zhong, J.J., 2004. Hyperproduction of cordycepin by two-stage dissolved oxygen control in submerged cultivation of medicinal mushroom *Cordyceps militaris* in bioreactors. Biotechnol. Progress 20, 1408–1413.

Mao, X.B., Zhong, J.J., 2006. Significant effect of NH_4^+ on cordycepin production by submerged cultivation of medicinal mushroom *C. militaris*. Enzyme Microb. Technol. 38, 343–350.

Mitchell, D.A., Berovic, M., 2014. Solid state bioprocessing. In: Berovic, M., Lübbert, A. (Eds.), Principles of Biochemical Engineering University of Ljubljana, FKKT Press, Ljubljana, Slovenia, pp. 244–277.

Park, J.P., Kim, S.W., Hwang, H.J., Yun, J.W., 2001. Optimization of submerged culture conditions for the mycelial growth and exo-biopolymer production by *Cordyceps militaris*. Lett. Appl. Microbiol. 33, 76–81.

Park, J.P., Kim, S.W., Hwang, H.J., Cho, Y.J., Yun, J.W., 2002. Stimulatory effect of plant oils and fatty acids on the exo-biopolymer production in *Cordyceps militaris*. Enzyme Microb. Technol. 31, 250–255.

Shih, I.L., Chou, B.W., Chen, C.C., Wu, J.Y., Hsieh, C., 2008. Study of mycelial growth and bioactive polysaccharide production in batch and fed-batch culture of *Grifola frondosa*. Bioresour. Technol. 99, 785–793.

Simonic, J., Stajic, M., Glamoclija, J., Vukojevic, J., Duletic Lausevic, S., Brceski, I., 2008. Optimization of submerged cultivation conditions for extra- and intracellular polysaccharide production by medicinal Ling Zhi or Reishi mushroom *G. lucidum* (W. Curt.: Fr.) P. Karst. (Aphyllophoromycetideae). Int. J. Med. Mushrooms 10, 351–360.

Song, C.H., Jeon, Y.J., Yang, B.K., Ra, K.S., Sung, J.M., 1998. Anti-complementary activity of exopolymers produced from submerged mycelial cultures of higher fungi with particular reference to *Cordyceps militaris*. J. Microbiol. Biotechnol. 8, 536–539.

Song, M., Kim, N., Lee, S., Hwang, S., 2007. Use of whey permeate for cultivating *Ganoderma lucidum* mycelia. J. Dairy Sci. 90, 2141–2146.

Stoilova, I., Krastanov, A., Stanchev, V., 2010. Properties of crude laccase from *T. versicolor* produced by solid-substrate cultivation. Adv. Biosci. Biotechnol. 1, 208–215.

Suzuki, I., Hashimoto, K., Oikawa, S., Sato, K., Osawa, M., Yadomae, T., 1989. Antitumor and immunomodulating activities of a beta-glucan obtained from liquid cultured *Grifola frondosa*. Chem. Pharm. Bull. 37, 410–413.

Svagelj, M., Berovic, M., Boh, B., Menard, A., Simcic, S., Wraber, B., 2008. Solid-state cultivation of *Grifola frondosa* (Dicks: Fr) S.F. Gray biomass and immunostimulatory effects of fungal intra- and extracellular ß-polysaccharides. New Biotechnol. 25, 150–155.

Tang, Y.J., Zhong, J.J., 2002. Fed-batch fermentation of *Ganoderma lucidum* for hyperproduction of polysaccharide and ganoderic acid. Enzyme Microb. Technol. 31(1), 20–28.

Tang, Y.J., Zhong, J.J., 2003. Role of oxygen supply in submerged cultivation of *Ganoderma lucidum* for production of *Ganoderma* polysaccharide and ganoderic acid. Enzyme Microb. Technol. 32, 478–484.

Tang, Y.J., Zhu, L.W., Li, H.M., Li, D.S., 2007. Submerged cultivation of mushrooms. Food Technol. Biotechnol. 45, 221–229.

Tisma, M., Znidarsic-Plazl, P., Vasic-Racki, D., Zelic, B., 2012. Optimization of laccase productivity by *Trametes versicolor* cultivation on industrial waste. Appl. Biochem. Biotechnol. 166, 36–46.

Wasser, S.P., 2002. Medicinal mushrooms as a source of antitumor and immunomodulating polysaccharides. Appl. Microbiol. Biotechnol. 60, 258–274.

Xing, Z.T., Cheng, J.H., Tan, Q., Pan, Y.J., 2006. Effect of nutritional parameters on laccase production by the culinary and medicinal mushroom, *Grifola frondosa*. World J. Microbiol. Biotechnol. 22, 799–806.

Xu, P., Ding, Z.Y., Qian, Z., Zhao, C.X., Zhang, K.C., 2008. Improved production of mycelial biomass and ganoderic acid by submerged culture of *Ganoderma lucidum* SB97 using complex media. Enzyme Microb. Technol. 42, 325–331.

Yang, F.C., Liau, C.B., 1998. The influence of environmental conditions on polysaccharide formation by *Ganoderma lucidum* in submerged cultures. Process Biochem. 33, 547–553.

Yang, F.C., Ke, Y.F., Kuo, S.S., 2000. Effect of fatty acids on the mycelial growth and polysaccharide formation by *Ganoderma lucidum* in shake-flask cultures. Enzyme Microbial. Technol. 27, 295–301.

Yang, F.C., Yang, M.-J., Cheng, S.H., 2009. A novel method to enhance the mycelia production of *Ganoderma lucidum* in submerged cultures by polymer additives and agitation strategies. J. Taiwan Inst. Chem. Eng. 40, 148–154.

Zhang, W.X., Zhong, J.J., 2010. Effect of oxygen concentration in gas phase on sporulation and individual ganoderic acids accumulation in liquid static culture of *Ganoderma lucidum*. J. Biosci. Bioeng. 109, 37–40.

Zhang, Z., Liu, R., Tang, Q., Zhang, J., Yang, Y., Shang, X., 2015. A new diterpene from the fungal mycelia of *Hericium erinaceus*. Phytochem. Lett. 11, 151–156.

Zhu, L.W., Zhong, J.J., Tang, Y.J., 2008. Significance of fungal elicitors on the production of ganoderic acid and *Ganoderma* polysaccharides by the submerged culture of medicinal mushroom *G. lucidum*. Process Biochem. 43, 1359–1370.

Zhu, Y., 2006. Health Promoting Dairy and Food Products Containing Mushroom Glucan Produced Through Cultivation of *Grifola frondosa*. European Patent EP 1709969 A1, October 11, 2006.

Zore, I., Berovic, M., Boh, B., Hodzar, D., Pohleven, F., 1998. Procedure for Preparation of Inoculum for Growing of Fungus *Ganoderma lucidum* by Submersion Cultivation. SI Patent SI 9700014, August 31, 1998.

CHAPTER

10 USE OF *ASPERGILLUS NIGER* EXTRACTS OBTAINED BY SOLID-STATE FERMENTATION

Noelia Pérez-Rodríguez[1,2], Ana Torrado-Agrasar[3] and José M. Domínguez[1,2]

[1]Department of Chemical Engineering, Faculty of Sciences, University of Vigo (Campus Ourense), Ourense, Spain
[2]Laboratory of Agro-food Biotechnology, CITI (University of Vigo)-Tecnópole, Technological Park of Galicia, San Cibrao das Viñas, Ourense, Spain [3]Bromatology Group, Department of Analytical and Food Chemistry, Faculty of Sciences, University of Vigo (Campus Ourense), As Lagoas, Ourense, Spain

10.1 AGRO-FOOD INDUSTRIAL WASTES AS RAW MATERIALS

Agricultural and agro-food by-products or residues such as bagasse, marc, bran, husk, shell, peel, pits, seeds, straw, leaves, or prunings from various crops (sugarcane, wheat, soy, rice, barley, corn, apple, orange, grapes, vine, treenuts, olives, cotton) are constantly generated as a result of industrial processing (Devesa-Rey et al., 2011; Dhillon et al., 2011; Mirabella et al., 2014; Sánchez, 2009).

These by-products have no major applications in the production line, being rarely employed as an alternative substrate for a second process or use. As a consequence, their disposal causes problems in various ways. Additionally, residues must be kept until their disposal, occupying space in the industrial enclosure or in fields, resulting in extra costs for companies. This drawback is usually avoided by burning some agro-food wastes, mainly those of a lignocellulosic nature, although this practice is tending to be banned since it leads to environmental pollution problems as well as a risk to human health, due to the formation of toxic compounds from lignin combustion (Bustos et al., 2004).

Another common treatment involves composting of organic matter, which could be applied in the ground as a natural fertilizer. It must be taken into account that composting generates leachates that can infiltrate the subsoil and reach groundwater, thus becoming an environmental problem if not correctly treated (García-López et al., 2014; Mirabella et al., 2014). Nevertheless, there are several disadvantages associated with this type of treatment, since attainment of compost requires considerable time, usually months, and it is an activity which does not get full use of the potential residue as a raw material. Once obtained, the use of compost from agroindustrial residues rarely meets nutritional soil requirements because of its unbalanced C/N composition and the possible presence of toxigenic compounds such as heavy metals (Cu, Ni, Zn, Pb, Cd) or herbicides that could destroy or pollute crops and soils. These aspects necessitate a strategy of adjusting compost composition to plant needs and removing such harmful compounds from composts to make them suitable for soil nourishment (de Alencar Guimaraes et al., 2013; Facchini et al., 2011; Özbaş and Balkaya, 2014; Sánchez, 2009; Santos et al., 2014).

Mushroom Biotechnology. DOI: http://dx.doi.org/10.1016/B978-0-12-802794-3.00010-2

Although some residues have been employed for animal feeding, there is no doubt that this alternative does not take advantage of the whole waste potential for the production of high value-added products, dismissing a good opportunity, considering that agricultural and food industry residues represent an abundant and inexpensive alternative source of valuable compounds.

Currently, a growing number of scientific works in the biorefinery field aim at "zero waste" disposal, where wastes are used as raw material for new products such as saccharides, organic acids, or phenolic compounds by such clean technologies as fermentation or enzymatic hydrolysis. In this way, an initial waste and environmental problem is turned into a cheap and readily available substrate. Consequently, these residues have attracted considerable interest because of their possible use in biotransformation to obtain "low volume-high cost" products or bioenergy (Mirabella et al., 2014). Because of all that, agricultural and food industry by-products and residues represent an inexpensive alternative source for microbial growth and enzyme production mainly using solid-state fermentation (SSF) (de Alencar Guimaraes et al., 2013; Haltrich et al., 1997).

SSF is useful for valorizing complex materials including agricultural, forestry, and food processing residues and wastes. Valorization as raw materials can be achieved, on the one hand, by using lignocellulosic materials as fungal mycelia support and a nutritional source for fungi growth and enzymatic production. On the other hand, different compounds can be enzymatically released from agriculture and food industry residues matrices, including saccharides, organic acids, or phenolic compounds among others, with many applications in sundry industrial fields.

10.2 LIGNOCELLULOSIC COMPOSITION OF AGROINDUSTRIAL WASTES

Most of the potentially valuable residues from agricultural, food, and feed industries have a lignocellulosic nature. The crops' plant cell wall is mainly composed of three polymers: cellulose, hemicellulose, and lignin. Cellulose and hemicellulose represent the two major quantitative polysaccharides in plant cell wall, and lignin is the most abundant noncarbohydrate constituent (Berlin et al., 2005; de Vries and Visser, 2001). The principal component of lignocellulosic biomass is cellulose, a linear homopolymer composed of D-glucose units linked by β-1,4 glucosidic bonds (de Vries and Visser, 2001). The second most abundant organic structure in the plant cell wall is hemicellulose, which is a heterogeneous polysaccharide. The predominant polymer in hardwood hemicelluloses fraction is xylan, comprising up to 20–35% dry weight of agricultural wastes. Xylan consists mainly of a backbone chain of β-1,4-linked D-xylopyranose residues, which are substituted by different side groups such as L-arabinose, D-galactose, acetyl, feruloyl, *p*-coumaroyl, and glucuronic acid residues, which represent a small fraction (de Vries and Visser, 2001; Pérez-Rodríguez et al., 2014). Finally, lignin has a complex, heterogeneous, polymeric structure derived from three monolignol monomers, methoxylated to various degrees: *p*-coumaryl alcohol, coniferyl alcohol, and sinapyl alcohol (Majumdar et al., 2014).

The hemicellulose chains coat the cellulose fibrils and interact with lignin, being all entangled, thereby ensuring the mechanical and structural stability of the plant cell wall, limiting cell growth, and supplying resistance to chemical and biological degradation agents (de Vries and Visser, 2001). Given that the already mentioned characteristics provided by hemicellulose, cellulose, and lignin help the plant to persist in natural environment, degradation of polysaccharides becomes a complex step in the industrial use of lignocellulosic materials.

Chemical processing for the fractionation and valorization of lignocellulosic materials has been employed for years in various industries. However, an increasing awareness of environmental pollution has forced industries to replace the use of harsh chemicals with environmentally friendly procedures (de Alencar Guimaraes et al., 2013). Enzymatic hydrolysis is a greener technology which provides a good alternative to preserve nature, especially when a subsequent food or feed use is required.

10.3 ENZYMES INVOLVED IN LIGNOCELLULOSE DEGRADATION

Several classes of enzymes are involved in the biodegradation of cellulose, hemicelluloses, and lignin fractions. The cellulose fraction can be hydrolyzed by four classes of enzymes, namely endoglucanases, cellobiohydrolases, β-glucosidases, and exoglucanases (de Vries and Visser, 2001). The hemicellulosic fraction presents a more heterogeneous polymeric composition, so that many diverse enzymes are involved in hemicellulose degradation. Although endoxylanases and β-xylosidase are the enzymes that hydrolyze the xylan backbone (Juturu and Wu, 2012), their activity is improved by the activity of so-called accessory enzymes such as α-L-arabinofuranosidases, arabinoxylan arabinofuranohydrolase, α-glucuronidases, β-xylosidases, arabinases, galactosidases, galactanases and feruloyl, coumaroyl, and acetyl and methylesterases, which are responsible for cleavage of the xylan side chains (de Vries et al., 2000). Although in low amounts, pectins are also components of the plant cell wall and, consequently, different enzymes are involved in the degradation of this polysaccharide with lyase and hydrolase activities. There are several studies reporting the ability of several *Aspergillus* species to produce a broad spectrum of most of these enzymes (de Vries and Visser, 2001), and the importance of synergism between them has also been described, which is in good agreement with the coordinated regulation of gene expression (de Vries et al., 2000). Finally, lignin's degrading enzymes are laccase, lignin peroxidase, and manganese peroxidase (Wang et al., 2013), which are mainly produced by white-rot fungi.

Hydrolysis of lignocellulosic materials has been traditionally focused on cellulose, which is the most abundant plant polysaccharide, to obtain glucose. However, the presence of hemicellulose and lignin can restrict cellulose hydrolysis. In particular, lignin appears to limit cellulose hydrolysis by forming a physical barrier that reduces the access of enzymes to their specific substrates (Berlin et al., 2005). Effective hydrolysis of lignocellulosic materials requires a synergistic action between the enzymes responsible for cleaving the different bonds, making possible the cooperation between the main-chain-cleaving enzymes with the so-called "accessory enzymes." Hemicellulases can enhance cellulose hydrolysis in terms of speed and hydrolysis yield by the removal of the noncellulosic polysaccharides or by enabling the release of other saccharides and other compounds from the recalcitrant substrate (Berlin et al., 2007; de Vries and Visser, 2001; Robl et al., 2013).

In particular, xylanases are hemicellulosic enzymes with accessory roles in the hydrolysis of cellulose in lignocellulose materials. Xylanases are involved in the depolymerization of xylan by releasing D-xylooligosaccharides (Collins et al., 2005). Filamentous fungi are particularly good producers of this enzyme from an industrial point of view, attracting greater attention than bacteria and yeasts due to the generally higher values of xylanolytic activity produced, and also because of fungal metabolism, having a natural capacity to grow and degrade lignocellulosic materials, allowing the production of related enzymes. Extracellular release of xylanases into the medium by fungi, avoiding the need for cell disruption, is another remarkable point for industrial purposes (de Alencar Guimaraes et al., 2013; Haltrich et al., 1997).

10.4 FUNGAL SSF

Fungi are saprotrophic organisms with a naturally adapted metabolism enabling them to synthesize and secrete enzymes having the capacity to degrade tissues. Thus, fungi are the spotlight microorganisms of many scientific assays to obtain hydrolytic enzymes. In this context, SSF is the most widespread technique for fungal growth and enzyme production, emerging as an appropriate technology for the management of agroindustrial residues. SSF can be defined as the culture of microorganisms on moist solids in the absence or near absence of free water. Solid culture processes present advantages over submerged fermentation, including higher concentration of products, higher volumetric productivity, better product recovery, higher oxygen availability, lower capital investment, simpler fermentation equipment and technique, and less generation of effluents, among others (Facchini et al., 2011; Mrudula and Murugammal, 2011; Pérez-Rodríguez et al., 2014). These cultivation conditions are particularly suitable for the growth of fungi, which are known to be able to colonize solid materials with relatively low water activity, unlike most bacteria and yeasts (Pérez-Rodríguez et al., 2014). Fermentation conditions in solid-state systems resemble those circumstances present in the natural habitats of fungi. Fungi's inherent mechanisms for enzyme production developed and improved over centuries of adaptation to these environments (Singhania et al., 2009), which are usually characterized by a high surface area in contact with air and reduced water activity. They have a significant influence on fungal metabolism since they favor oxygen availability for aerobic metabolism and can even affect gene expression (te Biesebeke et al., 2002).

The regulation of fungal synthesis of enzymes degrading polymeric substances, such as xylanases, is based on many feedback processes. First, filamentous fungi synthesize and release into the external environment low levels of certain hydrolases. These hydrolases act on the polymeric matrices, such as lignocellulose materials, releasing soluble monomers, which penetrate the cell membrane. Then the liberated compounds and other molecules present in the medium generate the appropriate cell signal that triggers the production of larger quantities of the enzymes necessary for its breakdown (Haltrich et al., 1997). Fungi produce specific enzymes to get nutrients from a particular medium, and in this way fungal metabolism induces the synthesis of hydrolytic enzymes to degrade polymers from biomass to digestible compounds.

According to the aforementioned, xylanolytic activity in fungi is inducible, and xylan is the best xylanolytic enzyme inducer for several filamentous fungi (Knob et al., 2009). Therefore, substrates containing large quantities of xylan, such as corn cob, are keys in xylanase induction. Considering that fungi seem to produce specific xylanases based on the chemical composition and structure of their substrate, the best way to obtain such enzymes for corn cob hydrolysis consists in producing them *ad hoc* by growing fungal cultures also on corn cob (Figure 10.1). In this way, fungi can upgrade xylanase secretion to the same substrate in a more specific process that can offer the prospect of higher hydrolysis yields.

Fungal enzyme production in SSF is affected by several environmental factors. The nature of solid substrates, temperature, pH, water activity and moisture, carbon and nitrogen sources, minerals, particle size, aeration, pretreatment of lignocellulosic residues, or bed-loading are parameters that can affect the metabolism of filamentous fungi, and many authors have studied several conditions to improve xylanase production (Ahmad et al., 2012b; Bakri et al., 2008; Benedetti et al., 2013; Dobrev et al., 2007; Pérez-Rodríguez et al., 2014; Singhania et al., 2009). Furthermore, during the scale-up of fungal

FIGURE 10.1

Aspergillus niger CECT 2700 grown on milled corn cob to produce xylanases by solid-state fermentation (Pérez-Rodríguez et al., 2014).

SSF systems by using bioreactors, the most important parameters that need consideration include agitation, aeration and oxygen transfer, temperature and moisture of the solid substrate, and air humidity (Pérez-Rodríguez et al., 2014).

10.5 *ASPERGILLUS NIGER* FOR THE PRODUCTION OF XYLANASES

The ability of the genus *Aspergillus* to produce a wide range of natural products such as enzymes justifies extensive research and the economical and industrial importance of biotechnological processes involving these fungi (Carvalho et al., 2012; de Vries and Visser, 2001; Knuf and Nielsen, 2012).

Aspergillus are filamentous fungi included in the Ascomycota phylum. Other renowned fungi such as *Penicillium* spp. or *Trichoderma* spp. are also part of this taxonomic group (Karpe et al., 2014). The genus *Aspergillus* consists of a large number of species, the most important of them being *Aspergillus awamori*, *Aspergillus flavus*, *Aspergillus fumigatus*, *Aspergillus japonicas*, *Aspergillus nidulans*, *Aspergillus niger*, and *Aspergillus terreus* (de Vries and Visser, 2001; Knuf and Nielsen, 2012). *Aspergillus* are saprotrophic fungi with a high tolerance to a wide range of pH, making them suitable for growing on a broad spectrum of substrates. *Aspergillus* species synthesize enzymes, including several hydrolytic enzymes useful for the polysaccharide decomposition of plant cell wall. Therefore, good fermentation capabilities and high levels of protein secretion are characteristics that make them ideal organisms for industrial applications (de Vries and Visser, 2001; Stroparo et al., 2012; Ward et al., 2005).

In particular, *A. niger* has been widely studied and employed in biotechnological processes due to its considerable extracellular enzyme production using different agricultural and food industry wastes (wheat bran, soybean bran, barley bran, rice straw, corn cob, apple pomace, orange peel, mango residue, sugarcane bagasse, oil, and wine residues) as substrates. In recent years, several scientific works have corroborated and supported the importance of *A. niger* for the production of xylanases (Ahmad et al., 2012b; Bakri et al., 2008; Benedetti et al., 2013; de Alencar Guimaraes et al., 2013; Dhillon et al., 2012; Dobrev et al., 2007; Ma et al., 2009; Okafor et al., 2007; Pal and Khanum, 2010; Pérez-Rodríguez et al., 2014; Takahashi et al., 2013), and for obtaining other lignocellulosic enzymes such as cellulase, β-glucosidase, pectinase, or feruloyl esterase (Bansal et al., 2012; Hegde and Muralikrishna, 2009; Mamma et al., 2008; Ou et al., 2011; Rodríguez-Zúñiga et al., 2014; Sánchez, 2009; Vitcosque et al., 2012).

On an industrial scale, xylanases are produced mainly by *Aspergillus* species (de Alencar Guimaraes et al., 2013; de Vries and Visser, 2001; Shekiro et al., 2012; Yoon et al., 2006). Among the *Aspergillus* genus, there are many species that are used to produce xylanases, including *A. niger* (Ahmad et al., 2012a,b; Bakri et al., 2008; Benedetti et al., 2013; de Alencar Guimaraes et al., 2013; de Vries and Visser, 2001; Dobrev et al., 2007; Pérez-Rodríguez et al., 2014).

10.6 CORN COB AS A CARBON SOURCE FOR XYLANASE PRODUCTION BY *A. NIGER*

According to the Food and Agriculture Organization of the United Nations, maize was the second most produced crop in the world in 2013, with a total production of 1.02×10^9 tons, preceded by sugarcane (1.88×10^9 tons) and followed by other important cereal crops, mainly rice (7.46×10^8 tons) and wheat (7.13×10^8 tons). The xylan content of corn cob is approximately 35% (Aachary and Prapulla, 2009). Corn cob xylan can be separated into two different structural types. One is a low-branched arabinoglucuronoxylan, which is mostly water insoluble, and the second is a highly branched, water-soluble heteroxylan. These two xylan structural types can be distinguished by the amount and proportion of non-regularly distributed side chains. These are mainly single α-(1→2)-linked 4-*O*-methyl-D-glucopyranosyl uronic acid units and α-(1→3)-linked L-arabinofuranose residues, as well as 2-*O*-β-D-xylopyranosyl-α-L-arabinofuranose. Thus, the difference is mainly due to the content of arabinose and uronic acid, both of which are higher in the water-soluble fraction (Ebringerová et al., 1992). To sum up, due to the notable percentage that corn cob makes up as a by-product of maize processing, the high worldwide production of this cereal and corn cob's notable xylan composition clearly indicate the potential utilization of corn cob substrate for fungal growth as well as xylanase and xylose production.

Corn cobs were found to be the best substrates for high yield of xylanases with poor cellulase production (Shah and Madamwar, 2005), so various authors have assayed corn cob as substrate for the synthesis of xylanases by *A. niger* (Ahmad et al., 2012b; Bakri et al., 2008; Benedetti et al., 2013; Dobrev et al., 2007; Pérez-Rodríguez et al., 2014; Ximenes et al., 2007). For instance, Bakri et al. (2008) evaluated the production of xylanases by *A. niger* SS7 using a submerged culture with a basal medium (yeast extract 2.0 g/L, peptone 2.0 g/L, K_2HPO_4 1.5 g/L, $MgSO_4 \cdot 7H_2O$ 0.5 g/L) supplemented with various agriculture residues (barley straw, wheat straw, wheat bran, corn hulls, olives pulp, apple pulp, and sawdust). Corn cob hulls were considered the best agricultural residue among all the studied wastes, obtaining a maximum of 293.82 U/mL after 5 days at 30°C, pH 7.0, 120 rpm, and 3.0% concentration of corn cob. Similarly, Benedetti et al. (2013) assayed the production of xylanases by *A. niger* FCUP1 in a medium made up of 0.1% $CaCO_3$, 0.5% NaCl, 0.1% NH_4Cl, 0.5% corn steep liquor, and 1% of different residues as carbon source (corn cob, sugarcane bagasse, wheat bran, and soy husks) at 40°C for 16 h, and proved that powdered corn cob was an excellent inducer for the *A. niger* xylanolytic complex.

Likewise, Ahmad et al. (2012b) tested the synthesis of xylanases by *A. niger* on three different concentration levels of corn cob (2.5%, 3.0%, and 3.5%), using four different pH levels (5.0, 5.5, 6.0, and 6.5) and four incubation temperatures (25.0°C, 27.5°C, 30.0°C, and 32.5°C). The highest xylanase activity (60.03 ± 1.83 IU/mL) was achieved at 30°C, pH 5.5, 3.0% concentration of corn cob, and an incubation period of 72 h. Meanwhile, Dobrev et al. (2007) optimized the composition of nutrients added to the culture broth ($(NH_4)_2HPO_4$ 2.6 g/L, urea 0.9 g/L, malt sprout 6.0 g/L, corn cob 24.0 g/L, and wheat bran 14.6 g/L), increasing by 33% the xylanase activity of xylanases secreted by *A. niger* B03 (996.30 U/mL) in comparison to the activity obtained with the basic medium, all in a submerged culture (Dobrev et al., 2007).

On the other hand, Pérez-Rodríguez et al. (2014) optimized the production of xylanases in solid-state culture by *A. niger* CECT 2700, enhancing the amount of xylanases from 504 ± 7 U/g dry corn cob (using the nonoptimized culture broth) to up to 2452.7 U/g dry corn cob after improving the operational conditions and supplementing the nutrients (corn cob without pretreatments moistened in a ratio 1:3.6 (w/v) with a solution containing 5.0 g/L $NaNO_3$, 1.3 g/L $(NH_4)_2SO_4$, 4.5 g/L KH_2PO_4, and 3 g/L yeast extract, after 8 days of fermentation at 30°C in Erlenmeyer flasks). This study was complemented by the scale-up of xylanase production, where the effects of various air flows was assayed. Successful results were achieved in a laboratory-scale horizontal tube bioreactor, reaching highest xylanase activity (2926 U/g dry corn cob) at a flow rate of 0.2 L/min.

Taking into account that substrates should not only provide nutrients for fungi but also support, facilitate microbial attachment, and allow oxygen transfer and heat dispersion among other conditions, various pretreatments have been investigated to improve the use of corn cob as a substrate for fungal xylanase production under SSF. Pretreatments of lignocellulosic substrates can modify the physical structure of lignin, increase the available surface area and pore size (Figure 10.2), cause

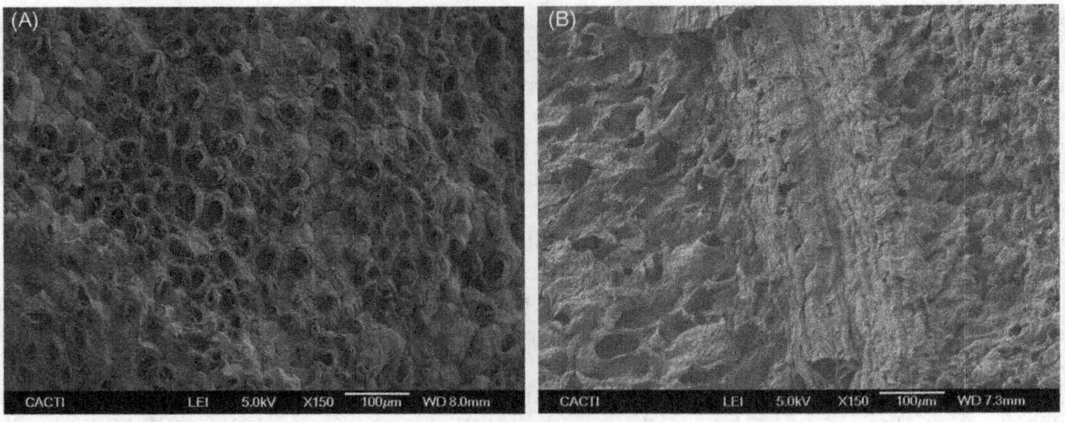

FIGURE 10.2

Scanning electron microscopy (SEM) photographs show the morphological changes of corn cob before (A) and after pretreatment (B) (destarched corn cob followed by NaOH treatment) (Pérez-Rodríguez et al., 2014).

partial depolymerization of hemicellulose, reduce the crystallinity of cellulose and lead to deacetylation of hemicellulose (Pérez-Rodríguez et al., 2014; Shah and Madamwar, 2005). Ferreira et al. (1999) evaluated the production of xylanase and β-xylosidase by *Aspergillus tamarii* in solid-state culture using corn cob subjected to three different pretreatments: alkali treatment (adding 1 mL of 0.3 N NaOH per g substrate), hydrogen peroxide containing $MnSO_4$ treatment, and heating (121°C for 1 h). However, they concluded that neither alkali, hydrogen peroxide, nor heat treatment of corn cob caused further improvements in enzyme production. In like manner, corn cob was enzymatically treated to remove the starch, pretreated with alkali (sodium hydroxide or ammonia solution), or subjected to physical pretreatments with autoclave and microwave by Pérez-Rodríguez et al. (2014) to assay xylanase production by *A. niger* CECT 2700. These treatments did not improve xylanase production, giving in all cases lower values compared to those obtained using the raw material. Similar results were obtained by Rahnama et al. (2013) working with untreated and alkali-pretreated rice straw. They justified that their best results regarding cellulase and xylanase production and those reported by researchers using untreated substrates could be related to the overall complexity of the untreated matrix, that could impact and induce greater enzyme production. Considering also the higher energy consumption and the additional economic costs related to the application of pretreatments, untreated corn cob appears to be the most environmentally friendly and sustainable technology for xylanolytic enzyme production by *A. niger* (Pérez-Rodríguez et al., 2014).

10.7 INDUSTRIAL APPLICATION OF FUNGAL XYLANASES

Fungal xylanases have gained importance in recent years due to their biotechnological application in diverse sectors such as paper, pulp, food, and animal feed industries, as well as for bioconversion of lignocellulosic waste into value-added compounds and to fermentative products (de Vries and Visser, 2001; Juturu and Wu, 2012; Pérez-Rodríguez et al., 2014; Polizeli et al., 2005).

In the pulp and paper industry enzymatic degradation focuses on the hemicellulose-lignin complexes, where the cellulose fibers are still intact and strong. Endoxylanase is the most remarkable enzyme used in bio-bleaching of kraft pulps. Nevertheless, the addition of other xylanolytic enzymes has also been shown to be effective (de Vries and Visser, 2001). Xylanolytic enzyme preparations lacking in cellulase can be of great value in the bio-bleaching of pulps, thus reducing the amount of bleaching chemicals, such as chlorine, and the mechanical methods required in the pretreatment of paper pulp. As a result, the cost of the process is decreased because of the diminution in the use of chemicals and energy consumption and the environmental impact of the paper and pulp industries is reduced since the pollution caused by the use of chlorine and toxic discharges is diminished (de Vries and Visser, 2001; Collins et al., 2005). Various studies describing the use for pulp bio-bleaching of cellulase-free xylanases produced by different *Aspergillus* species such as *terricola*, *ochraceus*, *niger*, *niveus*, and *oryzae* reflected the potential of these enzymes for this application (Szendefy et al., 2006; Betini et al., 2009; Michelin et al., 2010).

A second area in which xylanases find application is represented by the food and beverage industries. Products obtained microbiologically or enzymatically have been regarded as a GRAS (generally recognized as safe) status by the US Food and Drug Administration (USFDA). This regulation allows industries to use *A. niger* products in food and feed applications (Knuf and Nielsen, 2012), which yields an advantage in comparison to chemical and physicochemical procedures. Xylanolytic enzyme preparations are widely used in the production of bakery goods, coffee, starch, plant oil, and juice (Harris and Ramalingam, 2010). In the bakery industry, xylanases improve the elasticity and strength of the dough as a consequence of the arabinoxylan fraction. Thus, xylanolytic hydrolysis enables easier handling and larger loaf volumes, enhances bread texture, and improves dough quality (Collins et al., 2005; de Vries and Visser, 2001). *Aspergillus* xylanases, and especially *A. niger* xylanases, have shown interesting abilities for improving dough and bread quality attributes (Ahmad et al., 2012a; Shah et al., 2006; Romanowska et al., 2006; Camacho and Aguilar, 2003). Xylanases are also used in coffee extraction and in the preparation of instant coffee. The solubilization of cellulose, galactomannan, arabinogalactan, and xylan determines the yield of soluble solids used for the final products and their organoleptic characteristics. Thus, xylanolytic enzymes influence the properties of the coffee beverage (Jooste et al., 2013).

Arabinoxylans are cereal grain polysaccharides that appear in relatively small amounts. However, the unique physicochemical properties of arabinoxylans give them an outstanding significance, affecting grain functionality in biotechnological cereal-based processes such as separation of wheat flour into gluten and starch. The addition of xylanases to wheat flour changes the gluten agglomeration behavior during gluten–starch separation (Dornez et al., 2009). Hemicellulases have a considerable effect on the viscosity of the dough depending on their specificity toward the soluble and insoluble fractions of the polysaccharide and their dosage (Frederix et al., 2003). Low viscosity increases the mobility of the components during the batter stage and improves the rate of gluten agglomeration. Hence, an enzyme blend containing xylanases with selectivity toward water-extractable arabinoxylan or a high concentration mixture of both xylanases active with soluble and insoluble arabinoxylan improves gluten agglomeration behavior, starch yield, gluten–starch separation, and the global efficiency of cereal-based processes (Dornez et al., 2009). Studies using *Aspergillus aculeatus* xylanase with selectivity toward water-extractable arabinoxylan and no inhibition by wheat flour xylanase inhibitors, and *Bacillus subtilis* xylanase with selectivity toward water-unextractable arabinoxylan and inhibition by wheat flour xylanase inhibitors, have shown the better performance of the *Aspergillus* enzyme on gluten separation and the convenience of applying both enzymes together at high dosages to even further improve the process (Frederix et al., 2003). In fruit and vegetable processing, including juice, oil, brewing, and wine production, xylanases enhance extraction yield and filtration process performance, increasing must filterability and reducing final turbidity. Xylanases, acting in conjunction with other enzymes, yield improvements in consistency and palatability of drinks as well as increasing the recovery of essential oils, vitamins, mineral salts, and edible dyes or pigments from fruits and vegetables (Harris and Ramalingam, 2010), a necessity for processing industries.

Other applications in which xylanolytic enzymes are used include increasing feedstock nutritive value, digestibility, and conversion efficiency by cattle. Arabinoxylans are one of the major nonstarch polysaccharide fractions present in agricultural silage and grain feed. Arabinoxylans create a physical barrier to nutrient digestion and absorption by increasing gut viscosity, thus interfering in reducing the digestibility and assimilation of nutrients. The previously mentioned polysaccharides also produce an alteration in the secretion of endogenous digestive enzymes, water, and electrolytes, modifying the digestive tract functions and modulating gut microbiota (Zhang et al., 2014a). Addition of xylanases to feed enhances digestibility by leading to a decrease of digesta viscosity, an increase in the release of the enclosed nutrients from cell wall polysaccharides, and an improvement in protein and starch utilization. Nourishing diets containing feed treated with a cocktail of hydrolytic enzymes (mainly cellulase and xylanase, with or without feruloyl esterases) release monosaccharides due to two principal reasons: the breakdown of nonstarch polysaccharides and the release of their constituent monomers, and the subsequently release of the starch within the endosperm of grains exposed to the endogenous amylase, releasing more glucose. Thereby, ruminant feed enzyme additives boost animal growth performance and body weight gain, milk yield, and milk composition of dairy animals without changing feed intake (Dhiman et al., 2002; Phakachoed et al., 2012; Zhang et al., 2014a). Moreover, the enzyme supplementation of feed has a positive collateral impact on the environment because of the reduction of undesirable residues in excrements such as phosphorus, nitrogen, copper, and zinc, thus decreasing environmental pollution (Harris and Ramalingam, 2010; Polizeli et al., 2005). Crude xylanase from *A. niger* has been successfully assayed in pig diets, improving the *in vitro* feed digestibility and the average daily gain, as well as reducing the incidence of diarrhea (Tapingkae et al., 2008). A xylanase produced by a genetically modified *A. niger* was also assayed for chicken feeding, and good results were obtained for growth, which were related to the positive effect of the enzyme on digestion, absorption, metabolism, and immunity (Gao et al., 2007).

As an interesting alternative to the development and use of first-generation bioethanol obtained by fermentation of sugar and starchy materials, which has generated considerable controversy due to the competition between food or feed and fuel production, second-generation bioethanol produced from lignocellulosic biomass has emerged as an alternative renewable energy source to fuel petroleum. Agroindustrial wastes of a lignocellulosic nature are a good option to be used as raw materials for this purpose (Matsushika et al., 2009; Viikari et al., 2012). The conversion includes two steps: hydrolysis of cellulose in the lignocellulosic materials to fermentable reducing sugars, and fermentation of the sugars to ethanol. Bioconversion of the hemicelluloses by xylanases helps to achieve a reasonable commercial price for bioethanol by enhancing hydrolysis in two ways. On the one hand, xylanases and other hemicellulases expose the cellulose fibers to cellulase action, facilitating hydrolysis, and, on the other hand, releasing fermentable sugars from the hemicellulosic fraction. Regarding this, it must be considered that pentoses (mainly xylose and arabinose) account for the majority of hemicellulose monomers and that there are a limited number of microorganisms able to use xylose or arabinose (Kumar and Wyman, 2009), which do not include the *Saccharomyces* species usually employed for bioethanol fermentation (Hahn-Hägerdal et al., 2006). Because of this, great effort has been made in recent years to find wild or transformed efficient pentose-fermenting yeasts. In this way, the enzymatic hydrolysis of the cellulose and hemicellulose, which comprise up two-thirds of lignocellulosic materials (Gírio et al., 2010), results in sugar-rich liquid streams useful for the biological production of fuel ethanol, reducing the demand for and dependence on petroleum, mitigating global warming by reducing the amounts of greenhouse gas emissions into the atmosphere and also achieving at the same time an acceptable disposal of industrial residues (Matsushika et al., 2009; Viikari et al., 2012). The use of an *A. nidulans* endoxylanase in combination with a commercial cellulase preparation has allowed an improvement of more than 28% in ethanol production from wheat straw by a xylose-fermenting *Saccharomyces cerevisiae* strain (Alvira et al., 2011). In the case of *A. niger*, the particular ability of this fungus to produce significant levels of feruloyl esterase together with xylanases makes *A. niger* enzymatic extracts very suitable for this purpose, since xylan hydrolysis is synergistically enhanced by the previous cleavage of the ferulic side chains (Tabka et al., 2006).

10.8 CORN COB AS SUBSTRATE FOR THE ENZYMATIC PRODUCTION OF XYLOOLIGOSACCHARIDES AND XYLOSE

Xylooligosaccharides are getting increasing attention as functional food ingredients with beneficial properties for gastrointestinal health by their prebiotic action, increasing enteric microbiota stability. Some studies have found that these sugars may reduce the risk of colon cancer and suppress blood cholesterol levels, and are also capable of improving diabetic symptoms (Khandeparker and Numan, 2008). Consequently, xylooligosaccharides have proved to have potential uses in various fields including pharmaceuticals, feed formulations, and agricultural and food applications (de Vries and Visser, 2001; Pérez-Rodríguez et al., 2014).

Xylooligosaccharides are produced from corn cob by several methods, including hydrothermal treatment, chemical (acid) hydrolysis, enzymatic hydrolysis, or a combination of these methods (Chapla et al., 2012). Nevertheless, enzymatic hydrolysis is attractive because it does not produce undesirable by-products, generates less production of monosaccharides, and does not require special equipment (Chapla et al., 2012). Fungal xylanases are the preferred catalysts for the hydrolysis of xylan, mainly due to their higher specificity, milder reaction conditions, and negligible substrate losses (Pérez-Rodríguez et al., 2014). Xylanases release xylooligomers, xylobiose, and xylose. The main drawback of the enzymatic production of xylooligosaccharides is the accessibility of the enzymes to the substrate inside the lignocellulosic matrix. One method involves the extraction of xylan prior to the hydrolysis step. Chapla et al. (2012) carried out four different extraction protocols in order to recover the maximum amount of xylan from raw corn cobs. Treatments with sulfuric acid, sodium hypochlorite solution, sodium hydroxide, and autohydrolysis were applied. The best yield in corn cob xylan extraction (178.73 ± 5.8 g of xylan/kg of corn cob) was achieved using a mild alkali treatment. A partially purified xylanase from *Aspergillus foetidus* MTCC 4898 was used for the selective enzymatic production of xylooligosaccharides. Maximum production of xylooligosaccharides was 6.73 ± 0.23 mg/mL after 8 h of reaction time using 20 U of xylanase at a temperature of 45°C (Chapla et al., 2012).

An interesting method involves the application of different pretreatments to the lignocellulosic matrix for the purpose of making corn cob xylan more accessible for the enzymatic reaction instead of the complete extraction of xylan as substrate (Aachary and Prapulla, 2009). Aachary and Prapulla (2009) tested various pretreatment methods to evaluate the efficiency of xylooligosaccharide production of alkali- and acid-pretreated powdered corn cob and the combination of corn cob that was acid treated and cooked under pressure in xylooligosaccharide production. Compared with the xylan content of native corn cob (31.9 ± 2.3%), alkali-pretreated, acid-pretreated, and pressure-cooked corn cob powder showed greater percentages of xylan (40.8 ± 1.2%, 39.2 ± 1.1%, and 40.0 ± 0.8%, respectively). As a result, alkali pretreatment was found to be more efficient for xylooligosaccharide production. In the release of xylooligosaccharides studies, 200 U of endoxylanase (Bioxyl P40) per gram of material was employed. The percentage of xylooligosaccharides obtained by the enzymatic hydrolysis of raw corn cob was 89 ± 1.1%, whereas 81 ± 1.5%, 77 ± 1.8%, and 52 ± 3.2% of xylooligosaccharides were obtained by the enzymatic hydrolysis of alkali-pretreated, steam-cooked corn cob powder and acid pretreatment. Considering the percentages of both xylan and xylooligosaccharides released by endoxylanase Bioxyl P40, the alkali-pretreatment was the most efficient method, providing the highest xylooligosaccharides production (equivalent to 5.8 ± 0.14 mg/mL).

Zhu et al. (2006) also used pretreated corn cob as substrate for enzymatic release of xylooligosaccharides. Corn cob was pretreated in aqueous ammonia (15% ammonia, 60°C, L/S 2.8, and 48 h), which generated a delignified and xylan-rich substrate. Subsequently, both untreated and treated corn cobs were subjected to enzymatic hydrolysis using endoxylanase X2753 for production of xylooligosaccharides. Aqueous ammonia treatment of corn cobs yielded a superior substrate for saccharification (15.74 mg/mL) compared with untreated corn cob (4.11 mg/mL), with a total molar yield of 80.5% of xylan. Moreover, Zhu et al. (2006) studied corn stover as well as corn cob; they concluded that corn cobs are a more suitable feedstock than corn stover for xylooligosaccharide production, since corn cob has higher xylan content and greater bulk density.

A combined deacetylation and laboratory beating of pulp with a PFI mill as refining pretreatment was applied to corn cob by Zhang et al. (2014b) for the enhancement of a two-step enzymatic hydrolysis of both hemicellulose and cellulose to xylooligosaccharides and glucose, respectively. In the first step, the pretreated corn cob was hydrolyzed by xylanase (Habio Enzyme) to produce xylooligosaccharides. Further hydrolysis of the solid residue isolated from the first step was carried out by cellulase (Celluclast 1.5 L) and β-glucosidase (Novozyme 188). In deacetylation pretreatment, two weak bases, Na_2CO_3 and $Ca(OH)_2$, and one strong base, NaOH, were tested to remove acetyl groups. To further increase enzymatic digestibility, the deacetylated and washed corn cob was milled using a PFI mill, which applied a great proportion of compressive to shear forces on the corn cob. According to this work, the most suitable alkali for deacetylation of corn cob was $Ca(OH)_2$. It was also found that increasing the charge of the alkali the percentage of enzymatic hydrolysis and sugar yields were improved. Finally, 86.8% of xylan and 86.1% of glucan in the raw corn cob were hydrolyzed in the two-stage enzymatic hydrolysis with the combined refining pretreatments of deacetylation and PFI. The same authors also suggested that corn cob could be highly enzymatically hydrolyzed after mild deacetylation treatment compared to other cellulosic materials (such as corn stover and aspen wood).

Xylose is the final product of the complete hydrolysis of the xylan chain. It has applications in the food industry as a noncariogenic light sweetener, and it is the precursor for the biotechnological production of xylitol, which is based on the fermentation of xylose from the hemicellulosic fraction of agroindustrial wastes into xylitol by microorganisms (mainly yeasts) (Parajó et al., 1998; Rivas et al., 2006; Silva et al., 2007). Xylitol is a five-carbon sugar alcohol with sweetness similar to that of sucrose, which has many advantages as a food ingredient. It has anticariogenic properties, inhibiting the growth of oral bacteria and so reducing plaque formation and preventing caries. This polyol has a lower energy value than sucrose, so xylitol can be supplied in the diet, perhaps limiting the tendency to obesity. Xylitol can be used in diabetic food formulations because, with fructose, it is not insulin-mediated, which causes only limited increases in glucose and insulin blood levels. The incorporation of xylitol in formulations can reduce darkening and preserve the nutritional value of proteins because it does not undergo the Maillard reaction, and also produces an enhancement of color and taste without causing changes in properties during storage (Parajó et al., 1998; Rivas et al., 2006; Silva et al., 2005; Su et al., 2013). Due to all the mentioned desirable properties and vast potential uses of xylitol, industries are committed to xylitol incorporation as an ingredient in drugs, food, and dental products among others (Parajó et al., 1998; Rivas et al., 2006).

The generation of monomeric xylose by enzymatic processes using xylan from corn cob as substrate and *A. niger* xylanases as catalysts, was reported by Yoon et al. (2006). These authors evaluated three commercial xylanase preparations (Rapidase Pomaliqan (Gist-Brocades International), Clarex ML (Genenor), and Validase (Valley Research)) in the enzymatic production of pentoses and xylooligosaccharides from the hemicellulose fraction of corn husk and corn cob. Rapidase Pomaliqan enzyme preparation (1941 U/mL xylanase), derived from *A. niger* and *Trichoderma reesei*, was capable of increasing the concentration of sugars from an initial value of 106.5 to 210.6 g/kg dry matter corn husk and of 8.6 to 141.6 g/kg dry matter of corn cob, respectively, under favorable conditions (480 min of reaction at pH 5.0 and 50°C). Referring strictly to xylose, 83.7 ± 8.3 g/kg of dry matter from corn husks and 31.1 ± 6.1 g/kg of dry matter from corn cob were achieved using Rapidase Pomaliqan. Yields, including arabinose, xylose, xylobiose, and xylotriose, from the hemicellulose fraction of corn husks and corn cob with Rapidase Pomaliq xylanase were 75.2% and 65.6%, respectively.

However, the most common studies have been focused on the pretreatment of lignocellulosic materials in order to heighten xylose release. There are a large number of possible pretreatments (Banerjee et al., 2010). An effective pretreatment is characterized by several criteria. It must avoid the need for reducing the size of biomass particles, preserve the pentose (hemicellulose) fractions, limit formation of degradation products that inhibit growth of fermentative microorganisms, minimize energy demand, and limit cost (Mosier et al., 2005). For instance, Ximenes et al. (2007) employed corn spent from ethanol production, either used as delivered from the industry (untreated) or after being pretreated with hot water, and two fungal strains (*Trichoderma reesei* Rut C-30 and *A. niger* NRRL 2001) to first produce xylanases. *Aspergillus niger* NRRL 2001 secreted a maximum amount of xylanases (72.21 IU/mL) at 28°C after 132 h in flasks, thus improving the production after 264 h in a 4-L bioreactor (64.31 IU/mL). Then they evaluated the hydrolysis of xylan from corn waste pretreated with hot water to xylose. Although *A. niger* enzymes released more xylose (64%) compared to *T. reesei* (48%), when added separately, the mixture of *T. reesei* and *A. niger* enzymes increased the xylose content up to 71%. Weijuna et al. (2010) optimized alkali liquor pretreatment conditions of corn cob to remove lignin and improve the enzymatic hydrolysis yield of hemicellulose and cellulose. Under the optimal pretreatment conditions obtained (reaction time 12 h, temperature 70°C, liquid to solid ratio 23:1, aqueous ammonia concentration 2.5%(w/v), hydrogen peroxide 0.6%, sodium silicate 5%, magnesium sulfate 0.05% in reaction mixture), the yield of enzymatic hydrolysis of hemicellulose, cellulose, and corn cob residue was 41.36%, 88.09%, and 69.21%, respectively, by two steps of hydrolysis (xylanase and cellulose). These results mean an increase in hydrolysis yield compared with the untreated materials.

10.9 CONCLUSIONS

Lignocellulosic residues, such as corn cob, generated from agricultural activities and food or feed industries represent an inexpensive natural resource for fungal growth and production of enzymes or value-added compounds. Considering that the enzyme production cost is one of the main factors in determining the economic efficiency of a biotechnological process (Pérez-Rodríguez et al., 2014), its reduction by optimizing the fermentation substrates by incorporating lignocellulosic wastes as a carbon source must be a serious objective for industrial applications. SSF is an optimal method for waste valorization processes based on fungal culture. Considering the ability of several fungi to produce a broad spectrum of complementary and synergic enzymes, and taking into account that in many cases their regulatory mechanisms of gene expression are coordinated (de Vries et al., 2000), it is reasonable that the mixture of enzymes produced from a specific substrate could be the most appropriate for hydrolyzing it. Therefore, the use of the same substrate for fungal xylanase production and for enzymatic release of xylose and xylooligosaccharide seems to be an interesting methodology. Consequently, corn cob can be considered potentially recoverable as a raw material for a broad number of applications in industrial biotechnology processes.

REFERENCES

Aachary, A.A., Prapulla, S.G., 2009. Value addition to corncob: production and characterization of xylooligosaccharides from alkali pretreated lignin-saccharide complex using *Aspergillus oryzae* MTCC 5154. Bioresour. Technol. 100, 991–995.

Ahmad, Z., Butt, M.S., Ahmed, A., Riaz, M., Sabir, S.M., Farooq, U., et al., 2012a. Effect of *Aspergillus niger* xylanase on dough characteristics and bread quality attributes. J. Food Sci. Technol. 51, 2445–2453.

Ahmad, Z., Butt, M.S., Anjum, F.M., Siddique, M., Rathore, H.A., Nadeem, M.T., et al., 2012b. Effect of corn cobs concentration on xylanase biosynthesis by *Aspergillus niger*. Afr. J. Biotechnol. 11, 1674–1682.

Alvira, P., Tomás-Pejó, E., José Negro, M., Ballesteros, A.M., 2011. Strategies of xylanase supplementation for an efficient saccharification and cofermentation process from pretreated wheat straw. Biotechnol. Prog. 27, 944–950.

Bakri, Y., Al-jazairi, M., Al-kayat, G., 2008. Xylanase production by a newly isolated *Aspergillus niger* SS7 in submerged culture. Polish J. Microbiol. 57, 249–251.

Banerjee, G., Car, S., Scott-Craig, J.S., Borrusch, M.S., Walton, J.D., 2010. Rapid optimization of enzyme mixtures for deconstruction of diverse pretreatment/biomass feedstock combinations. Biotechnol. Biofuels 3, 22.

Bansal, N., Tewari, R., Soni, R., Kumar, S., 2012. Production of cellulases from *Aspergillus niger* NS-2 in solid-state fermentation on agricultural and kitchen waste residues. Waste Manag. 32, 1341–1346.

Benedetti, A.C.E.P., da Costa, E.D., Aragon, C.C., dos Santos, A.F., Goulart, A.J., Attilli-Angelis, D., et al., 2013. Low-cost carbon sources for the production of a thermostable xylanase by *Aspergillus niger*. Revista de Ciências Farmacêuticas Básica e Aplicada 34, 25–31.

Berlin, A., Gilkes, N., Kurabi, A., Bura, R., Tu, M., Kilburn, D., et al., 2005. Weak lignin-binding enzymes. Appl. Biochem. Biotechnol. 121, 163–170.

Berlin, A., Maximenko, V., Gilkes, N., Saddler, J., 2007. Optimization of enzyme complexes for lignocellulose hydrolysis. Biotechnol. Bioeng. 97, 287–296.

Betini, J.H.A., Michelin, M., Peixoto-Nogueira, S.C., Jorge, J.A., Terenzi, H.F., Polizeli, M.L.T.M., 2009. Xylanases from *Aspergillus niger*, *Aspergillus niveus* and *Aspergillus ochraceus* produced under solid-state fermentation and their application in cellulose pulp bleaching. Bioprocess Biosyst. Eng. 32, 819–824.

Bustos, G., Moldes, A.B., Cruz, J.M., Domínguez, J.M., 2004. Production of fermentable media from vine-trimming wastes and bioconversion into lactic acid by Lactobacillus pentosus. J. Sci. Food Agric. 84, 2105–2112.

Camacho, N.A., Aguilar, O.G., 2003. Production, purification, and characterization of a low-molecular-mass xylanase from *Aspergillus* sp. and its application in baking. Appl. Biochem. Biotechnol. 104, 159–171.

Carvalho, T., Paulo, D., Gomes, P., Cristina, R., Bonomo, F., Franco, M., 2012. Optimisation of solid state fermentation of potato peel for the production of cellulolytic enzymes. Food Chem. 133, 1299–1304.

Chapla, D., Pandit, P., Shah, A., 2012. Production of xylooligosaccharides from corncob xylan by fungal xylanase and their utilization by probiotics. Bioresour. Technol. 115, 215–221.

Collins, T., Gerday, C., Feller, G., 2005. Xylanases, xylanase families and extremophilic xylanases. FEMS Microbiol. Rev. 29, 3–23.

De Alencar Guimaraes, N.C., Sorgatto, M., Peixoto-Nogueira, S.D.C., Betini, J.H.A., Zanoelo, F.F., Marques, M.R., et al., 2013. Bioprocess and biotechnology: effect of xylanase from *Aspergillus niger* and *Aspergillus flavus* on pulp biobleaching and enzyme production using agroindustrial residues as substract. SpringerPlus 2, 380.

Devesa-Rey, R., Vecino, X., Varela-Alende, J.L., Barral, M.T., Cruz, J.M., Moldes, A.B., 2011. Valorization of winery waste vs. the costs of not recycling. Waste Manage. 31, 2327–2335.

De Vries, R.P., Visser, J., 2001. *Aspergillus* enzymes involved in degradation of plant cell wall polysaccharides. Microbiol. Mol. Biol. Rev. 65, 497–522.

De Vries, R.P., Kester, H.C.M., Poulsen, C.H., Benen, J.A.E., Visser, J., 2000. Synergy between enzymes from *Aspergillus* involved in the degradation of plant cell wall polysaccharides. Carbohydr. Res. 327, 401–410.

Dhillon, G.S., Oberoi, H.S., Kaur, S., Bansal, S., Brar, S.K., 2011. Value-addition of agricultural wastes for augmented cellulase and xylanase production through solid-state tray fermentation employing mixed-culture of fungi. Ind. Crops Prod. 34, 1160–1167.

Dhillon, G.S., Kaur, S., Brar, S.K., Gassara, F., Verma, M., 2012. Improved xylanase production using apple pomace waste by *Aspergillus niger* in koji fermentation. Eng. Life Sci. 12, 198–208.

Dhiman, T.R., Zaman, M.S., Gimenez, R.R., Walters, J.L., Treacher, R., 2002. Performance of dairy cows fed forage treated with fibrolytic enzymes prior to feeding. Anim. Feed Sci. Technol. 101, 115–125.

Dobrev, G.T., Pishtiyski, I.G., Stanchev, V.S., Mircheva, R., 2007. Optimization of nutrient medium containing agricultural wastes for xylanase production by *Aspergillus niger* B03 using optimal composite experimental design. Bioresour. Technol. 98, 2671–2678.

Dornez, E., Gebruers, K., Delcour, J.A., Courtin, C.M., 2009. Grain-associated xylanases: occurrence, variability, and implications for cereal processing. Trends Food Sci. Technol. 20, 495–510.

Ebringerová, A., Hromádková, Z., Alföldi, J., 1992. Structural and solution properties of corn cob heteroxylans. Carbohydr. Polym. 19, 99–105.

Facchini, F.D.A., Vici, A.C., Reis, V.R.A., Jorge, J.A., Terenzi, H.F., Reis, R.A., et al., 2011. Production of fibrolytic enzymes by *Aspergillus japonicus* C03 using agro-industrial residues with potential application as additives in animal feed. Bioprocess Biosyst. Eng. 34, 347–355.

Ferreira, G., Boer, C.G., Peralta, R.M., 1999. Production of xylanolytic enzymes by *Aspergillus tamarii* in solid state fermentation. FEMS Microbiol. Lett. 173, 335–339.

Frederix, S.A., Courtin, C.M., Delcour, J.A., 2003. Impact of xylanases with different substrate selectivity on gluten—starch separation of wheat flour. J. Agric. Food Chem. 51, 7338–7345.

Gao, F., Jiang, Y., Zhou, G.H., Han, Z.K., 2007. The effects of xylanase supplementation on growth, digestion, circulating hormone and metabolite levels, immunity and gut microflora in cockerels fed on wheat-based diets. Br. Poultry Sci. 48, 480–488.

García-López, J., Rad, C., Navarro, M., 2014. Strategies of management for the whole treatment of leachates generated in a landfill and in a composting plant. J. Environ. Sci. Health 49, 1520–1530.

Gírio, F.M., Fonseca, C., Carvalheiro, F., Duarte, L.C., Marques, S., Bogel-Łukasik, R., 2010. Hemicelluloses for fuel ethanol: a review. Bioresour. Technol. 101, 4775–4800.

Hahn-Hägerdal, B., Galbe, M., Gorwa-Grauslund, M.F., Lidén, G., Zacchi, G., 2006. Bio-ethanol—the fuel of tomorrow from the residues of today. Trends Biotechnol. 24, 549–556.

Haltrich, D., Nidetzky, B., Kulbe, K.D., Zupan, S., 1997. Production of fungal xylanases. Bioresour. Technol. 58, 137–161.

Harris, A.D., Ramalingam, C., 2010. Xylanases and its application in food industry : a review. J. Exp. Sci. 1, 1–11.

Hegde, S., Muralikrishna, G., 2009. Isolation and partial characterization of alkaline feruloyl esterases from *Aspergillus niger* CFR 1105 grown on wheat bran. World J. Microbiol. Biotechnol. 25, 1963–1969.

Jooste, T., García-Aparicio, M.P., Brienzo, M., van Zyl, W.H., Görgens, J.F., 2013. Enzymatic hydrolysis of spent coffee ground. Appl. Biochem. Biotechnol. 169, 2248–2262.

Juturu, V., Wu, J.C., 2012. Microbial xylanases: engineering, production and industrial applications. Biotechnol. Adv. 30, 1219–1227.

Karpe, A.V., Beale, D.J., Harding, I.H., Palombo, E.A., 2014. Optimization of degradation of winery-derived biomass waste by Ascomycetes. J. Chem. Technol. Biotechnol. (online volume) http://dx.doi.org/10.1002/jctb.4486.

Khandeparker, R., Numan, M.T., 2008. Bifunctional xylanases and their potential use in biotechnology. J. Ind. Microbiol. Biotechnol. 35, 635–644.

Knob, A., Terrasan, C.R.F., Carmona, E.C., 2009. β-Xylosidases from filamentous fungi: an overview. World J. Microbiol. Biotechnol. 26, 389–407.

Knuf, C., Nielsen, J., 2012. Aspergilli: systems biology and industrial applications. Biotechnol. J. 7, 1147–1155.

Kumar, R., Wyman, C.E., 2009. Effect of enzyme supplementation at moderate cellulase loadings on initial glucose and xylose release from corn stover solids pretreated by leading technologies. Biotechnol. Bioeng. 102, 457–467.

Ma, H., Liu, W.-W., Chen, X., Wu, Y.-J., Yu, Z.-L., 2009. Enhanced enzymatic saccharification of rice straw by microwave pretreatment. Bioresour. Technol. 100, 1279–1284.

Majumdar, S., Lukk, T., Solbiati, J.O., Bauer, S., Nair, S.K., Cronan, J.E., et al., 2014. Roles of small laccases from *Streptomyces* in lignin degradation. Biochemistry 53, 4047–4058.

Mamma, D., Kourtoglou, E., Christakopoulos, P., 2008. Fungal multienzyme production on industrial by-products of the citrus-processing industry. Bioresour. Technol. 99, 2373–2383.

Matsushika, A., Inoue, H., Kodaki, T., Sawayama, S., 2009. Ethanol production from xylose in engineered *Saccharomyces cerevisiae* strains: current state and perspectives. Appl. Microbiol. Biotechnol. 84, 37–53.

Michelin, M., Peixoto-Nogueira, S.C., Betini, J.H.A., da Silva, T.M., Jorge, J.A., Terenzi, H.F., et al., 2010. Production and properties of xylanases from *Aspergillus terricola* Marchal and *Aspergillus ochraceus* and their use in cellulose pulp bleaching. Bioproc. Biosyst. Eng. 33, 813–821.

Mirabella, N., Castellani, V., Sala, S., 2014. Current options for the valorization of food manufacturing waste: a review. J. Clean. Prod. 65, 28–41.

Mosier, N., Wyman, C., Dale, B., Elander, R., Lee, Y.Y., Holtzapple, M., et al., 2005. Features of promising technologies for pretreatment of lignocellulosic biomass. Bioresour. Technol. 96, 673–686.

Mrudula, S., Murugammal, R., 2011. Production of cellulase. Braz. J. Microbiol. 42, 1119–1127.

Okafor, U.A., Okochi, V.I., Onyegeme-okerenta, B.M., Nwodo-Chinedu, S., 2007. Xylanase production by *Aspergillus niger* ANL 301 using agro-wastes. Afr. J. Biotechnol. 6, 1710–1714.

Ou, S., Zhang, J., Wang, Y., Zhang, N., 2011. Production of feruloyl esterase from *Aspergillus niger* by solid-state fermentation on different carbon sources. Enzyme Res. 2011, 1–4.

Özbaş, E., Balkaya, N., 2014. Removal of heavy metals (Cu, Ni, Zn, Pb, Cd) from compost by molasses hydrolysate. J. Environ. Eng. Landsc. Manage. (in press).

Pal, A., Khanum, F., 2010. Production and extraction optimization of xylanase from *Aspergillus niger* DFR-5 through solid-state-fermentation. Bioresour. Technol. 101, 7563–7569.

Parajó, J.C., Domínguez, H., Domínguez, J.M., 1998. Biotechnological production of xylitol. Part 1: interest of xylitol and fundamentals of its biosynthesis. Bioresour. Technol. 65, 191–201.

Pérez-Rodríguez, N., Oliveira, F., Pérez-Bibbins, B., Belo, I., Torrado Agrasar, A., Domínguez, J.M., 2014. Optimization of xylanase production by filamentous fungi in solid-state fermentation and scale-up to horizontal tube bioreactor. Appl. Biochem. Biotechnol. 173, 803–825.

Phakachoed, N., Lounglawan, P., Suksombat, W., 2012. Effects of xylanase supplementation on ruminal digestibility in fistulated non-lactating dairy cows fed rice straw. Livestock Sci. 149, 104–108.

Polizeli, M.L.T.M., Rizzatti, A.C.S., Monti, R., Terenzi, H.F., Jorge, J.A., Amorim, D.S., 2005. Xylanases from fungi: properties and industrial applications. Appl. Microbiol. Biotechnol. 67, 577–591.

Rahnama, N., Mamat, S., Shah, U.K.M., Ling, F.H., Rahman, N.A.A., Ariff, A.B., 2013. Effect of alkali pretreatment of rice straw on cellulase and xylanase production by local trichoderma harzianum SNRS3 under solid state fermentation. BioResources 8, 2881–2896.

Rivas, B., Torre, P., Domínguez, J.M., Converti, A., Parajó, J.C., 2006. Purification of xylitol obtained by fermentation of corncob hydrolysates. J. Agric. Food Chem. 54, 4430–4435.

Robl, D., Delabona, P.D.S., Mergel, C.M., Rojas, J.D., Costa, P.D.S., Pimentel, I.C., et al., 2013. The capability of endophytic fungi for production of hemicellulases and related enzymes. BMC Biotechnol. 13, 94.

Rodríguez-Zúñiga, U.F., Bertucci Neto, V., Couri, S., Crestana, S., Farinas, C.S., 2014. Use of spectroscopic and imaging techniques to evaluate pretreated sugarcane bagasse as a substrate for cellulase production under solid-state fermentation. Appl. Biochem. Biotechnol. 172, 2348–2362.

Romanowska, I., Polak, J., Bielecki, S., 2006. Isolation and properties of *Aspergillus niger* IBT-90 xylanase for bakery. Appl. Microbiol. Biotechnol. 69, 665–671.

Sánchez, C., 2009. Lignocellulosic residues: biodegradation and bioconversion by fungi. Biotechnol. Adv. 27, 185–194.

Santos, R.C.A., Araújo, K.B., Zubiolo, C., Soares, C.M.F., Lima, A.S., Aquino, L.C.L.D.S., 2014. Microbial lipase obtained from the fermentation of pumpkin seeds: immobilization potential of hydrophobic matrices. Acta Sci. Technol. 36, 193–201.

Shah, A.R., Madamwar, D., 2005. Xylanase production under solid-state fermentation and its characterization by an isolated strain of *Aspergillus foetidus* in India. World J. Microbiol. Biotechnol. 21, 233–243.

Shah, A.R., Shah, R.K., Madamwar, D., 2006. Improvement of the quality of whole wheat bread by supplementation of xylanase from *Aspergillus foetidus*. Bioresour. Technol. 97, 2047–2053.

Shekiro, J., Kuhn, E.M., Selig, M.J., Nagle, N.J., Decker, S.R., Elander, R.T., 2012. Enzymatic conversion of xylan residues from dilute acid-pretreated corn stover. Appl. Biochem. Biotechnol. 168, 421–433.

Silva, R., Lago, E.S., Merheb, C.W., Macchione, M.M., Park, Y.K., Gomes, E., 2005. Production of xylanase and CMCase on solid state fermentation in different residues by *Thermoascus aurantiacus* Miehe. Braz. J. Microbiol. 36, 235–241.

Silva, S.S., Mussatto, S. i., Santos, J.C., Santos, D.T., Polizel, J., 2007. Cell immobilization and xylitol production using sugarcane bagasse as raw material. Appl. Biochem. Biotechnol. 141, 215–227.

Singhania, R.R., Patel, A.K., Soccol, C.R., Pandey, A., 2009. Recent advances in solid-state fermentation. Biochem. Eng. J. 44, 13–18.

Stroparo, E.C., Beitel, S.M., Resende, J.T.V., Knob, A., 2012. Filamentous fungi and agro-industrial residues selection for enzyme production of biotechnological interest. Semin. Ciênc. Agrár. 33, 2267–2278.

Su, B., Wu, M., Lin, J., Yang, L., 2013. Metabolic engineering strategies for improving xylitol production from hemicellulosic sugars. Biotechnol. Lett. 35, 1781–1789.

Szendefy, J., Szakacs, G., Christopher, L., 2006. Potential of solid-state fermentation enzymes of *Aspergillus oryzae* in biobleaching of paper pulp. Enzyme Microbial. Technol. 39, 1354–1360.

Tabka, M.G., Herpoël-Gimbert, I., Monod, F., Asther, M., Sigoillot, J.C., 2006. Enzymatic saccharification of wheat straw for bioethanol production by a combined cellulase xylanase and feruloyl esterase treatment. Enzyme Microb. Technol. 39, 897–902.

Takahashi, Y., Kawabata, H., Murakami, S., 2013. Analysis of functional xylanases in xylan degradation by *Aspergillus niger* E-1 and characterization of the GH family 10 xylanase XynVII. SpringerPlus 2, 447.

Tapingkae, W., Yachai, M., Visessanguan, W., Pongtanya, P., Pongpiachan, P., 2008. Influence of crude xylanase from *Aspergillus niger* FAS128 on the *in vitro* digestibility and production performance of piglets. Anim. Feed Sci. Technol. 140, 125–138.

te Biesebeke, R., Ruijter, G., Rahardjo, Y.S.P., Hoogschagen, M.J., Heerikkhuisen, M., Levin, A., et al., 2002. Different control mechanisms regulate glucoamylase and protease gene transcription in *Aspergillus oryzae* in solid-state and submerged fermentation. FEMS Yeast Res. 2, 245–248.

Viikari, L., Vehmaanperä, J., Koivula, A., 2012. Lignocellulosic ethanol: from science to industry. Biomass Bioenergy 46, 13–24.

Vitcosque, G.L., Fonseca, R.F., Rodríguez-Zúñiga, U.F., Bertucci Neto, V., Couri, S., Farinas, C.S., 2012. Production of biomass-degrading multienzyme complexes under solid-state fermentation of soybean meal using a bioreactor. Enzyme Res. 2012, 1–9.

Wang, F.-Q., Xie, H., Chen, W., Wang, E.-T., Du, F.-G., Song, A.-D., 2013. Biological pretreatment of corn stover with ligninolytic enzyme for high efficient enzymatic hydrolysis. Bioresour. Technol. 144, 572–578.

Ward, O.P., Qin, W.M., Dhanjoon, J., Ye, J., Singh, A., 2005. Physiology and biotechnology of *Aspergillus*. Adv. Appl. Microbiol. 58, 1–75.

Weijuna, Q., Yefub, C., Huanyingc, Z., Ruisheng, W., Dongguang, X., 2010. Optimization of pretreatment conditions for corn cob with alkali liquor. Trans. Chin. Soc. Agric. Eng. 4, 248–256.

Ximenes, E. a., Dien, B.S., Ladish, M.R., Mosier, N., Cotta, M.A., Li, X.-L., 2007. Enzyme production by industrially relevant fungi cultured on coproduct from corn dry grind ethanol plants. Appl. Biochem. Biotechnol. 136-140, 171–183.

Yoon, K.Y., Woodams, E.E., Hang, Y.D., 2006. Enzymatic production of pentoses from the hemicellulose fraction of corn residues. LWT - Food Sci. Technol. 39, 388–392.

Zhang, L., Xu, J., Lei, L., Jiang, Y., Gao, F., Zhou, G.H., 2014a. Effects of xylanase supplementation on growth performance, nutrient digestibility and non-starch polysaccharide degradation in different sections of the gastrointestinal tract of broilers fed wheat-based diets. Asian Australas. J. Anim. Sci. 27, 855–861.

Zhang, Y., Mu, X., Wang, H., Li, B., Peng, H., 2014b. Combined deacetylation and PFI refining pretreatment of corn cob for the improvement of a two-stage enzymatic hydrolysis. J. Agric. Food Chem. 62, 4661–4667.

Zhu, Y., Kim, T.H., Lee, Y.Y., Chen, R., Elander, R.T., 2006. Enzymatic production of xylooligosaccharides from corn stover and corn cobs treated with aqueous ammonia. Appl. Biochem. Biotechnol. 129, 586–598.

CHAPTER

11

IDENTIFICATION AND APPLICATION OF *VOLVARIELLA VOLVACEA* MATING TYPE GENES TO MUSHROOM BREEDING

Dapeng Bao and Hong Wang

Institute of Edible Fungi, Shanghai Academy of Agricultural Sciences, Shanghai City, People's Republic of China

11.1 INTRODUCTION

Volvariella volvacea, also known as the Chinese mushroom or straw mushroom, is an edible fungal species grown in tropical and subtropical regions. In the eighteenth century, Buddhist monks of Nanhua Temple located in the Chinese province of Guangdong, developed a primitive method of cultivating the straw mushroom to enrich their diet. Subsequently, *V. volvacea* was presented as tribute to China's royalty (Chang, 1969, 1977).

Although *V. volvacea* has been cultivated for approximately 300 years, there are still many associated production problems that greatly restrict the commercial exploitation of this mushroom. Mushroom yields in proportion to the dry weight of compost at spawning (biological efficiency) are only ~15% for straw substrates and 40% for cotton-waste "composts" (Chang, 1980), which are very low values compared with many of the other major cultivated edible mushroom species such as *Agaricus bisporus*, *Lentinula edodes*, and *Pleurotus* spp. (Chang, 1974). *V. volvacea* is also very sensitive to low temperatures, and the fungal mycelium will lose viability when exposed to temperatures below 15°C. Furthermore, mushroom fruit bodies suffer chilling damage and undergo autolysis when stored at low temperatures (4°C) (Chang, 1978), thereby restricting their shelf life.

V. volvacea is generally recognized as a primary homothallic Basidiomycete (Chang, 1969; Chang and Chu, 1969; Chang and Yan, 1971), although some single spore isolates are self-fertile (heterokaryotic) while others are self-sterile (homokaryotic) (Chiu and Moore, 1999). The heterokaryotic isolates complete the sexual cycle without mating, but homokaryotic isolates convert to the dikaryotic form with mating (Figure 11.1). Moreover, *V. volvacea* hyphae are multinucleate (Chang and Ling, 1970) and do not form clamp connections, which, in other fungi, serve as morphological markers to distinguish self-fertile from self-sterile mycelia (Chang, 1983). These traits are problematic for the development of a simple and efficient method for *V. volvacea* cross-breeding.

The sequencing of the genome of the monokaryotic *V. volvacea* strain V23-1 has been completed (Bao et al., 2013). These data will facilitate elucidation of the reproductive issues of the mushroom.

Mushroom Biotechnology. DOI: http://dx.doi.org/10.1016/B978-0-12-802794-3.00011-4

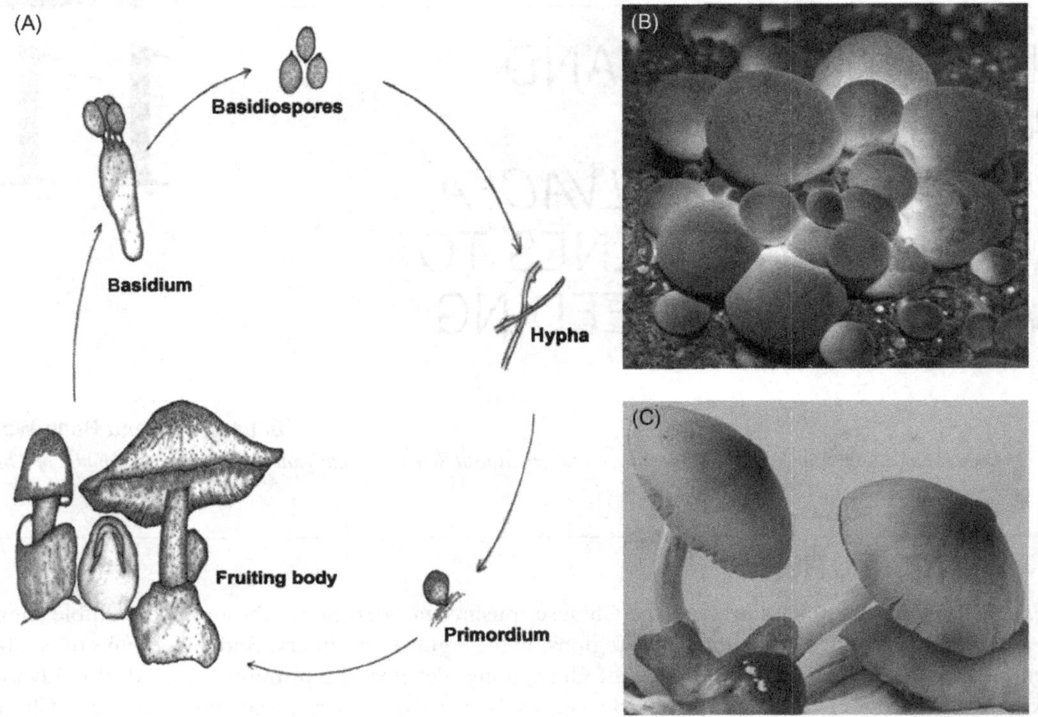

FIGURE 11.1

Life cycle and the fruiting body of *Volvariella volvacea*. (A) Basidiospores of *V. volvacea* are in tetrad and attached to a basidium by four sterigmata. Mature basidiospore germinates under favorable conditions to form primary mycelium. Secondary mycelia form by the fusion of primary mycelia, and give rise to fruiting bodies, which create another new generation of basidiospore. The fruiting mycelium has no clamp connections. The life cycle of *V. volvacea* in general is complete in 4–5 weeks. (B) "Button" stages in the development of *V. volvacea*, which are harvested and sold in the market. (C) A mature fruiting body.

From Bao et al. (2013).

The molecular genetic structure of the mating type locus and phylogenetic analysis of the mating type gene of *V. volvacea* will be helpful in identifying the sexual pattern of *V. volvacea* accurately, and also in finding the proper molecular markers for cross-breeding.

11.2 THE GENERAL FEATURES OF THE *V. VOLVACEA* GENOME

The *V. volvacea* whole genome sequence was assembled into 62 scaffolds with an N50 of 388 kb and a total size of 35.7 Mb, determined using combined Roche 454 GS FLX sequencing and Illumina Solexa sequencing (Bao et al., 2013) (Table 11.1). The size of the *V. volvacea* genome is similar to the genomes of several other species assigned to Agaricaceae, including *Schizophyllum commune* (38.5 M)

Table 11.1 Features of the *Volvariella volvacea* Genome

General Features	
Size of assembled genome (Mb)	36.45
GC content (%)	48.86%
Length of classified repeats (%)	2.25 Mb (6.18%)
Number of predicted gene models	11,084
Average gene length (with intron) (bp)	2,087
Average transcript length (bp)	1,572
Number of single-exon genes	1,066
Average number of exons per multiexon gene	7
Average exon size (bp)	229
Average intron size (bp)	88

(Ohm et al., 2010), *Coprinopsis cinerea* (37 M) (Stajich et al., 2010), and *Pleurotus ostreatus* (35 M) (Ramírez et al., 2011), but is bigger than that of another straw rotting fungus, *A. bisporus* (30.2 M) (Morin et al., 2012). Annotation of the assembled genome sequence generated 11,084 gene models, 76.43% of which were supported by EST data. The average transcript length was 1572 bp, with an average of six introns per multiexon gene. The average exon and intron sizes were 252.3 and 83.2 bp, respectively (Bao et al., 2013) (Table 11.1).

The data of the *V. volvacea* whole genome sequence have been deposited at DDBJ/EMBL/GenBank (http://www.ncbi.nlm.nih.gov/) under the accession number AMXZ01000000, and also at JGI (http://genome.jgi-psf.org/).

11.3 MATING TYPE LOCI AND MATING TYPE GENES OF *V. VOLVACEA*

In the genome sequence of the single spore-derived *V. volvacea*, strain V23-1, the homologous gene scanning revealed two homeodomain (HD) genes, *VVO_04854* and *VVO_05004* (designated *vv-HD1*$^{\text{-}A1\text{-}V23\text{-}1}$ and *vv-HD2*$^{\text{-}A1\text{-}V23\text{-}1}$), located 271 bp apart on scaffold 07 (Bao et al., 2013). Both encoded HD proteins have a high similarity with homologs from the bipolar Basidiomycetes *Pholiota microspora* (*P. nameko*) and the tetrapolar Basidiomycetes *C. cinerea* and *S. commune*. We suggested that *V. volvacea* has one *A* mating type locus containing *HD1* and *HD2* genes. Subsequently, we identified a pair of genes, *vv-HD1*$^{\text{-}A2\text{-}V23\text{-}18}$ and *vv-HD2*$^{\text{-}A2\text{-}V23\text{-}18}$ (accession number: JX157875), within the *A2* locus of another single spore-derived *V. volvacea*, strain V23-18, which is compatible with strain V23-1. The encoded HD1 and HD2 proteins from the spore monokaryons V23-1 and V23-18 showed 48% and 49% similarity, respectively (Bao et al., 2013). Studies on the molecular genetic structure of mating type *A* locus of *V. volvacea* showed that the flanking genes of the *A* mating type locus are the *mip* gene (coding mitochondrial intermediate peptidase, MIP) and *β-fg* gene (Bao et al., 2013) (Figure 11.2), which has highly conserved synteny with other Basidiomycetes, such as *C. cinerea* (Brown and Casselton, 2001), *P. djamor* (James et al., 2004), *Coprinellus disseminates* (James et al., 2006), *Phanerochaete chrysosporium* (James et al., 2011), *P. microspora* (*P. nameko*) (Yi et al., 2009), and *Postia placenta* (Martinez et al., 2009).

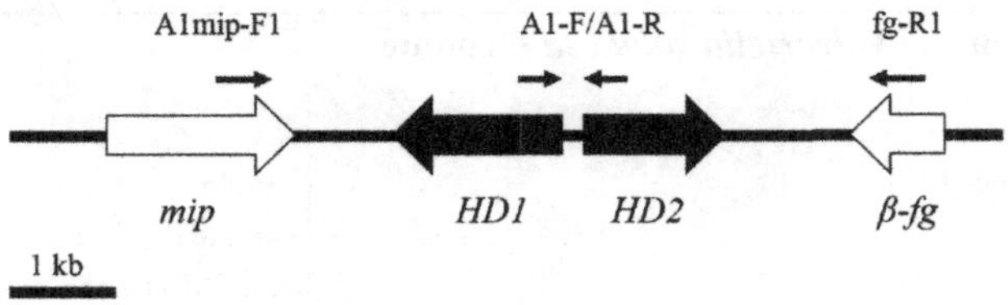

FIGURE 11.2

Relative position of the mitochondrial intermediate peptidase gene (*mip*), *HD* genes, and the *β-fg* gene, and the primers designed to produce the full length of the *A* mating type locus and the polymorphic markers in the *V. volvacea*.

We also found three pheromone-receptor-like genes (named *vv-rcb1*, *vv-rcb2*, and *vv-rcb3*) on scaffold 24 by blasting homologs in the *V. volvacea* genome, which are 72 and 4 kb apart, respectively (Bao et al., 2013). The deduced protein sequences of *vv-rcb1*, *vv-rcb2*, and *vv-rcb3* genes all have the seven-transmembrane structure, which is a classic trait of the pheromone receptor at mating type *B* locus of Basidiomycetes, and have high similarity with other pheromone receptors. We suggested *V. volvacea* also has a *B* mating type locus. But the *B* mating type locus of *V. volvacea* shows poorly conserved synteny with those of other Basidiomycetes. The genes encoding pheromone precursor, which is necessary for the function of *B* mating factor, failed to be found at the flanking region of the *vv-rcb1*, *vv-rcb2*, and *vv-rcb3* genes. Another three pheromone-receptor-like genes obtained from the *B* mating type locus of strain V23-18 by specific primes shared completely identical DNA sequences with that of strain V23-1. These results suggested that pheromone-receptor-like genes in *V. volvacea* might not regulate mating compatibility, which is the same as those in bipolar fungi *C. disseminatus* (James et al., 2006) and *P. nameko* (Yi et al., 2009).

Mating compatibility, leading to heterokaryosis and fruiting, exists only between homokaryotic mycelia carrying different mating type genes and is the basis for cross-breeding. The *V. volvacea* whole genome has greatly increased our understanding of the mating type system operative in this mushroom. *V. volvacea* strains V23-1 and V23-18 have the compatible bipolar mating type alleles *A1* and *A2*, respectively, each of which consists of a pair of mating type genes *HD1* and *HD2*, demonstrating the possibility of mating between homokaryotic hyphae. And the sequences of mating type genes *HD1* and *HD2* between strains V23-1 and V23-18 have been shown to be different, which provides the basis to design the molecular markers.

11.4 SETTING THE MOLECULAR MARKER-ASSISTED BREEDING TECHNIQUES OF *V. VOLVACEA*

V. volvacea have two kinds of single spore isolates: self-fertile (heterokaryotic) isolates and self-sterile (homokaryotic) isolates (Chiu and Moore, 1999). To date, many features relating to the sexual pattern of *V. volvacea* remain unclear, such as the origin of heterokaryotic isolates and the mechanism(s) involved in generating genetic variation among progeny; but the existence of the homokaryotic isolates raises the possibility of using self-sterile isolates as the crossing parent strains to create strains with desirable traits. However, due to the absence of clamp connections in heterokaryotic hyphae, it is problematic to simply

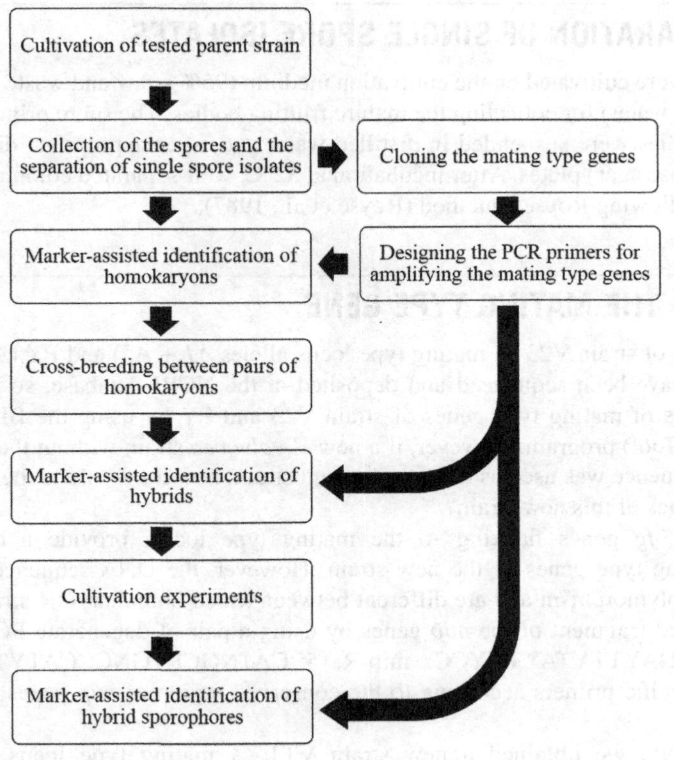

FIGURE 11.3

Process flow diagram regarding the steps of the molecular marker-assisted cross-breeding technique in *V. volvacea*.

and quickly identify the homokaryotic isolates from the heterokaryotic isolates according to morphological differentiations. Nevertheless, the identification of cross-bred *V. volvacea* hybrids (homokaryon × homokaryon crosses) has been reported using molecular screening based on RAPD, SRAP, ISSR, and SCAR markers (Chen, 2006; Li, 2008; Fu et al., 2010; Zhao, 2011). However, since compatible pairings between homokaryotic isolates are controlled by mating type locus, it should be possible to resolve the above-mentioned problem by using the mating type genes as the molecular markers. Guided by this information, we have identified homokaryotic, heterokaryotic, and cross-bred strains of *V. volvacea* using mating type molecular markers as auxiliary molecular markers, and established that the molecular marker-assisted cross-breeding technique represents a practical and highly efficient system for generating improved *V. volvacea* varieties with desirable characteristics. The technique consists of a series of sequential steps involving the separation of single spore isolates, cloning the mating type gene, designing the PCR primers for amplifying the mating type genes, marker-assisted identification of homokaryons, cross-breeding between pairs of homokaryons, marker-assisted identification of hybrids, cultivation experiments, and marker-assisted identification of hybrid sporophores (Xiong et al., 2014) (Figure 11.3).

11.5 THE SEPARATION OF SINGLE SPORE ISOLATES

The tested strains were cultivated on the cultivation medium (95% straw and waste cotton, 5% calcium carbonate, 65% tap water) for collecting the mature fruiting bodies. The spore prints obtained from the mature fruiting bodies were suspended in distilled water and, after appropriate dilution, spread onto PDA (potato dextrose agar) plates. After incubation at 32°C, well-separated colonies were subcultured onto PDA slants following Royse's method (Royse et al., 1987).

11.6 CLONING THE MATING TYPE GENE

The whole genome of strain V23 (*A* mating type locus alleles *A1* + *A2*) and PY (*A* mating type locus alleles *A3* + *A4*) have been sequenced and deposited at the NCBI database, so it is easy to obtain the DNA sequences of mating type genes of strain V23 and PY by using the BLAST (Basic Local Alignment Search Tool) program. However, if a new *V. volvacea* strain without the information of the whole genomic sequence was used as the tested parent strain, how do we clone the DNA sequences of the mating type genes of this new strain?

The *mip* and *β-fg* genes flanking *A*, the mating type locus, provide a convenient method to clone new mating type genes of the new strain. However, the DNA sequences of *mip* genes of *V. volvacea* have polymorphism and are different between tested homokaryotic strains. So we usually obtain the conserved fragment of the *mip* genes by using a pair of degenerate PCR primers (mip-F: 5′-TGGGAYMGNGAYTTYTAYTGYCC; mip-R: 5′-CATNGCRTGNCCCATYTCRTGRAA), and then design the specific primers according to the conserved region of *mip* genes for different tested strains.

In our laboratory, we obtained a new strain VT1 (*A* mating type locus alleles *A5* + *A6*) derived from Thailand and isolated the single spore isolates VT1-1 and VT1-2. We designed the upstream specific primer A5mip-F1 (5′-ATACAGAGTGAGGGAGTCATTTGG) for strain VT1-1 and A6mip-F1 (5′-GGTGGAGGTTAAATACCAAGTGAG) for strain VT1-2, according to the conserved DNA sequence of *mip* gene. The downstream specific primer fg-R1 (5′-TCGGAGGAAGCGGGTCCACTACA) is designed for both strains based on the conserved DNA sequence of *β-fg* gene (Figure 11.2).

The approximately 5-kb DNA sequence containing the *A5* mating type locus was obtained from *V. volvacea* strain VT-1 by long-chain PCR using a pair of primers, A5mip-F1 and fg-R1. Further analysis showed that the *A5* mating type locus of strain VT-1 contained *vv-HD1*$^{\text{-}A5\text{-}VT1\text{-}1}$ and *vv-HD2*$^{\text{-}A5\text{-}VT1\text{-}1}$ genes (GenBank accession numbers: KF702360 and KF702361), 162 bp apart in the opposite direction, which were 1341 and 1450 bp in length and encoded HD1$^{\text{-A5-VT1-1}}$ (446 amino acids) and HD2$^{\text{-A5-VT1-1}}$ (429 amino acids), respectively. The similarity of HD1$^{\text{-A5-VT1-1}}$ and HD2$^{\text{-A5-VT1-1}}$ with HD1$^{\text{-A1-V23-1}}$ (GeneBank accession number: AEO99207.1) and HD2$^{\text{-A1-V23-1}}$ (GenBank accession number: AEO99208.1) are 51% and 49%, and with HD1$^{\text{-A2-V23-18}}$ and HD2$^{\text{-A2-V23-18}}$ (GenBank accession number: JX157875) are 56% and 45%, respectively. The *vv-HD1*$^{\text{-}A5\text{-}VT1\text{-}1}$ gene is 531 bp apart from the *mip* gene.

The *A6* mating type locus of strain VT-2 was amplified using primers A6mip-F1 and fg-R1, and contained *vv-HD1*$^{\text{-}A6\text{-}VT1\text{-}2}$ and *vv-HD2*$^{\text{-}A6\text{-}VT1\text{-}2}$ genes (GeneBank accession numbers: KJ396286 and KJ396287), 161 bp apart in the opposite direction, which were 1442 and 1501 bp in length and encoded

Table 11.2 Primers Used for Identification of Mating Type of *V. volvacea*

Primer	Sequence (5′–3′)	Annealing Temp. (°C)	Mating Type Allele	Parent Strains
A1-F:	AGGGCATTCCAACCTATTCGCTTTC	51	*A1*	V23
A1-R:	AATGTGAACAGTTTGAGCGGAGT			
A2-F:	GTGGTTGGGATGGAAGGTTGTGA	51	*A2*	V23
A2-R:	CTGTGAGGGTTTGTGGTGGATA			
A3-F:	GCTGGTTTGATGTAAGCAGAGGG	53	*A3*	PY
A3-R:	GTGTTGGAAAGACGCTCTGCTGT			
A4-F:	GGCTTTGAATGGCAATCGCTCCT	53	*A4*	PY
A4-R:	CGTGAAAGGCGTCCAGTTTATGC			

HD1$^{\text{-A6-VT1-1}}$ (463 amino acids) and HD2$^{\text{-A6-VT1-2}}$ (443 amino acids), respectively. The similarity of HD1$^{\text{-A6-VT1-2}}$ and HD2$^{\text{-A6-VT1-2}}$ with HD1$^{\text{-A1-V23-1}}$ and HD2$^{\text{-A1-V23-1}}$ are 41% and 46%, and with HD1$^{\text{-A2-V23-18}}$ and HD2$^{\text{-A2-V23-18}}$ are 45% and 48%, respectively. The *vv-HD1*$^{\text{-A6-T1}}$ is 818 bp apart from the *mip* gene.

The above information on the *A* mating type loci of *V. volvacea* strain VT-1 will be helpful in designing the specific PCR primers for amplifying the mating type genes.

11.7 DESIGNING THE PCR PRIMERS FOR AMPLIFYING THE MATING TYPE GENES

The heterokaryon *V. volvacea* has two different mating type loci, each of which contains two mating type genes, *HD1* and *HD2*, usually having ~1700 bp length, respectively, and being <500 bp apart. A pair of specific PCR primers must be designed for each mating type locus to check the specificity of the mating type gene. Usually, the upstream primer was designed to locate at the HD1 gene, and the downstream one was within the HD2 gene. So the interval of *HD1* and *HD2* genes was spanned by the pair of PCR primers. The length of the amplified PCR fragment has usually <1000 bp, which is easy and quick to detect by electrophoresis. In our previous study, the primer pairs A1-F/A1-R and A2-F/A2-R (Figure 11.2, Table 11.2), which produced 655 and 265 bp amplified DNA fragments, respectively, were designed based on the *HD1* and *HD2* gene sequences (HQ343318 and JX157875) of the *A1* and *A2* alleles in the *A* mating type locus of strain V23 (Xiong et al., 2014). The primer pairs A3-F/A3-R and A4-F/A4-R, which produced 748 and 693 bp amplified DNA fragments, respectively, were designed according to the *HD1* and *HD2* gene sequences (JN578700 and JN578701) of the *A3* and *A4* alleles in the *A* mating type locus of strain PY (Xiong et al., 2014).

11.8 THE MARKER-ASSISTED IDENTIFICATION OF HOMOKARYONS

The single spore isolates of *V. volvacea* contained the self-fertile (heterokaryotic) strains and the self-sterile (homokaryotic) strains (Chiu and Moore, 1999). Only the homokaryon is a useful candidate for cross-breeding; the heterokaryon will disturb the results of the cross-breeding test. So the key step for the cross-breeding of *V. volvacea* is to select out the homokaryotic strains from the single spore isolates. Due to the absence of clamp connection formation in the heterokaryotic strain of *V. volvacea* (Chang, 1983), it is difficult to establish the difference between homokaryotic and heterokaryotic strains according to morphological trait as in the case of other mushrooms, such as *L. edodes* or *P. ostreates*.

Fortunately, the homokaryon of the single spore isolates has only one kind of the two mating type loci (*Ax* or *Ay*) of the parent strain, but the heterokaryon has both mating type loci (*Ax* and *Ay*) of the parent strain. Thus, it is possible to identify the homokaryon and heterokaryon by detecting the existence of the mating type loci according to the specificities of the mating type genes. For example, we obtained 124 single spore isolates at random isolated from *V. volvacea* strain V23 (*A* mating type locus alleles *A1* + *A2*). Mating types *A1* and *A2* of the single spore isolate were identified by DNA fragments amplified using the primer pairs A1-F/A1-R and A2-F/A2-R, respectively. By using the molecular screening, a total of 101 single spore isolates were putatively identified as homokaryons, of which 35 were mating type *A1* and 66 mating type *A2*. The remaining 23 were heterokaryons carrying both *A1* and *A2* mating type loci (Xiong et al., 2014). Of the 88 single spore isolates isolated at random from strain PY (*A* mating type locus alleles *A3* + *A4*), by detecting the DNA fragments amplified using the primer pairs A3-F/A3-R and A4-F/A4-R, 72 were putatively identified as homokaryons, of which 41 were mating type *A3* and 31 mating type *A4*. The remaining 16 were heterokaryons carrying both *A3* and *A4* mating type loci (Xiong et al., 2014). Among 62 single spore isolates at random isolated from strain No. 9715 (*A* mating type locus alleles *A7* + *A8*), there were 21 homokaryons carrying *A7* and 23 homokaryons carrying *A8*, and the remaining 20 were heterokaryons carrying both *A7* and *A8*.

11.9 CROSS-BREEDING BETWEEN PAIRS OF HOMOKARYONS

In the last step, the heterokaryotic strains could be picked out and not used as parent strains for the cross-breeding test. This will avoid disturbance of the heterokaryotic strains for pairing the cross strains. On the other hand, the homokaryotic strains will be accurately selected out with the information on the carried mating type genes, which will be helpful in designing compatible cross-pairings to be tested. In this step, the mycelia of two self-sterile isolates (homokaryons) with compatible mating types were inoculated 1.0 cm apart on to PDA plates and incubated at 32°C until they merged. Mycelium from the area of contact was subcultured onto fresh PDA medium and purified by repeated subculture from the edge of the fungal colony. In our study, we paired the 72 homokaryons (41 strains carrying *A3* mating type locus and 31 strains with *A4*) derived from strain PY with the homokaryotic strain V23-1 carrying the *A1* mating type locus (Xiong et al., 2014).

11.10 MARKER-ASSISTED IDENTIFICATION OF HYBRIDS

The putative hybrids had been purified by the repeated subculture in last step, but this operation still couldn't ensure all assumed hybrids are true ones. So we further used the molecular markers of the mating type genes to identify the realness of the putative hybrids. If the putative hybrid has only one of the mating type loci of the parent strains, this means the cross is incompatible and we will delete it.

If the putative hybrid has both mating type loci of the parent strains, we will treat it as assumed compatible crosses (the mixed culture of the homokaryotic parent strains cannot be found in this way, but the rate of this happening is usually very low) and use them for the cultivation test. In our laboratory, we identified the above-mentioned 72 crosses between the homokaryons derived from strain PY and the homokaryotic strain V23-1; the results revealed that the putative compatible crosses were 58 pairs, 33 of which derived from the crosses between 41 *A3* strains and the *A1* strain V23-1 (compatible mating rate: 80.5%), and 25 of which came from the crosses between 31 *A4* strains and the *A1* strain V23-1 (compatible mating rate: 80.6%) (Xiong et al., 2014). In our study, we deleted about 20% of candidate hybrids based on molecular marker-assisted identification in this step, which eliminated a lot of labor in the further experiment. In other laboratory work, 106 crosses between the single spore isolates of strain PY and strain No. 9715 were all identified as heterokaryons and shown to be compatible.

11.11 CULTIVATION EXPERIMENTS

All putative hybrids were to be cultivated for checking the ability of fruiting body production and selecting for biological traits such as: fruiting body yields, average weight of fruit bodies, time between substrate inoculation and primordia formation, fruiting body color, and tendency of fruit bodies to open (parachuting). In our laboratory, the cultivation test was carried out by the following process. The cultivation medium usually consisted of 95% straw and waste cotton, 5% calcium carbonate, and 65% tap water. The compost was turned each time the temperature at a depth of 10 cm reached 55°C, and the pH was regulated to 8.0–8.5 with lime. After uniformly mixing, the compost was spread over the cultivation beds to a thickness of approximately 12 cm and compost temperature was maintained at 65°C for 12 h by steam heating. The surface temperature was then allowed to fall to 30–32°C by natural cooling before inoculating with spawn, after which the temperature of the cultivating room was maintained at 30–32°C by air-conditioning. Tap water was sprayed on the compost at 500 g per square meter five days after inoculation, primordia appeared 2–3 days later, and harvesting commenced after a further 3–5 days. Each putative hybrid strain was cultivated in triplicate, and the cultivation data were subjected to statistical analysis using SAS software.

In our study, the cultivation tests revealed that all 58 putative hybrids produced fruiting bodies, which means that all crossed pairs were indeed compatible and produced true heterokaryotic cross-bred strains. The cultivation experiments undertaken to determine selected biological traits associated with the cross-bred strains revealed that highest yields (1735 g per 10 kg wet substrate) were obtained with hybrid PV26, while the lowest-yielding hybrid, PV5, produced only 135 g per 10 kg wet substrate. In the initial cultivation experiments, among 58 putative hybrids, only 18 strains producing high-quality fruiting bodies and exhibiting a range of desirable biological traits were selected for a second screening test using the parent strains (V23 and PY) as controls, three of which were designated as candidate hybrids for further selected cultivation experiments.

11.12 MARKER-ASSISTED IDENTIFICATION OF HYBRID SPOROPHORES

This step is usually carried out to confirm that the fruiting body derived from the hybrid is a true heterokaryon, using the molecular markers of the mating type genes. DNA was extracted from the mature

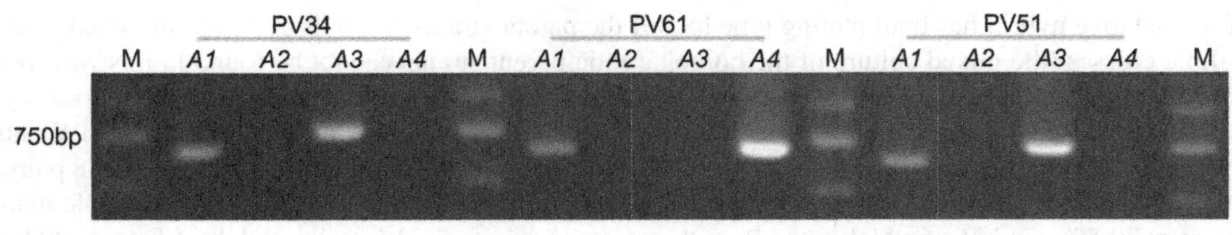

FIGURE 11.4

Identification of the mating type genes of the fruiting bodies derived from cross-bred strains PV34, PV61, and PV51. M: markers.

fruiting body using an improved CTAB method (Zhang et al., 2006). Mating types *A1* and *A2* were identified by the DNA fragments amplified using primer pairs A1-F/A1-R and A2-F/A2-R, respectively (Table 11.2). Mating types *A3* and *A4* were identified by the DNA fragments amplified using primer pairs A3-F/A3-R and A4-F/A4-R, respectively (Table 11.2). PCR reaction mixtures contained ddH_2O (18.25 μL); 2.5 μL 10 × buffer solution; 2.0 μL dNTP (each 2.5 mmol/L); primer, 1 μL (10 μmol/L); 0.25 μL Taq DNA polymerase (5 U/μL), 1 μL DNA template. Amplification conditions were: 94°C for 5 min; 35 cycles of 94°C for 30 s, 51–53°C (see Table 11.1) for 30 s, and 72°C for 1 min; followed by a final extension at 72°C for 10 min. PCR products were separated by electrophoresis on 1.5% sepharose (Xiong et al., 2014). The above-mentioned 18 selected strains were confirmed as true hybrids by validation of the mating type using DNA extracted from the fruiting bodies (Figure 11.4).

To date, the sexual pattern of *V. volvacea* still remains unclear, but the *V. volvacea* genome has facilitated the establishment of an efficient and facile technique for *V. volvacea* cross-breeding using mating type genes as molecular markers. The above-described methodology employing a series of sequential steps reduces the workload and uncertainty inherent in cross-breeding for generating improved *V. volvacea* cultivars, and further application of the technique will provide an important stimulus to the *V. volvacea* cultivation industry.

REFERENCES

Bao, D.B., Gong, M., Zheng, H.J., et al., 2013. Sequencing and comparative analysis of the straw mushroom (*Volvariella volvacea*) genome. PLoS One, http://dx.doi.org/10.1371/journal.pone.0058294.

Brown, A.J., Casselton, L.A., 2001. Mating in mushrooms: increasing the chances but prolonging the affair. Trends Genet. 17, 393–400.

Chang, S.T., 1969. A cytological study of spore germination of *Volvariella volvacea*. Bot. Mag. 82, 102–109.

Chang, S.T., 1974. Production of the straw mushroom (*Volvariella volvacea*) from cotton wastes. Mushroom J. 21, 348–354.

Chang, S.T., 1977. The origin and early development of straw mushroom cultivation. Econ. Bot. 31, 374–376.

Chang, S.T., 1978. *Volvariella volvacea*. In: Chang, S.T., Hayes, W.A. (Eds.), The Biology and Cultivation of Edible Mushroom Academic Press, New York, London, pp. 573–603.

Chang, S.T., 1980. Cultivation of *Volvariella* mushroom in southeast Asia. Mushroom Newslett. Tropics 1, 5–10.

Chang, S.T., 1983. A morfological study of *Volvariella volvacea*. Chung Chi J. 22, 91–103.

Chang, S.T., Chu, S.S., 1969. Nuclear behaviour in the basidium of Volvariella *volvacea*. Cytologia 34, 293–299.

Chang, S.T., Ling, K.Y., 1970. Nuclear behavior in the Basidiomycete, *Volvariella volvacea*. Am. J. Bot. 57, 165–171.

Chang, S.T., Yan, C.K., 1971. *Volvariella volvacea* and its life history. Am. J. Bot. 58, 552–561.

Chen, J., 2006. Researching and Hybridizing of Breed Resource Through Molecule Sign About *Volvariella volvacea*. Master's Thesis, Fujian Agriculture and Forestry University, 57pp.

Chiu, S.W., Moore, D., 1999. Segregation of genotypically diverse progeny from self-fertilized haploids of the Chinese straw mushroom, *Volvariella volvacea*. Mycol. Res. 103, 1335–1345.

Fu, J.S., Zhu, J., Xie, B.G., et al., 2010. Identification of hybrid strain 2628 of *Volvariella volvacea* and its variety test. Chin. Agric. Sci. Bull. 26, 48–53.

James, T.Y., Liou, S.R., Vilgalys, R., 2004. The genetic structure and diversity of the *A* and *B* mating-type genes from the tropical oyster mushroom, *Pleurotus djamor*. Fungal Genet. Biol. 41, 813–825.

James, T.Y., Srivilai, P., Kues, U., et al., 2006. Evolution of the bipolar mating system of the mushroom *Coprinellus disseminatus* from its tetrapolar ancestors involves loss of mating-type-specific pheromone receptor function. Genetics 172, 1877–1891.

James, T.Y., Lee, M., van Diepen, L.T.A., 2011. A single matingtype locus comprised of homeodomain genes promotes nuclear migration and heterokaryosis in the white-rot fungus *Phanerochaete chrysosporium*. Eukaryotic Cell 10, 249–261.

Li, G., 2008. The Research of Molecular Marker-Assisted Breeding About *Volvariella volvacea*. Master's Thesis, Fujian Agriculture and Forestry University, 80pp.

Martinez, D., Challacome, J., Morgenstern, I., et al., 2009. Genome, transcriptome, and secretome of the wood decay fungus *Postia placenta* supports unique mechanisms of lignocellulose conversion. Proc. Natl. Acad. Sci. USA 106, 1954–1959.

Morin, E., Kohler, A., Baker, A.R., et al., 2012. Genome sequence of the button mushroom *Agaricus bisporus* reveals mechanisms governing adaptation to a humic-rich ecological niche. Proc. Natl. Acad. Sci. USA 109, 17501–17506.

Ohm, R.A., de Jong, J.F., Lugones, L.G., et al., 2010. Genome sequence of the model mushroom *Schizophyllum commune*. Nat. Biotechnol. 28, 957–963.

Ramírez, L., Oguiza, J.A., Pérez, G., et al., 2011. Genomics and transcriptomics characterization of genes expressed during postharvest at 4°C by the edible Basidiomycete *Pleurotus ostreatus*. Int. Microbiol. 14, 111–120.

Royse, D.J., Jodon, M.H., Antonio, G.G., et al., 1987. Confirmation of intraspecific crossing and single and joint segregation of biochemical loci of *Volvariella volvacea*. Exp. Mycol. 11, 11–18.

Stajich, J.E., Wilke, S.K., Ahren, D., et al., 2010. Insights into evolution of multicellular fungi from the assembled chromosomes of the mushroom *Coprinopsis cinerea* (*Coprinus cinereus*). Proc. Natl. Acad. Sci. USA 107, 11889–11894.

Xiong, D., Wang, H., Chen, M.J., et al., 2014. Application of mating type genes in molecular marker-assisted breeding of the edible straw mushroom *Volvariella volvacea*. Sci. Hort. 180, 59–62.

Yi, R., Tachikawa, T., Ishikawa, M., et al., 2009. Genomic structure of the *A* mating-type locus in a bipolar Basidiomycete, *Pholiota nameko*. Mycol. Res. 113, 240–248.

Zhang, H., Qin, L.H., Tan, Q., et al., 2006. Extraction of genomic DNA from *Lentinula edodes* using the CTAB method. J. Shanghai Univ. (Natural Science Edition) 12, 547–550.

Zhao, G.H., 2011. *Volvariella volvacea* Breeding Research. Master's Thesis, Fujian Agriculture and Forestry University, 122pp.

CHAPTER

BIOTECHNOLOGICAL USE OF FUNGI FOR THE DEGRADATION OF RECALCITRANT AGRO-PESTICIDES

12

Reyna L. Camacho-Morales and José E. Sánchez
El Colegio de la Frontera Sur, Tapachula, Chiapas, México

12.1 INTRODUCTION

Agriculture is one of the main activities that provides food for the world population. One of the greatest challenges faced by agriculturalists is their ability to control pathogens, insect pests and weeds that destroy crops in which time and financial resources have been invested.

Prior to modern agriculture, appropriate land use, crop rotation, and recycling of organic matter were effective in preventing pest proliferation. Generally, these methods avoided alteration of the ecosystems that were exploited for agriculture (Altieri, 1995). However, as time progressed, the introduction and use of agrochemical compounds increased, permitting the total or partial elimination of insect pests and weeds in crop areas. Despite the obvious efficacy and benefits of these types of compounds, a series of restrictions and negative effects became evident, namely damage to human health, the elimination of beneficial insects, and the general modification of ecosystems. The World Health Organization has calculated that approximately 220,000 people die as a consequence of exposure to different insecticides each year (Eddleston et al., 2002). Pesticides present a serious predicament in that the use of agrochemical compounds protects production, investment, and public health through the eradication of hunger, but at the same time they are extremely harmful to living organisms. Many of these compounds, including glyphosate, chlorothalonil, and paraquat, have successfully eradicated various natural agricultural pests (Matlock and de la Cruz, 2002; Boza, 1972); however, their bioaccumulation and persistence in the environment represent serious problems.

The onset of World War II saw the development of several organic insecticides such as dichlorodiphenyltrichloroethane (DDT), which experienced a boom between 1942 and 1950. Subsequently, it was removed from the majority of countries due to its easy accumulation in organism tissues and high persistence in the soil (Longnecker et al., 1997), although it is still employed in vector control strategies against mosquitoes that transmit dengue and malaria (WHO, 2006, 2009).

Mushroom Biotechnology. DOI: http://dx.doi.org/10.1016/B978-0-12-802794-3.00012-6

The problem of persistence associated with insecticides is not limited to developing countries: in California, more than 6 million hectares of crop land were treated with pesticides between 1990 and 2003 (Epstein and Bassein, 2003). One of the solutions to this problem is the use of integrated pest management (IPM) (Kogan, 1998). IPM refers to a decision-making system on the use of multiple pest control strategies, resulting in a management strategy based on cost analysis, the impact on society and the environment, and, most importantly, a reduction in the use of chemical products (Metcalf, 1980; Allen and Rajotte, 1990; Grewell et al., 2003).

There is a long history of organochloride pesticide (OCP) use in Mexico. Between 1974 and 1991, close to 60,000 tons of OCP, in particular DDT, were released into the environment; however, the use of this compound and chlordane was prohibited in 2000 and 2003, respectively (López-Carrillo et al., 1996; Moody, 2003). Despite these measures, concentrations of these compounds in the soil, air, and blood of people who had been in contact with them were much higher than permitted levels (Wong et al., 2008; Alegría et al., 2006).

One of the most important aspects of pesticides is their high prevalence index in the soil and the risk of these compounds reaching underground water tables. The persistence of these compounds in the soil is a result of many factors, including physical, chemical, and biological processes such as absorption, volatilization, chemical degradation, and accumulation in plants. These processes control the transport and form in which the pesticide moves in the soil, air, or water (Linn et al., 1993; Moorman et al., 2001), and are associated with many soil characteristics including pH, salinity, porosity, and the quantity of organic matter (Boivin et al., 2004; Clausen and Fabricius, 2002).

Residues of pesticide compounds have been found in the air, water, and soil; furthermore, they have been identified in several geographic regions, including some that are far from the original application site, such as deserts, oceans, and even in the polar regions. Many organisms such as whales and other arctic animals have accumulated pesticides in their tissues, and these compounds are magnified as they pass from one trophic level to another within food chains (biomagnification). Humans are not exempt from exposure to these contaminants, and pesticide residues have been found in a variety of tissues and secretions (Asita and Hatane, 2012; Miranda-Contreras et al., 2013; Ansari et al., 2014).

The residues that persist in soil or water can be degraded by microorganisms that are naturally present in the contaminated environment (Guo et al., 2000). However, in many cases this tends to be a slow process due to the high concentrations of the compounds or the incapacity of the microorganisms to incorporate them into their metabolic pathways (Rao et al., 1983).

12.2 BIOREMEDIATION OF XENOBIOTICS

Due to the aforementioned factors, the use of technologies such as bioremediation have increased over the last few years, and the search for appropriate strategies that permit the reduction of recalcitrant compounds in the environment in a safe and natural way has been one of the main objectives of research centers.

Bioremediation exploits the capacity of microorganisms to degrade organic compounds, using those that are native to the contaminated sites or others that are genetically improved (Poindexter and Miller, 1994; Nocentini et al., 2000; McGuinness and Dowling, 2009). There are several bioremediation processes that are capable of biodegrading persistent compounds, the following being the most important: phytoremediation, algae treatment, and bacterial accumulation of xenobiotics.

12.2.1 PHYTOREMEDIATION

Phytoremediation consists of using of plants to accumulate, remove, and neutralize organic compounds, metals, or radioactive compounds. This method takes advantage of the associations between plants and endophytic bacteria such as *Acetobacter* and *Bacillus* (McGuinness and Dowling, 2009). Similarly, numerous studies have focused on the improved resistance of genetically modified plants when faced with an accumulation of toxic compounds (Gisbert et al., 2003; Krämer and Chardonnens, 2001).

Another system that has proved to be very efficient in removing contaminants is algae treatments. These treatments are based on the capacity of algae to accumulate large quantities of metals and several other compounds that are water contaminants (De Godos et al., 2009). An example is the use of the alga *Fucus vesiculosus*, which tolerates high concentrations of crude oil and its derivatives, which accumulate within its tissues (Wrabel and Peckol, 2000).

Another promising option for removing contaminants is the use of microorganisms such as bacteria, which possess metabolic pathways that enable the accumulation or metabolization of different types of xenobiotics. Research on these techniques has increased in recent years, particularly as it appears that they have less impact on the natural environment and can be applied simply and safely. *Pseudomonas putida* has been one of the most frequently used microorganisms for this type of bioremediation, and to date many research papers have been published, not only on the level of biodegradation of various compounds but on the specific mechanisms that carry out this process along the biodegradation metabolic pathways (Reardon et al., 2000; Van Beilen et al., 2001; Kumar et al., 2005; El-Naas et al., 2009).

12.2.2 BIOREMEDIATION BY FUNGI

One of the main groups of organisms used for bioremediation is white rot fungi. These living systems use a complex enzymatic system to degrade lignin, a highly complex polymer that is present in wood and is one of the main components in nature (Pointing, 2001). The enzymes that carry out this type of biodegradation are composed of extracellular oxidases and peroxidases, including laccases, manganese peroxidases, aryl alcohol oxidases, and lignin peroxidase, among others (Paszczynski and Crawford, 2000; Noyotný et al., 2004).

Many fungal species have been extensively studied in relation to biodegradation. *Phanerochaete chrysosporium*, a resupinate or crust fungus belonging to the white rot fungi group, has been used as a model for the biodegradation of aromatic hydrocarbons, colorants, and pesticides (Kullman and Matsumura, 1996; Fournier et al., 2004; Lopera et al., 2005).

Laccase is one of the most studied enzymes for biodegradation of xenobiotic compounds; however, its capacity for degrading only phenolic compounds, which are rarely found in lignin, places it at a disadvantage compared with other enzymes such as lignin peroxidase, which are capable of degrading both phenolic and nonphenolic compounds (Higuchi, 2004). The use of mediators, substances of low molecular weight that act as electron acceptors during the oxide reduction reaction, favor the oxidation of those nonphenolic substrates that have a high redox potential. There are many compounds used as mediators; these promote the biodegradation of recalcitrant compounds in the environment and are also used in the paper-making industry and for processing fruit juices. These include 2,2′-azino-*bis*(3-ethylbenzthiazoline-6-sulfonic acid) (ABTS), violuric acid, 1-hydroxybenzotriazole, *N*-hydroxyphthalimide, syringaldehyde, and 1-nitro-2-naphthol-3,6-disulfonic acid, among others (Moldes and Sanroman, 2006; Camarero et al., 2007; Baiocco et al., 2003).

The use of brown rot fungi as mediators for contaminant biodegradation has also been suggested. These types of mushrooms are capable of degrading cellulose and hemicelluloses, but not the lignin present in wood. Bioremediation is carried out using free radicals produced by the Fenton reaction. This mechanism has been reported to play a specific role in the biodegradation of many xenobiotic compounds such as enrofloxacin, chlorophenol, and 2,4,6 trinitrotoluene (Wetzstein et al., 1997; Schlosser et al., 2000; Newcombe et al., 2002). Other research includes the degradation of DDT by brown rot mushrooms such as *Fomitopsis pinicola* and *Daedalea dickinsii*, which can transform DDT to DDE 1,1-dichloro-2,2-*bis*(4-chlorophenyl) ethylene and DDD (1,1-dichloro-2,2-*bis* (4-chlorophenyl) ethane) via the Fenton reaction (Purnomo et al., 2010, 2011).

Due to their importance, use, and distribution, pesticides have been the center of many biodegradation studies. In this respect, the biodegradation of pentachlorophenol has been tested using *P. chrysosporium*, *Berjkandera adusta*, and *Pleurotus ostreatus*. The biodegradation percentages vary, with *P. chrysosporium* recording the highest with 96% after 64 days in a liquid culture medium (Lamar et al., 1990; Ruttimann and Lamar, 1997). Biodegradation of the insecticide lindane by the fungal species *Trametes hirsuta*, *Pleurotus eryngii*, and *P. chrysosporium* has been extensively studied. Liquid cultures as well as soil contaminated by these agrochemicals have been studied, resulting in biodegradation percentages ranging from 10.6% to 96%, demonstrating that *T. hirsuta* is the most suitable biodegrader (Arisoy, 1998; Quintero et al., 2007; Singh and Kuhad, 1999).

Another compound used as a study model is heptachlor; the use of this pesticide is currently prohibited, but residues that are highly toxic to living systems are still found in areas where it was previously applied. The species *P. chrysosporium* and two species from the *Pleurotus* genus, *Pleurotus sajor-caju* and *P. eryngii*, have been used in studies of heptachlor biodegradation (Arisoy, 1998; Kennedy et al., 1990).

Finally, a number of studies have been done on the biodegradation of glyphosate, one of the most widely used herbicides worldwide. In this case, the process was carried out using purified enzymes. Total degradation of this compound was achieved when laccase and manganese peroxide were incubated, using ABTS as a mediator for the former and manganese sulfate and Tween 80 for the latter (Pizzul et al., 2009).

Important advances have been made regarding the biodegradation of three model compounds: endosulfan (insecticide), chlorothalonil (fungicide), and paraquat (herbicide). These three compounds present aromatic structures that can be used as a target for nonspecific extracellular enzymes of the ligninolytic type that possess several macromycetes (Figure 12.1).

12.2.2.1 Endosulfan biodegradation by fungi

Endosulfan is an insecticide that is applied extensively to coffee, tea, and cotton crops, among others. In its commercial form, endosulfan is composed of two stereoisomers, α and β endosulfan, in a 70:30 ratio. This compound is especially toxic for aquatic organisms (Broomhall, 2002; Capkin et al., 2006). In mammals, it can induce reproductive toxicity (Saiyed et al., 2003; Bharath et al., 2011) and neurotoxicity (Ravi and Varma, 1998). In the soil, endosulfan is converted into endosulfan sulfate by oxidation and diol endosulfan by hydrolysis. Both isomers present different levels of persistence in the environment. The half-life of total endosulfan and α and β endosulfan is 1336 and 27.5 and 157 days, respectively, under aerobic conditions (GFEA-U, 2007; US EPA, 2007a). Endosulfan sulfate is a more persistent and toxic residue than endosulfan isomers (US EPA, 2007b).

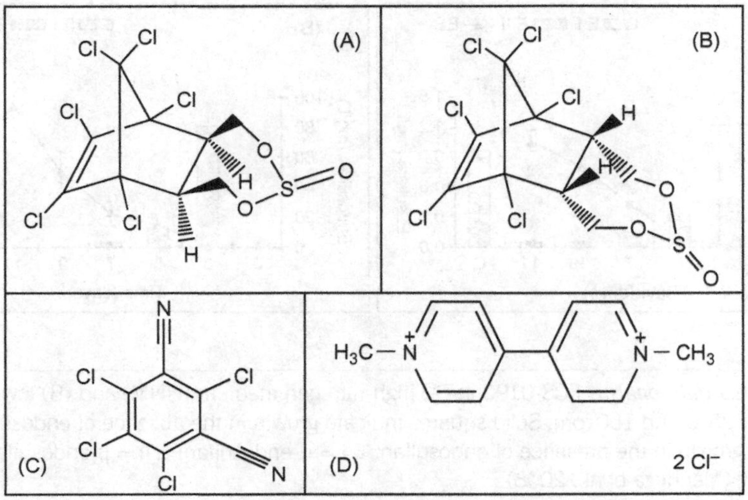

FIGURE 12.1

Structures of (A) α-endosulfan, (B) β-endosulfan, (C) chlorothalonil, and (D) paraquat (Kegley et al., 2014; Schmidt et al., 2001; Wales et al., 2003; Shia et al., 2013).

Various studies have been conducted on the biodegradation of the insecticide endosulfan. The capacity of *Aspergillus niger* to metabolize this compound was tested, revealing that this species is capable of eliminating a concentration of 400 ppm endosulfan after 12 days of incubation (Tejomyee and Pravin, 2007). Another study demonstrated the biodegradation of endosulfan and endosulfan sulfate by the fungi *T. hirsuta*, achieving 90% removal after 14 days incubation (Kamei et al., 2011).

The ability of fungal strains from the *Pleurotus* genus and the species *Auricularia fuscosuccinea* to degrade endosulfan was demonstrated. Eighty *Pleurotus* strains including *Pleurotus djamor*, *Pleurotus ostreatus*, *Pleurotus cornucopiae*, and *Pleurotus cystidiosus* were studied. The results show that all the evaluated strains were capable of growth in a concentration of 100 ppm endosulfan. The growth medium was deprived of glucose, indicating that these fungi were able to degrade the contaminant and use it as a source of carbon. One of the *Pleurotus pulmonarius* strains grown in a liquid culture achieved 96.98% and 93.62% biodegradation of α endosulfan and β endosulfan, respectively, after 8 days of growth. Figure 12.1 shows the influence of culture medium composition on biodegradation rates (Kegley et al., 2014; Schmidt et al., 2001; Wales et al., 2003; Shia et al., 2013). A medium with high nitrogen concentration (HNC) and another with low nitrogen concentration (LNC) were used; the LNC degraded compounds at a faster rate than the HNC (Mendoza et al., 2008) (Figure 12.2).

During fungal development, the growth substrate may contain considerable quantities of endosulfan; thus, it is important to know how the pasteurization and composting processes affect compound degradation and also compare these processes using biodegradation by the fungi *P. pulmonarius*. During the composting process of the grass *Digitaria decumbens*, in the presence of $Ca(OH)_2$, a reduction in α and β endosulfan concentration by 61.4% and 49.5%, respectively, was achieved; this was considerably more than was attained by the control (38.5%), which did not contain $Ca(OH)_2$.

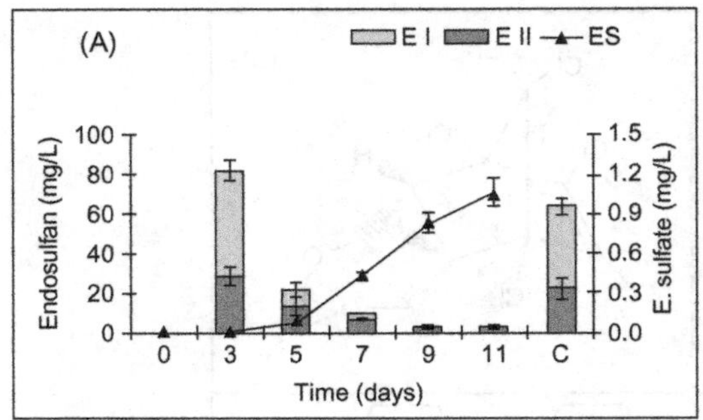

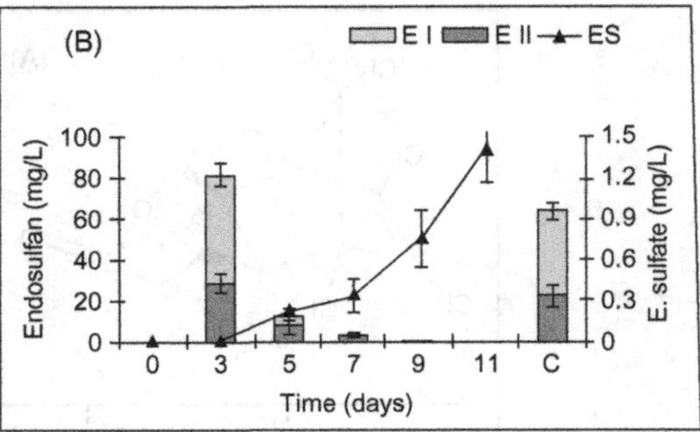

FIGURE 12.2

Growth of *Pleurotus pulmonarius* ECS-0190 in (A) high-nitrogen medium (HNM) and (B) low-nitrogen medium (LNM) at 25°C and 150 rpm. Solid squares indicate growth in the absence of endosulfan. Solid triangles indicate growth in the presence of endosulfan. E I = α-endosulfan; E II = β-endosulfan; ES = endosulfan sulfate (Mendoza et al., 2008).

This indicates that the process of substrate alkalinization is fundamental to the degradation process. In the case of sterilization, 84.8% of α endosulfan and 87.5% of β endosulfan was removed. Finally, subsequent to substrate colonization by *P. pulmonarius*, α and β endosulfan were degraded by 99%. These results demonstrated that although considerable removal of endosulfan is achieved during the composting and sterilization processes, much higher levels of degradation (99%) are achieved as a result of fungal growth (Hernández-Rodríguez et al., 2006).

In the case of *A. fuscosuccinea*, 100 ppm of endosulfan was removed over a period of 8 days using a medium with a limited source of carbon; however, biomass production was much lower when compared to the control, demonstrating that endosulfan is capable of inhibiting fungal growth (Escobar et al., 2002). Studies on endosulfan biodegradation by *A. fuscosuccinea* strains have continued; in one study, biodegradation was carried out using four species of *A. fuscosuccinea*. Their biodegradation capacity was evaluated after 4 and 8 days of growth; it was determined that biodegradation reached approximately 100% after 8 days of growth for all the evaluated strains; furthermore, the cell-free culture mix was capable of degrading the insecticide, indicating that an extracellular enzyme may be responsible for biodegradation. An analysis of the enzymes that participated in the biodegradation process revealed only laccase and phenol oxidase activities; thus, it can be inferred that these could be involved in endosulfan biodegradation (Yanez-Montalvo et al., 2015).

12.2.2.2 Chlorothalonil biodegradation by fungi

Around 45 species of microorganisms have been described that are capable of degrading the fungicide chlorothalonil; bacteria and actinomycetes comprise the large majority of the microorganisms studied, the most important being *Bacillus cereus*, *Moraxella* spp., and *Agrobacterium* spp., which are capable of degrading a concentration of 50 ppm chlorothalonil (Wang et al., 2011).

The bacterial species *Ochrobactrum lupine* can degrade 90.4% and 99.7% chlorothalonil after 4 and 7 days of incubation, respectively, using a mineral salt medium and a 50 ppm contaminant concentration (Shi et al., 2011). The assays conducted with this fungicide refer to biodegradation using spent mushroom substrate (SMS) from the strain *P. pulmonarius* ECS-0190. These experiments demonstrated that after two mushroom harvests, the remaining substrate could be used for degradation of an aqueous solution containing the fungicide (2 mg/L chlorothalonil). Freshly obtained SMS extract was able to reduce 100% of the initial concentration of chlorothalonil (2 mg/L) in a liquid effluent after 45 min of contact. SMS storage time had a negative effect on the stability of enzymatic activity: using spent substrate stored for a week, chlorothalonil concentration was reduced by 49.5% after 1 h reaction, while with substrate stored for 2 and 3 weeks, biodegradation efficiency decreased to 9.15% and 0%, respectively. Cooling and freezing the spent substrate extract also had a negative effect on chlorothalonil biodegradation (Córdova-Juárez et al., 2011). However, in these studies, the authors could not determine if biodegradation was caused by ligninolytic enzymes (essentially phenol oxidase, laccase, and MN-peroxidase) from the cultivated fungi. This suggests the influence of other factors such as the presence of mediators or alternative systems that enable degradation.

12.2.2.3 Biodegradation of paraquat

Paraquat is a nonsystemic contact herbicide that is applied to coffee, banana, mango, and sugar crops. Its use has been prohibited in many countries; however, it is still distributed in some countries, including Mexico (Wesseling et al., 2001). This highly toxic herbicide is still extensively used, mainly due to its low price, rapid action, and environmental characteristics. A large number of cases of poisoning and death from ingestion by humans have been reported. The compound is not absorbed into the intestinal tract and therefore tends to accumulate in the kidneys and lungs (Shimada et al., 2002; Murray and Gibson, 1974). Paraquat also causes the degeneration of dopaminergic neurons (Liou et al., 1996; Chanyachukul et al., 2004). Due to its high persistence in the soil, it is difficult to eliminate and therefore presents severe problems of contamination and accumulation (Smith and Mayfield, 1978). In the soil, paraquat can be removed by implementing two techniques. The first is the photolytic technique, using low wavelengths (300–200 nm). In this method the main metabolites produced from paraquat are monoquat and piridone; however, this only occurs in the soil surface layers and is very inefficient as most of the paraquat remains in the soil layers below the surface. The second method is the microbial technique, producing mainly monoquat, which can subsequently be metabolized into ammonia, carbon dioxide, and water. This type of biodegradation depends largely on soil type and the quantity of paraquat (Roberts et al., 2002).

There are few reports on paraquat degradation using microorganisms. The bacterial species *Pseudomonas putida* has been used as a model for research on the biodegradation of diverse environmental contaminants. One such study on paraquat biodegradation resulted in 90% removal of the contaminant in 24 h, using activated carbon (Kopytko et al., 2002). Other studies have analyzed photolytic degradation of this herbicide; for example, the effectiveness of X-ray diffraction, UV diffuse reflectance spectroscopy, and X-ray adsorption spectroscopy have been demonstrated. These methods have attained 100% contaminant reduction using a support of copper and titanium mesoporous material (Sorolla et al., 2012).

A large number of fungal species have not yet been studied, and as many of these grow in areas that have suffered from indiscriminate use of harmful substances, they can tolerate or even biodegrade large quantities of contaminants. To increase knowledge on this type of fungi, native strains from the southeast region of the state of Chiapas, Mexico, were isolated and characterized. These strains were collected from coffee farms and crop fields that had been susceptible to paraquat application. Some 105 strains were analyzed to determine the biodegradation capacity in a liquid medium with 200 ppm of paraquat. The experiment was carried out over a period of 15 days, after which the residual paraquat was extracted and the levels of biodegradation were quantified. Approximately 10 strains were capable of biodegrading between 30% and 90% of paraquat present in the growth medium.

12.3 PERSPECTIVES

Without doubt, the indiscriminate use of chemical products in agriculture has generated a huge amount of contamination in the air, water, and soil. Currently, the use of alternative technologies against pest species, such as bioinsecticides, integral pest management, and genetically modified plants that can exploit soil nutrients more efficiently, or with specific insect defense characteristics; together, these provide an encouraging picture. However, more work is needed on bioremediation treatments that include the use of ligninolytic enzymes, intracellular metabolic pathways, and biodegradation by native soil microorganisms, the latter considered as one of the best techniques in counteracting high pollution levels that persist in the soil.

Currently, studies and processes are being developed to improve the use of the enzymatic systems that only microorganisms possess. An alternative that offers great potential for the development of innovative applications refers to the use of self-propelled nanomotors (Soler and Sánchez, 2014). These systems are already being tested with ligninases. Orozco et al. (2014) report using a self-propelled tubular motor that releases an enzyme for efficient biocatalytic degradation of chemical pollutants. These processes are based on the Marangoni effect, involving the simultaneous release of an SDS surfactant and the enzyme remediation agent laccase in the polluted sample. The movement induces fluid convection and leads to the rapid dispersion of laccase into the contaminated solution and to a dramatically accelerated biocatalytic decontamination process.

These new alternatives, together with those that already exist, are potential tools that can be used to reduce or alleviate levels of agrochemical contamination in the environment. Essentially, they are still under development, and it will probably take more than 10 years until crops can be grown on soils that are free from agrochemical residues, or before the new treatments can be used to treat effluent waters from agroindustry and agriculture.

Despite having been little studied, the enzymatic systems of macromycetes appear to be very powerful, and therefore further research is needed, particularly when applications for the biodegradation of other contaminants are found, such as newly emerging contaminants for which there is currently no treatment method (De Morais et al., 2012; Santos et al., 2012).

REFERENCES

Alegría, H., Bidleman, T.F., Figueroa, M.S., Lopez-Carrillo, L., Torres-Arreola, L., Torres-Sanchez, L., et al., 2006. Organochlorine pesticides in the ambient air of Chiapas, Mexico. Environ. Pollut. 140, 483–491.

Allen, W.A., Rajotte, E.G., 1990. The changing role of extension entomology in the IPM era. Annu. Rev. Entomol. 35, 379–397.

Altieri, M.A., 1995. Agroecology: The Science of Sustainable Agriculture. Westview Press, Boulder, CO.

Ansari, M.S., Moraiet, M.A., Ahmad, S., 2014. Insecticides: impact on the environment and human health. Environmental Deterioration and Human Health. Springer, Netherlands, pp. 99–123.

Arisoy, M., 1998. Biodegradation of chlorinated organic compounds by white-rot fungi. Bull. Environ. Contam. Toxicol. 60 (6), 872–876.

Asita, A.O., Hatane, B.H., 2012. Cytotoxicity and genotoxicity of some agropesticides used in Southern Africa. J. Toxicol. Environ. Health Sci. 4 (10), 175–184.

Baiocco, P., Barreca, A.M., Fabbrini, M., Galli, C., Gentili, P., 2003. Promoting laccase activity towards non-phenolic substrates: a mechanistic investigation with some laccase–mediator systems. Org. Biomol. Chem. 1, 191–197.

Bharath, B.K., Srilatha, C., Anjaneyulu, Y., 2011. Reproductive toxicity of endosulfan on male rats. Int. J. Pharma Bio Sci. 2 (3), 508–512.

Boivin, A., Cherrier, R., Perrin-Ganier, C., Schiavon, M., 2004. Time effect on bentazone sorption and degradation in soil. Pesticide Manage. Sci. 60 (8), 809–814.

Boza, B.T., 1972. Ecological consequences of pesticides used for the control of cotton insects in the Cañete Valley, Peru. In: Farvar, M.Y., Milton, J.P. (Eds.), The Careless Technology: Ecology and International Development The Natural History Press, New York, NY.

Broomhall, S., 2002. The effects of endosulfan and variable water temperature on survivorship and subsequent vulnerability to predation in *Litoria citropa* tadpoles. Aquat. Toxicol. 61 (3–4), 243–250.

Camarero, S., Ibarra, D., Martínez, A., Romero, J., Gutierrez, A., del Río, J.C., 2007. Paper pulp delignification using laccase and natural mediators. Enzyme Microbial. Technol. 40, 1264–1271.

Capkin, E., Altinok, I., Karahan, S., 2006. Water quality and fish size affect toxicity of endosulfan, an organochlorine pesticide to rainbow trout. Chemosphere 64 (10), 1793–1800.

Chanyachukul, T., Yoovathaworn, K., Thongsaard, W., Chongthammakun, S., Navasumrit, P., Satayavivad, J., 2004. Attenuation of paraquat-induced motor behavior and neurochemical disturbances by L-valine *in vivo*. Toxicol. Lett. 150, 259–269.

Clausen, L., Fabricius, I., 2002. Atrazine, isoproturon, mecoprop, 2,4-D, and bentazone adsorption onto iron oxides. J. Environ. Qual. 30 (3), 858–869.

Córdova-Juárez, R.A., Gordillo-Dorry, L.L., Bello-Mendoza, R., Sánchez, J.E., 2011. Use of spent substrate after *Pleurotus pulmonarius* cultivation for treatment of chlorothalonil containing wastewater. J. Environ. Manage. 92, 948–952.

De Godos, I., Blanco, S., García-Encina, P.A., Becares, E., Muñoz, R., 2009. Long-term operation of high rate algal ponds for the bioremediation of piggery wastewaters at high loading rates. Bioresour. Technol. 100 (19), 4332–4339.

De Morais, R., Ferreira, P., Cintra, L., Alves, T., Campos, M., Schimidt, F., et al., 2012. Evaluation of the use of *Pycnoporus sanguineus* fungus for phenolics and genotoxicity decay of a pharmaceutical effluent treatment. Rev. Ambi. Água 7 (3), 41–50.

Eddleston, M., Karalliedde, L., Buckley, N., Fernando, R., Hutchinson, G., Isbister, G., et al., 2002. Pesticide poisoning in the developing world–a minimum pesticide list. Lancet 360, 1163–1167.

El-Naas, M.H., Al-Muhtaseb, S.A., Makhlouf, S., 2009. Biodegradation of phenol by *Pseudomonas putida* immobilized in polyvinyl alcohol (PVA) gel. J. Hazard. Mater. 164 (2–3), 720–725.

Epstein, L., Bassein, S., 2003. Patterns of pesticide use in California and the implications for strategies for reduction of pesticides. Annu. Rev. Phytopathol. 41, 351–375.

Escobar, V., Nieto, M., Sánchez, J., Cruz, L., 2002. Effect of endosulfan on mycelial growth of *Pleurotus ostreatus* and *Auricularia fuscosuccinea* in liquid culture. In: Proceedings of IV International Conference on Mushroom Biology and Mushroom Products, Cuernavaca, Mexico, pp. 399–408.

Fournier, D., Halasz, A., Spain, J., Spanggord, R., Bottaro, J.C., Hawari, J., 2004. Biodegradation of hexahydro-1,3,5-trinitro-1,3,5-triazine ring cleavage product 4-nitro-2,4-diazabutanal by *Phanerochaete chrysosporium*. Appl. Environ. Microbiol. 71 (2), 1123–1128.

GFEA-U, 2007. Endosulfan Draft Dossier Prepared in Support of a Proposal of Endosulfan to be Considered as a Candidate for Inclusion in the CLRTAP Protocol on Persistent Organic Pollutants. German Federal Environment Agency, Berlin.

Gisbert, C., Ros, R., De Haro, A., Walker, D.J., Bernal, M.P., Serrano, R., et al., 2003. A plant genetically modified that accumulates Pb is especially promising for phytoremediation. Biochem. Biophys. Res. Commun. 303 (2), 440–445.

Grewell, J.B., Landry, C.J., Conko, G., 2003. Ecological Agrarian: Agriculture's First Evolution in 10,000 Years. Purdue University Press, Indiana.

Guo, L., Jury, W.A., Wagenet, R.J., Flury, M., 2000. Dependence of pesticide degradation on sorption: nonequilibrium model and application to soil reactors. J. Contam. Hydrol. 43 (1), 45–62.

Hernández-Rodríguez, D., Sánchez, J.E., Nieto, M.G., Márquez, F.J., 2006. Degradation of endosulfan during substrate preparation and cultivation of *Pleurotus pulmonarius*. World J. Microbiol. Biotechnol. 22, 753–760.

Higuchi, T., 2004. Microbial degradation of lignin role of lignin peroxidase, manganese peroxidase, and laccase. Proc. Jpn. Acad. B 80, 204–214.

Kamei, I., Takagi, K., Kondo, R., 2011. Degradation of endosulfan and endosulfan sulfate by white-rot fungus *Trametes hirsuta*. J. Wood Sci. 57 (4), 317–322.

Kegley, S.E., Hill, B.R., Orme S., Choi A.H., 2014. PAN Pesticide Database. Pesticide Action Network, North America, Oakland, CA. <http://www.pesticideinfo.org/Detail_Chemical.jsp?Rec_Id=PC34550>.

Kennedy, D., Aust, S., Bumpus, A., 1990. Comparative biodegradation of alkyl halide insecticides by the white rot fungus, *Phanerochaete chrysosporium* BKM-F-1767. Appl. Environ. Microbiol. 56 (8), 2347–2353.

Kogan, M., 1998. Integrated pest management: historical perspectives and contemporary developments. Annu. Rev. Entomol. 43, 243–270.

Kopytko, M., Chalela, G., Zauscher, F., 2002. Biodegradation of two commercial herbicides (Gramoxone and Matancha) by the bacteria *Pseudomonas putida*. Electron. J. Biotechnol. 5 (2), 1.

Krämer, U., Chardonnens, A.N., 2001. The use of transgenic plants in the bioremediation of soils contaminated with trace elements. Appl. Microbiol. Biotechnol. 55, 661–672.

Kullman, S.W., Matsumura, F., 1996. Metabolic pathways utilized by *Phanerochaete chrysosporium* for degradation of the cyclodiene pesticide endosulfan. Appl. Environ. Microbiol. 62 (2), 593–600.

Kumar, A., Kumar, S., Kumar, S., 2005. Biodegradation kinetics of phenol and catechol using *Pseudomonas putida* MTCC 1194. Biochem. Eng. J. 22 (2), 151–159.

Lamar, R.T., Larsen, M.J., Kirk, T.K., 1990. Sensitivity to and degradation of pentachlorophenol PCP by *Phanerochaete chrysosporium*. Appl. Environ. Microbiol. 56 (11), 3519–3526.

Linn, D.M., Carski, T.H., Brusseau, M.L., Chang, F.H., 1993. Sorption and Degradation of Pesticides and Organic Chemicals in Soil. Soil Science Society of America, Madison, WI, p. 260.

Liou, H.H., Chen, R.C., Tsai, Y.F., Chen, W.P., Chang, Y.C., Tsai, M.C., 1996. Effects of paraquat on the substantia nigra of the Wistar rats: neurochemical, histological, and behavioral studies. Toxicol. Appl. Pharmacol. 137, 34–41.

Longnecker, M.P., Rogan, W.J., Lucier, G., 1997. The human health effects of DDT (dichlorodiphenyltrichloroethane) and PCBS (polychlorinated biphenyls) and an overview of organochlorines in public health. Annu. Rev. Public. Health 18, 211–244.

Lopera, M.M.M., Peñuela, M.G.A., Dominguez, G.M.C., Mejía, Z.G.M., 2005. Evaluación de la degradación del plaguicida clorpirifos en muestras de suelo utilizando el hongo *Phanerochaete chrysosporium*. Rev. Fac. Ing. 33, 58–69.

López-Carrillo, L., Torres-Arreola, L., Torres-Sánchez, L., Espinosa-Torres, F., Jimenez, C., Cebrian, M., et al., 1996. Is DDT use a public health program in Mexico. Environ. Health Perspect. 104, 584–588.

Matlock, R.B., de la Cruz, R., 2002. An inventory of parasitic *Hymenoptera* in banana plantations under two pesticide regimes. Agric. Ecosyst. Environ. 93 (1), 147–164.

McGuinness, M., Dowling, D., 2009. Plant-Associated bacterial degradation of toxic organic compounds in soil. Int. J. Environ. Res. Public Health 6, 2226–2247.

Mendoza, G.M., Sánchez, J.E., Nieto, M.G., Márquez Rocha, J., 2008. Degradation of endosulfan by *Pleurotus* spp. In: Lelley, J.I. (Ed.), Proceedings of 6th International Conference on Mushroom Biology and Mushroom Products World Society for Mushroom Biology and Mushroom Products, Bonn, pp. 36–47.

Metcalf, R.L., 1980. Changing role of insecticides in crop protection. Ann. Rev. Entomol. 25 (1), 219–256.

Miranda-Contreras, L., Gómez-Pérez, R.R.G., Cruz, I.B.L., Salmen, S., Colmenares, M., Barreto, S., et al., 2013. Occupational exposure to organophosphate and carbamate pesticides affects sperm chromatin integrity and reproductive hormone levels among Venezuelan farm workers. J. Occup. Health 55, 195–203.

Moldes, D., Sanroman, M.A., 2006. Amelioration of the ability to decolorize dyes by laccase: relationship between redox mediators and laccase isoenzymes in *Trametes versicolor*. World J. Microbiol. Biotechnol. 22, 1197–1204.

Moody, J., 2003. North America eliminates use of chlordane. Trio Newslett., 4. North American Commission for Environmental Cooperation, Montreal. <http://www.cec.org/trio/stories/index.cfm?edZ9&IDZ111&varlanZenglishO>.

Moorman, T.B., Jayachandran, K., Reungsang, A., 2001. Adsorption and desorption of atrazine in soils and subsurface sediments. Soil Sci. 166 (12), 921–929.

Murray, R.E., Gibson, J.E., 1974. Paraquat disposition in rats, guinea pigs and monkeys. Toxicol. Appl. Pharmacol. 27, 283–291.

Newcombe, D., Paszcynsky, A., Gajewska, W., Kroger, M., Feis, G., Crawford, R., 2002. Production of small molecular weight catalyst and the mechanism of trinitrotoluene degradation by several *Gloeophyllum* species. Enzyme Microbial. Technol. 30, 506–517.

Nocentini, M., Pinelli, D., Fava, F., 2000. Bioremediation of a soil contaminated by hydrocarbon mixtures: the residual concentration problem. Chemosphere 41, 1115–1123.

Noyotný, C., Svobodová, K., Erbanová, P., Cajthaml, T., Kasinath, A., Lang, E., et al., 2004. Ligninolytic fungi in bioremediation: extracellular enzyme production and degradation rate. Soil Biol. Biochem. 36, 1545–1551.

Orozco, J., Vilela, D., Valdés-Ramírez, G., Fedorak, Y., Escarpa, A., Vazquez-Duhalt, R., et al., 2014. Efficient biocatalytic degradation of pollutants by enzyme-releasing self-propelled motors. Chem. Eur. J. 20, 2866–2871.

Paszczynski, A., Crawford, R., 2000. Recent advances in the use of fungi in environmental remediation and biotechnology. Soil Biochem. 10, 379–422.

Pizzul, L., Castillo, M.P., Stenström, J., 2009. Degradation of glyphosate and other pesticides by ligninolytic enzymes. Biodegradation 20 (6), 751–759.

Poindexter, L., Miller, P., 1994. Predictability of Bioremediation Performance Cannot Be Made with a High Level of Confidence. American Academic of Microbiology, Washington, DC.

Pointing, S.B., 2001. Feasibility of bioremediation by white-rot fungi. Appl. Microbiol. Biotechnol. 57, 20–33.

Purnomo, A.S., Mori, T., Ryuichiro, K., 2010. Involvement of Fenton reaction in DDT degradation by brown-rot fungi. Int. Biodeterior. Biodegrad. 64, 560–565.

Purnomo, A.S., Morib, T.K., Ichiro, K.R., 2011. Basic studies and applications on bioremediation of DDT: a review. Int. Biodeterior. Biodegrad. 65, 921–930.

Quintero, J.C., Lú-Chau, T.L., Moreira, M.T., Feijoo, G., Lema, J.M., 2007. Bioremediation of HCH present in soil by the white-rot fungus *Bjerkandera adusta* in a slurry batch bioreactor. Int. Biodeterior. Biodegrad. 60 (4), 319–326.

Rao, P.S.C., Mansell, R.S., Baldwin, L.B., Laurent, M.F., 1983. Pesticides and their behavior in soil and water. Soil Science Fact Sheet SL 40 (Revised). Florida Cooperative Extension Service, Gainesville, p. 4.

Ravi, K., Varma, M.N., 1998. Biochemical studies on endosulfan toxicity in different age groups of rats. Toxicol. Lett. 44 (3), 247–252.

Reardon, K.F., Mosteller, D.C., Bull Rogers, J.D., 2000. Biodegradation kinetics of benzene, toluene, and phenol as single and mixed substrates for *Pseudomonas putida*. Biotechnol. Microbiol. 70 (2), 1123–1128.

Roberts, T.R., Dyson, J.S., Lane, M.C.G., 2002. Deactivation of the biological activity of paraquat in the soil environment: a review of long-term environmental fate. J. Agric. Food Chem. 50 (13), 3623–3631.

Ruttimann, C., Lamar, R., 1997. Binding of substances in pentachlorophenol to humic soil by the action of white rot fungi. Soil Biol. Biochem. 9 (7), 1143–1148.

Saiyed, H., Dewan, A., Bhatnagar, V., Shenoy, U., Shenoy, R., Rajmohan, H., et al., 2003. Effect of endosulfan on male reproductive development. Environ. Health Perspect. 111 (16), 1958–1962.

Santos, I., Grossman, M., Sartoratto, A., Ponezi, A., Durrant, L., 2012. Degradation of the recalcitrant pharmaceuticals carbamazepine and 17α-ethinylestradiol by ligninolytic fungi. Chem. Eng. Trans. 27, 169–174.

Schlosser, D., Fahr, K., Karl, W., Wetzstein, H.G., 2000. Hydroxylated metabolites of 2,4-dichlorophenol imply a Fenton-type reaction in *Gloeophyllum striatum*. Appl. Environ. Microbiol. 66, 2479–2483.

Schmidt, W.F., Bilboulian, S., Rice, C.P., Fettinger, J.C., McConnell, L.L., Hapeman, C.J., 2001. Thermodynamic, spectroscopic, and computational evidence for the irreversible conversion of β- to α-endosulfan. J. Agric. Food Chem. 49, 53–72.

Shi, X.-Z., Guo, R.-J., Takagi, K., Miao, Z.-Q., Li, S.-D., 2011. Chlorothalonil degradation by *Ochrobactrum lupini* strain TP-D1 and identification of its metabolites. World J. Microbiol. Biotechnol. 27 (8), 1755–1764.

Shia, J.C., Yabu, M., Van Tran, K., Young, M.S., 2013. Improved Resolution for Paraquat and Diquat: Drinking Water Analysis Using the CORTECS UPLC HILIC Column. Waters Application Note 720004732en.

Shimada, H., Furuno, H., Hirai, K., Koyama, J., Ariyama, J., Simamura, E., 2002. Paraquat detoxicative system in the mouse liver postmitochondrial fraction. Arch. Biochem. Biophys. 402, 149–157.

Singh, B.K., Kuhad, R.C., 1999. Biodegradation of lindane gamma-hexachlorocyclohexane by the white-rot fungus *Trametes hirsutus*. Lett. Appl. Microbiol. 28 (3), 238–241.

Smith, E.A., Mayfield, C.I., 1978. Paraquat: determination, degradation and mobility in soil. Water Air Soil Pollut. 9, 439–452.

Soler, L., Sánchez, S., 2014. Catalytic nanomotors for environmental monitoring and water remediation. Nanoscale 6, 7175–7182.

Sorolla, M.G., Dalida, M.L., Khemthong, P., Grisdanurak, N., 2012. Photocatalytic degradation of paraquat using nano-sized Cu-TiO_2/SBA-15 under UV and visible light. J. Environ. Sci. 24 (6), 1125–1132.

Tejomyee, S.B., Pravin, R.P., 2007. Biodegradation of organochlorine pesticide, endosulfan, by a fungal soil isolate, *Aspergillus niger*. Int. Biodeterior. Biodegrad. 59 (4), 315–321.

US EPA, 2007a. Note to Reader. Endosulfan Readers Guide. November 16. EPA-HQ-OPP-2002-0262-0057.

US EPA, 2007b. Addendum to the Ecological Risk Assessment for Endosulfan, Memorandum to Special Review and Reregistration branch. October 31. EPA-HQ-QPP-202-0262-0063.

Van Beilen, J.B., Pankea, S., Lucchinib, S., Franchini, A., Rothlisberger, M., Witholt, B., 2001. Analysis of *Pseudomonas putida* alkane-degradation gene clusters and flanking insertion sequences: evolution and regulation of the alk genes. Microbiology 147 (6), 1621–1630.

Wales, S.S., Scott, G.I., Ferry, J.L., 2003. Stereo-selective degradation of aqueous endosulfan in modular estuarine mesocosms: formation of endosulfan γ-hydroxycarboxylate. J. Environ. Monit. 5, 373–379.

Wang, G., Liang, B., Li, F., Li, S., 2011. Recent advances in the biodegradation of chlorothalonil. Curr. Microbiol. 63, 450–457.

Wesseling, C., Van Wendel, B.J., Ruepert, C., León, C., Monge, P., Hermosillo, H., et al., 2001. Paraquat in developing countries. J. Occup. Environ. Health 7, 275–286.

Wetzstein, H.G., Schemer, N., Karl, W., 1997. Degradation of fluoroquinolone enrofloxacin by the brown rot fungus *Gloeophyllum striatum*: identification of metabolites. Appl. Environ. Microbiol. 63, 4272–4281.

Wong, F., Alegria, H.A., Jantunen, L.M., Bidleman, T.F., Figueroa, M.S., Gold-Bouchot, G., et al., 2008. Organochlorine pesticides in soils and air of southern Mexico: chemical profiles and potential for soil emissions. Atmos. Environ. 42, 7737–7745.

World Health Organization (WHO), 2006. WHO Gives Indoor Use of DDT a Clean Bill of Health for Controlling Malaria. <http://www.who.int/mediacentre/news/releases/2006/pr50/en/>.

World Health Organization, 2009. Countries Move Toward More Sustainable Ways to Roll Back Malaria. <http://www.who.int/mediacentre/news/releases/2009/malaria_ddt_20090506/en>.

Wrabel, M.L., Peckol, P., 2000. Effects of bioremediation on toxicity and chemical composition of No. 2 fuel oil: growth responses of the brown alga *Fucus vesiculosus*. Mar. Pollut. Bull. 40 (2), 135–139.

Yanez-Montalvo, A., Vazquez-Duhalt, R., Cruz-l6pez, L., Calixto, M.A., Sanchez, J.E., 2015. Purification and partial characterization of a phenol oxidase from the Edible mushroom *Auricularia fuscosuccinea*. JJEnzyme 2 (1), 006.

Index

Note: Page numbers followed by "*f*" and "*t*" refer to figures and tables, respectively.

C

D

X

Printed in the United States
By Bookmasters